Sustainable Civil Infrastructures

Sustainable Civil Infrastructures (SUCI) is a series of peer-reviewed books and proceedings based on the best studies on emerging research from all fields related to sustainable infrastructures and aiming at improving our well-being and day-to-day lives. The infrastructures we are building today will shape our lives tomorrow. The complex and diverse nature of the impacts due to weather extremes on transportation and civil infrastructures can be seen in our roadways, bridges, and buildings. Extreme summer temperatures, droughts, flash floods, and rising numbers of freeze-thaw cycles pose challenges for civil infrastructure and can endanger public safety.

We constantly hear how civil infrastructures need constant attention, preservation, and upgrading. Such improvements and developments would obviously benefit from our desired book series that provides sustainable engineering materials and designs. The economic impact is huge, and much research has been conducted worldwide. The future holds many opportunities, not only for researchers in a given country, but also for the worldwide field engineers who apply and implement these technologies. We believe that no approach can succeed if it does not unite the efforts of various engineering disciplines from all over the world under one umbrella to offer a beacon of modern solutions to the global infrastructure. Experts from the various engineering disciplines around the globe participate in this series, including: Geotechnical, Geological, Geoscience, Petroleum, Structural, Transportation, Bridge, Infrastructure, Energy, Architectural, Chemical and Materials, and other related Engineering disciplines.

The SUCI series publishes the best research papers from various IEREK international conferences, while external conference proceedings proposals are no longer accepted as direct publications. The series does invite significant expansions of selected papers previously submitted to conferences. The Editors' role should be evident in the reviews through an intensive **double-blind peer review** process, helping authors with edits, and collaborating with them to demonstrate **great academic significance** through clear research findings and innovative solutions that go well beyond a typical conference submission.

With the commitment to the above instructions, the SUCI series welcomes external proposals and requests to publish **edited books, contributed volumes, monographs, and textbooks** that align with the series' scope and fit with its policy.

SUCI series is now indexed in SCOPUS and EI Compendex.

Editor-in-ChiefSeries Editor Hany Farouk Shehata Mourad Amer SUCI.series@ierek.com

Lokesh Choudhary · Vaishali Sahu · Aman Garg
Editors

Integrated Approaches for Smart and Resilient Urban Infrastructure

Springer

Editors
Lokesh Choudhary
Department of Civil Engineering, School of Engineering and Technology
Manav Rachna International Institute of Research and Studies
Faridabad, Haryana, India

Vaishali Sahu
Department of Multidisciplinary Engineering, School of Engineering and Technology
The NorthCap University
Gurugram, Haryana, India

Aman Garg
Department of Civil Engineering and Smart Cities
Shantou University
Shantou, Guangdong, China

ISSN 2366-3405 ISSN 2366-3413 (electronic)
Sustainable Civil Infrastructures
ISBN 978-3-032-34635-3 ISBN 978-3-032-34636-0 (eBook)
https://doi.org/10.1007/978-3-032-34636-0

This Springer imprint is published by the registered company Springer Nature Switzerland AG
The registered company address is: Gewerbestrasse 11, 6330 Cham, Switzerland

Foreword

Urban infrastructure systems are facing unprecedented challenges arising from rapid urbanisation, climate change, resource scarcity, and technological transformation. Meeting these challenges requires infrastructure solutions that are not only efficient and intelligent but also sustainable, resilient, and adaptable to future needs. Consequently, there is an increasing need for integrated approaches that combine advances in engineering, environmental sustainability, digital technologies, and urban planning.

The volume Integrated Approaches for Smart and Resilient Urban Infrastructure addresses this need by bringing together contemporary research and innovations from diverse domains of infrastructure development. The contributions explore emerging solutions that support the transition toward sustainable, technology-enabled, and resilient urban systems, reflecting the growing convergence of advanced materials, smart technologies, environmental stewardship, and data-driven decision-making.

The book is organized into six thematic sections covering sustainable construction materials and advanced concrete technologies; energy-efficient buildings and net-zero infrastructure; smart mobility and transportation systems; water, wastewater, and environmental sustainability; digital technologies, IoT, and urban connectivity; and sustainable urban systems for smart cities. Together, these chapters provide valuable insights into the challenges and opportunities associated with creating future-ready urban infrastructure.

Aligned with the objectives of the Sustainable Civil Infrastructures series, this volume highlights innovative approaches that enhance the long-term performance, sustainability, and resilience of infrastructure systems. The contributions demonstrate how interdisciplinary research and integrated solutions can support the development of smarter, greener, and more livable cities.

It is hoped that this book will serve as a useful reference for researchers, academicians, graduate students, policymakers, and practicing engineers working in the fields of civil engineering, urban development, and sustainable infrastructure. By encouraging holistic and forward-looking perspectives, the volume aims to contribute to the advancement of knowledge and practice in resilient urban infrastructure development.

Lokesh Choudhary

Preface

Urban infrastructure systems worldwide are undergoing a paradigm shift driven by rapid technological advancements, increasing environmental concerns, and the pressing need for resilience against natural and anthropogenic hazards. Cities today are expected not only to function efficiently but also to adapt to climate variability, resource constraints, and evolving societal needs. Against this backdrop, the integration of smart technologies with resilient and sustainable infrastructure design has become an essential focus area for researchers and practitioners alike.

The present volume, *Integrated Approaches for Smart and Resilient Urban Infrastructure*, is conceived with the objective of providing a comprehensive platform that captures recent advances and multidisciplinary perspectives in the field of civil and infrastructure engineering. The book brings together 22 carefully selected chapters that address a wide spectrum of themes including smart transportation and traffic management, digital and IoT-enabled urban systems, sustainable and advanced construction materials, energy-efficient buildings, renewable energy integration, water resources and wastewater management, environmental protection, urban ecology, and regulatory and governance frameworks.

All chapters in this edited volume are based on original research contributions presented at the *International Conference on Technological Innovations in Multidisciplinary Engineering and Sciences (TIMES 2025)*. The conference attracted high-quality submissions from researchers across academia, industry, and research organisations, reflecting the multidisciplinary nature of modern engineering challenges. Following the conference, selected manuscripts were invited for inclusion in this book and were subjected to a thorough peer-review and revision process to meet the academic and publication standards of Springer and the *Sustainable Civil Infrastructures* series.

A distinctive feature of this book is its emphasis on integrated approaches that bridge theory and practice. Several chapters focus on the application of artificial intelligence, machine learning, and IoT technologies for smart city development, while others address sustainable material innovations such as waste-derived binders, self-healing concrete, and low-carbon construction practices. The volume also highlights emerging concerns related to climate change, environmental pollution, water security, and urban resilience, offering both technical insights and policy-oriented perspectives.

This book is intended to serve as a reference resource for postgraduate students, doctoral researchers, academicians, and professionals working in the areas of civil engineering, urban infrastructure, sustainability, and smart city development. It is our hope that the ideas, methodologies, and case studies presented in this volume will stimulate further research, encourage interdisciplinary collaboration, and contribute to the development of smarter, greener, and more resilient urban infrastructure systems.

Acknowledgements

The editors would like to express sincere gratitude to all contributing authors for their high-quality research contributions and for their continued cooperation throughout the manuscript preparation, review, and revision process. The authors' commitment to addressing reviewer comments and adhering to the editorial guidelines has been instrumental in maintaining the academic rigor and coherence of this volume. Their collective efforts have significantly enriched the technical depth and multidisciplinary character of the book.

The editors also extend appreciation to the reviewers for their thoughtful and constructive evaluations, which helped strengthen the quality, clarity, and relevance of the individual chapters.

Sincere thanks are due to the Springer editorial and production team for their professional guidance, timely support, and assistance at various stages of the publication process. Finally, acknowledgment is made to the organising committee of the *International Conference on Technological Innovations in Multidisciplinary Engineering and Sciences (TIMES 2025)* for providing a vibrant academic platform that facilitated meaningful technical exchange and served as the foundation for this scholarly compilation.

Contents

Section 1: Sustainable Construction Materials and Advanced Concrete Technologies

Toward Carbon-Neutral Concrete: Insights on Waste-Derived Binders and Bio-Enzymatic Approaches

Diparshan Rowort Dikhalipa Rai[1(✉)], Arjun Shahi[1], and Mitun Chakraborty[2]

[1] Department of Civil Engineering, Manav Rachna International Institute of Research and Studies, Faridabad, Haryana 121004, India
rdiparshan@gmail.com

[2] Department of Biotechnology, Parul Institute of Technology, Parul University, Vadodara, Gujarat 391760, India
mitun.chakraborty34470@paruluniversity.ac.in

Abstract. The pressure is on the construction industry to minimize its carbon footprint, and cement production contributes about 8% of global greenhouse gases. This review analyses the new technology of capturing CO_2 in concrete that involves use of waste-based binders and bio-enzymatic additives as a way of making construction sustainable (Sustainable Development Goal (SDG) 9, 11). The paper summarizes the existing knowledge on the effects of using carbonation, the use of industrial waste materials such as fly ash and ground granulated blast furnace slag, and the new use of biological agents to improve CO_2 fixation and self-healing capabilities. The combination of these technologies offers great potential in the realization of net-zero carbon concrete in the process of controlling waste management issues (SDG 12). Significant results show that optimized mixtures can be used to obtain CO_2 sequestration efficiencies of more than 40%, embodied carbon reductions as much as 56%, with good mechanical performance and durability at the same time. But such issues as scalability, long-lasting viability, and standardization are still significant impediments to large-scale industrial use. This review presents some of the research gaps that are critical and gives recommendations on how to develop this technology to commercial viability.

Keywords: Bio-enzymatic additives · CO_2 sequestration · geopolymers · mineralization · sustainable concrete · waste-derived binders

1 Introduction

A never witnessed sustainability issue exists in the global concrete industry. The production of concrete is responsible for approximately 5–8% of global anthropogenic CO_2 with the energy-intensive process of producing Ordinary Portland Cement (OPC) and calcining limestone in clinker production being the key drivers (Antunes et al., 2021). With today's demand for concrete at more than 14 billion tonnes a year with existing and predicted growth of over 30–50% by 2050, traditional construction methods cannot be accommodated within any realistic climate change targets to limit warming to 1.5 °C (Mundaca et al., 2019).

L. Choudhary et al. (Eds.): TIMES 2025, SCI, pp. 3–16, 2027.
https://doi.org/10.1007/978-3-032-34636-0_1

It is therefore, increasingly necessary to shift the paradigm to a carbon-neutral or even carbon-negative construction material. CO_2 capturing concrete is a game-changing method in which hardening composite actively captures atmospheric carbon dioxide, together with waste-derived binders i.e. industrial by-products which are usually sent to landfills. Apart from the mentioned waste valorisation, as developing topics of the bio-enzymatic process were gradually being looked upon, innovative environmentally friendly concretes with improved carbonation processes or with self-repairing ability guarantee further benefits for environment (Adebanjo et al., 2025).

The present review bridges together and synthesizes the understanding on three linked developments: (1) accelerated carbonation and direct CO_2 mineralisation in concrete, (2) waste-derived binders derived from industrial residues as precursors, and (3) the role of biological and enzymatic additives that accelerate this development. There is a high level of expectation of low-carbon or carbon-negative construction materials. In the next 10 years the convergence of these technologies would prove to be a boon for the Construction and Infrastructure industry (Adebanjo et al., 2025).

Figure 1 presents schematically the progressive carbon lock-in process in concrete and how physical capture, and chemical mineralization, waste-derived binder synergy, and bio-enzymatic acceleration are all part of the process of permanently capturing CO_2.

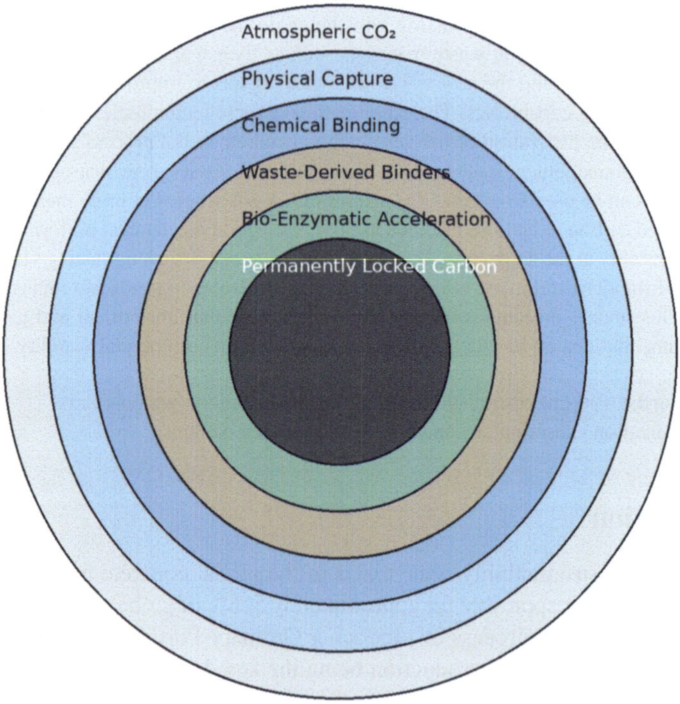

Fig. 1. Carbon Lock-in Mechanism Map (Source: Author's contribution)

2 Carbon Dioxide Sequestration Mechanisms in Concrete

2.1 Fundamental Chemistry of Carbonation

Carbonation in cementitious materials represents the reaction of CO_2 with calcium-bearing phases to form calcium carbonate minerals. The fundamental reaction equation is:

$$Ca(OH)_2 + CO_2 \rightarrow CaCO_3 + H_2O \quad (1)$$

Hydration of concrete produces around 15–20% of calcium hydroxide ($Ca(OH)_2$), a reactive phase which can enhance CO_2 sequestration (Teune et al., 2023). Old-style passive carbonation takes place over many decades at the surface of a concrete perhaps just a few millimetres (mm) in depth to begin with. This is because it is exposed to air and its combination which includes oxygen, water vapor and CO_2. It is important to note that this approach can be followed in a few hours up to a couple of days by performing the carbonation mechanism with either pressurized CO_2 injection or gas rich environment, therefore becoming commercially feasible (Antunes et al., 2021).

Recent research demonstrates that cement suspension carbonation represents an alternative pathway with exceptional efficiency. When CO_2 is added directly into a cement slurry during mixing (prior to normal setting), fast reactions of aqueous ions take over from the classical diffusion-controlled processes (Adebanjo et al., 2025). This method could achieve CO_2 sequestration ratio of 45% at most, a significant breakthrough in practical carbonation rate as well as uncompromised strength of the concrete.

2.2 Carbonation Efficiency and Optimization

The capture efficiency of CO_2 is dramatically changed under different reaction conditions, material composition and processing parameters. Early studies on practical applicability revealed capacities of 9.8–18.9% for stationary and 6.3–8.1% for direct flue gas-based systems kept at adsorption conditions along circulation lines with flowing particles, respectively (Lai et al., 2021). Newer optimized systems have much higher efficiencies. For instance, the foam concrete made with slag as the raw materials under the optimal condition had 91.59 kg/m^3 of CO_2 sequestration (Wen et al., 2025).

Some highly important sequestration efficiencies are related to moisture content, sample size, ambient pressure and temperature (Antunes et al., 2021). Kinetics is especially affected by water content because carbonation reactions proceed via aqueous ionic pathways. Enriched pressures promote reaction rates; however, the benefits of higher pressures start to diminish at above 1000 psi. Temperature effects are opposite, due to the fact that increased temperatures also decrease CO_2 solubility in aqueous solutions while they enhance reaction kinetics (Wen et al., 2025).

2.3 Advanced Carbonation Technologies

Three-dimensional concrete printing (3DCP) and in-situ CO_2 sequestration is an untapped frontier of advanced carbonation. During printing, purging with pressurized CO_2 gas resulted in a 15% increase in uptake of CO_2 compared to the control specimens, and its effectiveness was highly sensitive to adequate spraying parameters and curing conditions (Tay et al., 2024). The benefits of 3DCP implementation are no more constraints on chamber size as well as the potential for large scale sequestration without specific facilities like in traditional accelerated carbonation.

The sequestration of CO_2 through the carbonated mixing water is another feasible way (Motie and Assareh, 2020). This method stably captures an amount of CO_2 up to 0.84% by mass of CO_2 in the cement within the cement matrix as calcium carbonate. Further this process reduces embodied CO_2 by approximately 47% in ready-mix concrete mixtures. Poor workability resulting from this approach was successfully addressed through inclusion of recycled concrete aggregate as external alkalinity source and lubrication agent (Motie and Assareh, 2020).

3 Waste-Derived Binders: Valorisation and Performance

3.1 Industrial Waste Streams as Precursors

The worldwide amounts of industrial residues with potential application in concrete mixes are overwhelming: Fly Ash (FA) is produced in excess of 800 million tonnes annually, Ground Granulated Blast Furnace Slag (GGBS) production exceeds 350 million tonnes a year, and Construction and Demolition (C&D) waste also approaches 600 million tonnes in a year (Wu et al., 2019). However, only a marginal quantity of these materials is used as fuel and most of them are disposed in landfill at large environmental and economic costs.

Fly Ash, which is a major waste material from coal-fired electricity production process, have pozzolanic properties that allow it to react in cementitious systems once activated. Compositional analysis demonstrates that FA is generally composed of 40–60% SiO_2, 15–35% Al_2O_3 and 5–15% Fe_2O_3. Thus, it is a good precursor for alkali-activation (Ismail et al., 2014). GGBS (which is formed by quenching of molten iron slag) consists of the same silicate and aluminate crystal phases but with CaO content between 30–50%. This provides hybrid properties between pure geopolymeric and Portland cement-like behaviour (Natarajan et al., 2022).

Other waste sources of attention are red mud (by-product from alumina production), copper-nickel slag, rice husk ash, steel slag and calcined marine clay (Balakrishnan et al., 2011). Different materials provide different phase assemblages and reactivity. For example, steel slag consists of olivine and wüstite phases promoting the reactivity with CO_2 and Rice Husk Ash (RHA) contains a highly amorphous silica favourable for alkali-activation (Ponomar et al., 2022).

3.2 Alkali-Activation Chemistry and Binder Formation

Alkali-activation is the major process to transform waste-based aluminosilicate materials into binders for use. The reactions are at their core unlike those of Portland cement chemistry, requiring highly alkaline media (pH 12–14) to dissolve the aluminosilicate sources and stimulate polycondensation reactions leading to a 3D gel network (Ponomar et al., 2022).

When fly ash and slag blend together, synergistic effects emerge: slag-rich systems start calcium silicate hydrate (C-S-H) phases which are typical as in the hydration of Portland cement, while fly ash favour geopolymeric gels (N-A-S-H) generation (Albitar et al., 2015). Binary and ternary combinations can thus take advantage of both routes. Optimum GGBS:FA ratio is, however, found to be 7:3–8:2 with 28-day ambient cured compressive strength of 49–62 MPa (Kumar et al., 2010). These properties are comparable to or better than those of OPC counterparts, with 40–56% lower carbon footprint.

The composition of the activator solution has a significant impact on reaction mechanisms. Sodium hydroxide (NaOH) (with usually 8–14 mol/L concentration) controls dissolution rates and pH, whereas the addition of sodium silicate (Na_2SiO_3) provides activation and further source of silica (Parveen and Singhal, 2017). A molar ratio of Na_2SiO_3/NaOH (alkali activator) range between 1.5 to 2.5, however, the best value depends on the composition of the material mix and applications targeted (Aldred and Day, 2012).

3.3 Performance Characteristics of Waste-Derived Binders

Waste-based alkali-activated materials (AAMs) exhibit performance typically equal to or better than OPC concrete in many demanding applications (Natarajan et al., 2022). The strength growth is composition, activation parameters and curing conditions dependent. The early-age strength gain rate is generally slower than that of OPC mixes for the ambient-cured systems (30–50% development of 28-day strength at 7 days) whereas, late-age strength gain is even higher than that of OPC (Jindal et al., 2022).

Durability characteristics in particular emphasize the AAMs advantage. Chloride penetration resistance is significantly higher than OPC concrete, due to denser microstructure and lower permeability (Balakrishnan et al., 2011). With repeated wetting and drying cycles in a saline environment, there is no appreciable drop of the compressive strength or its value even increased (to 87.9 MPa from the initial 66.5 MPa) when compared to control OPC cubes which exhibited significant loss (Balakrishnan et al., 2011). Similarly, resistance to sulfuric acid attack and thermal degradation is greater for AAMs surpassing conventional cements (Aliques-Granero et al., 2017).

However, challenges remain. AAMs exhibit higher drying shrinkage than OPC concrete, with values reaching 2000–2500 micro strain without mitigation measures (Messina et al., 2015). Creep behaviour differs drastically, with basic creep comprising approximately 70% of total drying creep at 90 days (compared to 30–40% for OPC) (Sagoe-Crentsil et al., 2013). Carbonation resistance also requires attention, though several strategies including limestone additions and optimized gel composition have shown promising performance (Feng et al., 2024).

4 Bio-Enzymatic Additives for Enhanced CO_2 Fixation

4.1 Microbially-Induced Calcite Precipitation (MICP)

The biological pathways of CO_2 storage in concrete include Microbial Induced Calcite Precipitation (MICP), which was studied to catalyse the formation of Calcium Carbonate ($CaCO_3$) by microorganisms through metabolic reactions (Zamba et al., 2023). Two main enzymatic pathways contribute to this process, (1) Ureolytic bacteria generating the hydrolysis of urea into ammonia and CO_2 by the enzyme urease, and (2) organisms producing carbonic anhydrase (CA), catalysing hydration of membrane-permeable CO_2 to bicarbonate (HCO_3^-).

Urease is the enzyme responsible for urea decomposition, and bacteria which yield this enzyme are known as ureolytic bacteria such as Bacillus and Sporosarcina (Mora and Arioli, 2014). Once urea and calcium chloride (or the other sources of Ca) are added into concrete where the bacterial spore is mixed in, ammonia along with CO_2 is produced at the local site due to enzymatic reaction. This increases pH locally and subsequently activates precipitation of $CaCO_3$.

$$2NH_3 + CO_2 + Ca^{2+} \rightarrow CaCO_3 + 2NH_{4+} \quad (2)$$

This mechanism has been found extremely effective in crack healing for cracks as wide as 2 mm with healing efficiency of 80–90% under optimum conditions (Adebanjo et al., 2025). The bacterial cell walls serve as nucleation sites for this process and $CaCO_3$ before growing in crack fissures precipitate initially on the cellular surface. Formation of calcite ($CaCO_3$) and vaterite polymorphs can further be confirmed by performing scanning electron microscopy (SEM) (Zailani et al., 2020).

4.2 Carbonic Anhydrase-Based Biocementation

Carbonic anhydrase (CA) also offers an enzymatic pathway with possibly better CO_2 sequestration kinetics. This omnipresent enzyme increases the rate of CO_2 hydration by orders of magnitudes to rates as high as 10^6 of that obtained in uncatalyzed reactions (Shekh et al., 2012):

$$CA : CO_2 + H_2O \leftrightarrow HCO_{3-} + H^+ \quad (3)$$

Such CA-based biocementation is fundamentally different from ureolytic processes: instead of depending on bacterial metabolism and the production of ammonia leading to pH increase, CA can directly catalyse bicarbonate formation (Ferrer et al., 2020). The pathway is adaptable to a number of implementation modalities from purified enzyme to thermophilic bacterial strains derived from submarine hydrothermal systems or fungi producing CA enzymes.

Fungal species, especially those that are originated from high-pH habitats could secrete both urease and carbonic anhydrase enzymes which can exert a dual enzymatic CO_2-fixation acceleration (Abubakari et al., 2025). Recent studies indicated that fungi might hasten the carbonation process and also provide strength to concrete through precipitation of $CaCO_3$ (Abubakari et al., 2025).

4.3 Enzyme Immobilization and Protection

A significant barrier to the introduction of biological agents into concrete is how to protect microbial cells and enzymes from the alkaline liquids and mechanical load in a setting concrete (Singh and Mishra, 2025). Various protection approaches are known, including encapsulation of the bacterial spore within natural minerals (bentonite, diatomaceous earth, sepiolite and pumice), microencapsulation in modified alginate hydrogels and enzyme immobilized in metal-carbonic hybrid carriers (Singh and Mishra, 2025).

Tailored alginate hydrogels are particularly promising and have been shown to exhibit microbial viability (Johnston et al., 2020). The encapsulated Bacillus sphaericus remained viable and formed large amounts of $CaCO_3$ (70% by weight of the matrix) in situ as recorded by its ability to consume oxygen. This protective effect was corroborated when comparing the respiratory activity of encapsulated and unbound bacterial samples, revealing that only encapsulated variants were metabolically active (Johnston et al., 2020).

Inorganic carriers with copper phosphates developed for carbonic anhydrase immobilization remained stable for 2 months over a broad pH (including acidic and alkaline) and temperature spectrum, featuring an increased enzymatic activity with respect to free enzyme (Ali et al., 2023). This method offers a combined mechanical protection, and the possibility of recovery and re-use of the biocatalyst.

5 Integration of Technologies: Synergistic Advantages

5.1 Bio-Enzymatic Enhancement of Waste-Derived Binder Systems

Synergistic effects greater of each technology alone are achieved by combining bio-enzymatic additives with alkali activated waste-based binders (Alshalif et al., 2021). The strengthening effect of Bacillus tequilensis on the slag/fly ash-based bio-foamed concrete brick was 35.5% compared with that for plain foamed concrete (reaching 8.22 MPa under optimal conditions) (Alshalif et al., 2021). The mechanisms rely on the concomitant activities of urease and carbonic anhydrase enzymes, such that they catalyse CO_2 fixation into calcium carbonate, while waste derived binders serve as a matrix of the structure and source of calcium.

The coupling of biological and chemical processes works as follows: The activation of bacterial spores and conditions that support enzymatic activity are supported by alkaline environment of AAMs (pH 12–14). The calcium carbonate precipitation is triggered when the local pH rises. This happens when the bacteria consume urea or catalyses bicarbonate formation. A dense microstructure, filled pores and added binding strength is the result of $CaCO_3$ formation (Alshalif et al., 2021).

5.2 Crack Healing and Durability Enhancement

Auto-healing occurs when bacterial spores in a dormant state inside the concrete during setting are activated when water penetrates into the cracks (Zamba et al., 2023). This self-healing (passive) behaviour increases the lifetime, without any external intervention. Studies on Ureolytic Bacteria (Bacillus sp.) added to high-performance concrete indicated a crack healing efficiency between 77–89% of fresh cracks having an initial width of 0.4–0.6 mm (Zamba et al., 2023).

The heal and grow mechanism includes (1) dormant spore germination due to the presence of moisture and nutrients, (2) enzymatic urea hydrolysis that leads to Ca^{2+} precipitation conditions, and (3) $CaCO_3$ deposition clogging cracks (Khushnood et al., 2019). Post-healing mechanical characterization revealed an improvement of 6.74% in flexural tensile strength when compared to non-treated specimens and crack closure was achieved 28 days after the first cracking appearance (Schwarz, 2024).

5.3 Embodied Carbon Reduction and LCA Perspectives

LCA research uses economic metrics to measure the environmental advantages of the waste-based binder and CO_2 capture technologies. Replacement of OPC with GGBS abates carbon emission by about 56% per cubic meter of concrete, whereas fly ash replacement abates carbon emission by 40–68% with respect to substitution rate (Verma et al., 2024). Combined with the in-situ CO_2 mineralization, the process also leads to further carbon reductions of 0.5–1.5 tonnes per tonne of concrete (Balakrishnan et al., 2011).

The analysis of building works including sustainable materials revealed overall embodied carbon savings of 20–31% at the project level when 50% OPC replacement with waste-based binders were associated with CO_2 sequestration technologies (Balakrishnan et al., 2011). The economic studies show that initial cost premiums of 10–20% are paid on waste-derived binder concrete, but lifetime cost models (including lower maintenance expenses through the use of improved durability) show 15–30% cost savings after 50 years of service (Fernando et al., 2021).

6 Challenges and Limitations

6.1 Scalability and Industrial Implementation

However positive the laboratory results are, the large-scale implementation has always been an issue. In accelerated carbonation plants, the supply system of CO_2, the pressure system and environment control needs a huge amount of money [14]. Industrial carbonation imposes processing time and energy, which have to be compensated against embodied carbon advantages. This is currently unfavourable in most climatic settings where grid electricity is based on fossil fuels (Kumar and Majid, 2020).

The performance of carbonation reduces with the increase in concrete density. Porous foam concretes have better CO_2 uptake than dense structural concrete, but concrete density must be high for durability requirements (Antunes et al., 2021). Such tension in the carbon capture performance and the structural performance necessitates new mix designs to balance such competing requirements.

6.2 Biological Agent Viability and Long-Term Performance

It is difficult to preserve bacterial spore viability in the process of concrete mixing, setting and service life. Although spores resist severe pH and mechanical stress levels compared to vegetative cells, survival characteristics under prolonged service are also very different in spores (Mora and Arioli, 2014). Certain studies show that loss of spores in mixing and setting is 10–20% which reduces cumulative healing capacity (Ulrich et al., 2018).

Passive healing is still probabilistic because the dormant spores will only be activated by certain trigger conditions (moisture, source of calcium, pH conditions, etc.), otherwise the cracks will not be healed. Also, it is challenging to measure the healing effectiveness in real structures, because it is hardly observable what is happening inside the structure, in terms of internal geometry of cracks and the distribution of moisture (Zhang et al., 2020).

6.3 Standardization and Performance Verification

The absence of standard testing protocols regarding the CO_2-capturing concrete is a setback in developing markets. The approaches to measuring CO_2 sequestration, assessment of the depth of carbonation, and long-term stability are also different among groups of research (Mwenketishi et al., 2023). The adoption at the industry level needs consensus standards to cover: (1) carbonation verification, (2) definition of the CO_2 source and traceability, (3) the minimum level of sequestration to obtain a carbon credit, (4) the long-term durability assessment protocols.

The adoption of carbon credits counting and verification will not be full proof unless regulations are standardised. Does sequestered CO_2 using carbonation qualify as corporate carbon neutral? Is biological CO_2 sequestration equivalent to engineered carbonation? It is these policy questions that are yet to be answered in the majority of jurisdictions (Mwenketishi et al., 2023).

7 New Opportunities and Future Prospects

The history and comparative maturity of the main technologies that are associated with carbon-neutral concrete have not continued to develop in a consistent manner. Figure 2 presents the comparative level of maturity of key research areas in early-stage research (2011–2014) and the current level of research (2023–2025) with its gaps.

Fig. 2. Technology Maturity Radar (2011–2025) (Source: Author's contribution)

7.1 Advanced Enzymatic Systems

The studies on the new sources of enzymes will yield increased efficiency of CO_2 fixation. Submarine hydrothermal organisms have thermophilic carbonic anhydrase with better stability, activity in high temperature and active in highly alkaline environments (Shekh et al., 2012). Marine bioprospecting is still discovering calcium carbonate precipitating organisms in new environmental habitats.

Experiments in the evolution of carbonic anhydrase in laboratories have created carbonic anhydrase variants with significantly increased catalytic rates and thermal stability (Ferrer et al., 2020). Synthetic biology solutions although still in initial stages of their use in construction may eventually provide the means of custom-made enzyme systems being optimized to work with concrete chemistry. The enzyme engineering and its directed evolution is at the forefront of the areas to be explored.

7.2 Multifunctional Waste-Derived Binder Systems

Studies are also examining ternary and quaternary binder system that combines several streams of waste in order to generate synergies in the properties (Kumar et al., 2021). Combination of slag, fly ash, and new types of wastes (copper-nickel slag, spent catalysts, marine clay) can be tuned to meet particular performance goals: rapid development of strength, increased durability, decreased shrinkage, or maximized CO_2 reactivity (Aliques-Granero et al., 2017; Jaime-Vargas et al., 2025; Sahu et al., 2025).

7.3 Alternatives with Additive Manufacturing

The Additive Manufacturing can be integrated with other processes including 3D printing which is commonly described as a production process. In-situ CO_2 sequestration coupled with three-dimensional concrete printing is one of the most promising frontiers. The process of layer-by-layer construction allows the carbonation of successive layers of deposition in real time, potentially with a more effective capture of CO_2 than post-casting carbonation and without imposing porosity penalties (Katariya et al., 2025).

8 Conclusions and Recommendations

- CO_2 capturing concrete based on waste-based binders and bio-enzyme additives is a novel direction to carbon-negative construction materials. Recent studies show technological feasibility: High CO_2 sequestration efficiency, embodied CO_2 reductions and comparable or better mechanical/durability capability than conventional concrete. Nevertheless, shift in research curiosity to industrial standard should be the concerted action in several spheres.
- The focus of research must be on: (1) Service life considerations while analysing interaction of enzymes and concrete, (2) Development of standardised procedures for quantification and verification, (3) long-term studies to assess the stability of CO_2 sequestered during carbonation (specifically chances of re-release into the environment) (4) Optimal development of ternary binders which can function well for myriad climatic and structural requirement.
- To develop the industry, it is needed: (1) to invest in a large-scale demonstration project that will demonstrate the technical and economic feasibility, (2) to develop standard testing structures that allow regulatory approval, and (3) policy frameworks that explain the eligibility of carbon credits.
- Policies and government should cover: (1) equalization of methods to quantify CO_2, which would allow reasonable accounting of carbon, (2) incentive schemes (carbon pricing, tax credits) that would allow economic competitiveness, and (3) connection with circular economy models where waste valorisation is valued.
- Combination of waste-generated binders, CO_2 sequestration, and bio-enzymatic catalysis provide tangible industry solutions to meet 2050 net-zero goals. These results can only be attained through the dedication of researchers, industry practitioners, and policymakers in their effort to achieve a common objective in regard to sustainability.

References

Abubakari, S., Helal, A., Khalil, A.B.: A review on CO2 mitigation by modified microbial carbonic anhydrase enzyme (CA). Arab. J. Sci. Eng. (2025). https://doi.org/10.1007/s13369-025-10852-0

Adebanjo, A.U., et al.: Carbon-capturing concrete: current status and future directions for development. Innov. Infrastruct. Solut. **10**, 566 (2025). https://doi.org/10.1007/s41062-025-02358-3

Ahmad Zailani, W.W., et al.: Bonding strength characteristics of FA-based geopolymer paste as a repair material when applied on OPC substrate. Appl. Sci. **10**, 3321 (2020). https://doi.org/10.3390/app10093321

Albitar, M., Visintin, P., Mohamed Ali, M.S., Drechsler, M.: Assessing behaviour of fresh and hardened geopolymer concrete mixed with class-F fly ash. KSCE J. Civ. Eng. **19**, 1445–1455 (2015). https://doi.org/10.1007/s12205-014-1254-z

Aldred, J., Day, J.: Is geopolymer concrete a suitable alternative to traditional concrete ? In: 37th Conference on Our World in Concrete & Structures, pp. 1–14. Singapore (2012)

Ali, J., Faridi, S., Sardar, M.: Carbonic anhydrase as a tool to mitigate global warming. Environ. Sci. Pollut. Res. **30**, 83093–83112 (2023). https://doi.org/10.1007/s11356-023-28122-7

Aliques-Granero, J., Tognonvi, T.M., Tagnit-Hamou, A.: Durability test methods and their application to AAMs: case of sulfuric-acid resistance. Mater. Struct. **50**, 36 (2017). https://doi.org/10.1617/s11527-016-0904-7

Alshalif, A.F., et al.: Optimisation of carbon dioxide sequestration into bio-foamed concrete bricks pores using Bacillus tequilensis. J. CO2 Util. **44**, 101412 (2021). https://doi.org/10.1016/j.jcou.2020.101412

Antunes, M., Santos, R.L., Pereira, J., Rocha, P., Horta, R.B., Colaço, R.: Alternative clinker technologies for reducing carbon emissions in cement industry: a critical review. Materials (Basel). **15**, 209 (2021). https://doi.org/10.3390/ma15010209

Balakrishnan, M., Batra, V.S., Hargreaves, J.S.J., Pulford, I.D.: Waste materials – catalytic opportunities: an overview of the application of large scale waste materials as resources for catalytic applications. Green Chem. **13**, 16–24 (2011). https://doi.org/10.1039/C0GC00685H

Feng, H., Wen, J., Shao, Q., Yang, Y., Yao, X.: Carbonation resistance of fly ash/slag based engineering geopolymer composites. Constr. Build. Mater. **449**, 138471 (2024). https://doi.org/10.1016/j.conbuildmat.2024.138471

Fernando, S., Gunasekara, C., Law, D.W., Nasvi, M.C.M., Setunge, S., Dissanayake, R.: Life cycle assessment and cost analysis of fly ash–rice husk ash blended alkali-activated concrete. J. Environ. Manage. **295**, 113140 (2021). https://doi.org/10.1016/j.jenvman.2021.113140

Ferrer, F.M., Hobart, K., Bailey, J.V.: Field detection of urease and carbonic anhydrase activity using rapid and economical tests to assess microbially induced carbonate precipitation. Microb. Biotechnol. **13**, 1877–1888 (2020). https://doi.org/10.1111/1751-7915.13630

Ismail, I., Bernal, S.A., Provis, J.L., San Nicolas, R., Hamdan, S., van Deventer, J.S.J.: Modification of phase evolution in alkali-activated blast furnace slag by the incorporation of fly ash. Cem. Concr. Compos. **45**, 125–135 (2014). https://doi.org/10.1016/j.cemconcomp.2013.09.006

Jaime-Vargas, J.A., Sahu, V., Ruiz, D.E.G., Mathur, A.: Carbon capture and storage in the cement industry: a comparative scenario of india and mexico. In: Innovations in the Development of Sustainable Infrastructure, pp. 3–14 (2025). https://doi.org/10.1007/978-3-031-83750-0_1

Jindal, B.B., Alomayri, T., Hasan, A., Kaze, C.R.: Geopolymer concrete with metakaolin for sustainability: a comprehensive review on raw material's properties, synthesis, performance, and potential application. Environ. Sci. Pollut. Res. **30**, 25299–25324 (2022). https://doi.org/10.1007/s11356-021-17849-w

Johnston, T.G., et al.: Compartmentalized microbes and co-cultures in hydrogels for on-demand bioproduction and preservation. Nat. Commun. **11**, 563 (2020). https://doi.org/10.1038/s41467-020-14371-4

Katariya, M., Mathur, A., Singh, S.: Role of intelligent manufacturing in achieving SDGs. In: Smart Cities and Sustainable Manufacturing, pp. 9–28. Elsevier (2025). https://doi.org/10.1016/B978-0-443-26474-0.00016-6

Khushnood, R.A., ud din, S., Shaheen, N., Ahmad, S., Zarrar, F.: Bio-inspired self-healing cementitious mortar using Bacillus subtilis immobilized on nano-/micro-additives. J. Intell. Mater. Syst. Struct. **30**, 3–15 (2019). https://doi.org/10.1177/1045389X18806401

Kumar, J.C.R., Majid, M.A.: Renewable energy for sustainable development in India: current status, future prospects, challenges, employment, and investment opportunities. Energy, Sustainability and Society **10**(1), 1–36 (2020). https://doi.org/10.1186/s13705-019-0232-1

Kumar, S., Kumar, R., Mehrotra, S.P.: Influence of granulated blast furnace slag on the reaction, structure and properties of fly ash based geopolymer. J. Mater. Sci. **45**, 607–615 (2010). https://doi.org/10.1007/s10853-009-3934-5

Kumar, V.S., Ganesan, N., Indira, P.V.: Engineering properties of hybrid fibre reinforced ternary blend geopolymer concrete. J. Compos. Sci. **5**, 203 (2021). https://doi.org/10.3390/jcs5080203

Lai, J.Y., Ngu, L.H., Hashim, S.S.: A review of CO2 adsorbents performance for different carbon capture technology processes conditions. Greenh. Gases Sci. Technol. **11**, 1076–1117 (2021). https://doi.org/10.1002/ghg.2112

Messina, F., et al.: Fiber Bragg grating sensors: a reliable method for the measurement of early age properties of traditional and innovative cements. In: 2015 Fotonica AEIT Italian Conference on Photonics Technologies, pp. 1–10. Institution of Engineering and Technology, Turin, Italy (2015). https://doi.org/10.1049/cp.2015.0162

Mora, D., Arioli, S.: Microbial urease in health and disease. PLoS Pathog. **10**, e1004472 (2014). https://doi.org/10.1371/journal.ppat.1004472

Motie, M., Assareh, M.: CO2 sequestration using carbonated water injection in depleted naturally fractured reservoirs: a simulation study. Int. J. Greenh. Gas Control **93**, 102893 (2020). https://doi.org/10.1016/j.ijggc.2019.102893

Mundaca, L., Ürge-Vorsatz, D., Wilson, C.: Demand-side approaches for limiting global warming to 1.5 °C. Energy Effic. **12**, 343–362 (2019). https://doi.org/10.1007/s12053-018-9722-9

Mwenketishi, G.T., Benkreira, H., Rahmanian, N.: A comprehensive review on carbon dioxide sequestration methods. Energies **16**, 7971 (2023). https://doi.org/10.3390/en16247971

Natarajan, P., Sivasakthi, M., Revathi, T., Jeyalakshmi, R.: A comparative study on fly ash pozzolanic cement mortar and ambient-cured alkali-activated fly ash–GGBS cement mortar after exposure to elevated temperature. Innov. Infrastruct. Solut. **7**, 30 (2022). https://doi.org/10.1007/s41062-021-00635-5

Parveen, Singhal, D.: Development of mix design method for geopolymer concrete. Adv. Concr. Constr. **5**, 377–390 (2017). https://doi.org/10.12989/acc.2017.5.4.377

Ponomar, V., Adesanya, E., Ohenoja, K., Illikainen, M.: High-temperature performance of slag-based Fe-rich alkali-activated materials. Cem. Concr. Res. **161**, 106960 (2022). https://doi.org/10.1016/j.cemconres.2022.106960

Sagoe-Crentsil, K., Brown, T., Taylor, A.: Drying shrinkage and creep performance of geopolymer concrete. J. Sustain. Cem. Mater. **2**, 35–42 (2013). https://doi.org/10.1080/21650373.2013.764963

Sahu, V., Mathur, A., Chundawat, T.S., Vargas, J.A.J., Ruiz, D.E.G.: Advancements and challenges in carbon capture and storage technologies: Review for smart city infrastructure. In: Smart Cities and Sustainable Manufacturing, pp. 295–306. Elsevier (2025). https://doi.org/10.1016/B978-0-443-26474-0.00009-9

Schwarz, K.: Bio-concrete and biogenic construction materials with cyanobacteria. Fraunhofer Inst. Ceram. Technol. Syst. IKTS 1–2 (2024)

Shekh, A.Y., et al.: Recent Advancements in Carbonic Anhydrase-Driven Processes for CO 2 Sequestration: Minireview. Crit. Rev. Environ. Sci. Technol. **42**, 1419–1440 (2012). https://doi.org/10.1080/10643389.2011.556884

Singh, S.K., Mishra, S.S.: Unlocking the potential of bacterial self-healing in concrete: applications, and challenges for sustainable infrastructure. Indian Chem. Eng. 1–38 (2025). https://doi.org/10.1080/00194506.2025.2524383

Tay, Y.W.D., et al.: Potential of carbon dioxide spraying on the properties of 3D concrete printed structures. Carbon Capture Sci. Technol. **13**, 100256 (2024). https://doi.org/10.1016/j.ccst.2024.100256

Teune, I.E., Schollbach, K., Florea, M.V.A., Brouwers, H.J.H.: Carbonation of hydrated cement: the impact of carbonation conditions on CO2 sequestration, phase formation, and reactivity. J. Build. Eng. **79**, 107785 (2023). https://doi.org/10.1016/j.jobe.2023.107785

Ulrich, N., Nagler, K., Laue, M., Cockell, C.S., Setlow, P., Moeller, R.: Experimental studies addressing the longevity of Bacillus subtilis spores – The first data from a 500-year experiment. PLoS ONE **13**, e0208425 (2018). https://doi.org/10.1371/journal.pone.0208425

Verma, N., Eswari, J.S., Mahapatra, C.: Determination of CO2 Sequestration into Bio-Concrete Bricks Pores using Fungi. J. Environ. Nanotechnol. **13**, 351–357 (2024). https://doi.org/10.13074/jent.2024.12.243865

Wen, C., et al.: Study of macro-/micro-properties and in-situ mineralization mechanism of CO2 foam concrete based on solid waste-derived SAC-gypsum composite. Constr. Build. Mater. **486**, 141911 (2025). https://doi.org/10.1016/j.conbuildmat.2025.141911

Wu, H., Zuo, J., Zillante, G., Wang, J., Yuan, H.: Status quo and future directions of construction and demolition waste research: A critical review. J. Clean. Prod. **240**, 118163 (2019). https://doi.org/10.1016/j.jclepro.2019.118163

Zamba, D.D., Mohammed, T.A., Getu, Y.M., Tufa, D.H., Demiss, B.A.: Experimental Investigation on Self-Healing Efficiency of Mortar with Bacillus subtilis and Bacillus cereus. Adv. Mater. Sci. Eng. **2023**, 1–11 (2023). https://doi.org/10.1155/2023/9399101

Zhang, X., Al-Dossary, A., Hussain, M., Setlow, P., Li, J.: Applications of bacillus subtilis spores in biotechnology and advanced materials. Appl. Environ. Microbiol. **86**, e01096-e1120 (2020). https://doi.org/10.1128/AEM.01096-20

Parametric Strength Study of Sustainable Steel Fibre Reinforced Concrete Using Taguchi's Method

Yaman Hooda(✉)

Department of Civil Engineering, Manav Rachna International Institute of Research and Studies, Sector – 43, Aravalli Hills, Delhi – Surajkund Road, Faridabad 121004, India
yaman.set@mriu.edu.in

Abstract. With the burgeoning construction industry, the rate of production and usage of concrete is also increasing with an alarming rate. One of the main concerns of the concrete production is generation of carbon dioxide, which is one of the main cause of "greenhouse effect" phenomenon. In regards of this concern, researchers are being using different materials in order to replacing the core components of the concrete composite partially so that the rate of carbon dioxide production will decrease. Since concrete is weak in tension, reinforcements are provided. With advancements in the technology, the method of reinforcements provided into the concrete are changing from steel bars to nanomaterials. This study will be focusing on making a sustainable reinforced concrete mix. The two – dimensional steel fibres are used as a reinforcing material with its addition of two percent by weight of cement. The sustainability approach is introduced into the concrete mix by partially replacing the concrete ingredients with industrial by – wastes; cement, fine aggregates and coarse aggregates are replaced partly with fly ash, GGBS & concrete aggregates obtained by recycling of old concrete composite. The proportions added into the concrete mix are decided based on Taguchi's Approach. Laboratory experiments were conducted as per the Indian Standard Codal provisions for the determination of mechanical properties of the sustainable concrete mix.

Keywords: Fibre Reinforced Concrete · Steel Fibre Reinforced Concrete · Sustainable Composite · Mechanical Strength · Taguchi Method · SDG 11

1 Introduction

Water is the most used substance on Earth, yet concrete being most often used man-made construction substance worldwide. It is made by combining the necessary amounts of cementing materials, water, aggregates, and admixtures. Concrete is the mixture, which becomes a rock-like mass after being poured into forms and given time to cure. Concrete gets older and becomes foster because of the enduring chemical reaction among water and cement that causes the hardening. The voids created by the bigger coarse aggregates are occupied with the presence of finer aggregate, & the voids formed by the

L. Choudhary et al. (Eds.): TIMES 2025, SCI, pp. 17–24, 2027.
https://doi.org/10.1007/978-3-032-34636-0_2

mixing of fine aggregates are occupied with the hardened concrete. In a concrete mixture, the cementitious substance and water come together to generate a paste recognized as "Cement – Water Paste" that adding up towards occupying the finer aggregates voids, act as a coating on the surfaces of coarse & fine aggregates, thereby acts as a binder for both aggregates as it cures and harden and densified. The innate fragility of plain concrete is caused by the occurrence of microcracks at the mortar-aggregate contact. Fibres can be added to the mixture to eliminate the weakness. At the internal micro-cracks, the fibres assist in the load transfer. Fiber-reinforced concrete is the name given to such a concrete; made primarily with regular concrete components with the ideation of reinforcement with the presence of fibres. In this system, short, irregular, and discrete fine fibres with a particular geometry are randomly distributed throughout the concrete to act as reinforcement. The mix becomes more cohesive and less likely to segregate but the workability is significantly reduced by the fibres' interlocking and entanglement with the aggregate particles. Steel, glass, and organic polymers were used to generate the fibres that can be used to reinforce concrete. In lieu of the fraction of the steel fibres employed in the concrete composite by volume, the composite of SFRC yields greater strength as compared to the conventional concrete composite in compression [1–4]. The presence of the fibres in the concrete composite enhances the property of toughness, which primarily depends on the availability of the fibres in the concrete composite; along with their proportion and type. The results were more impressive in case of employment of hooked – end steel fibres as compared to the addition of other domains of steel fibres. The steel fibers with hook ends were recommended for further research due to improved performance and is said to raise flexibility [5–7]. The addition of steel fibres enhances the overall stiffness, durability and plasticity of the concrete composite significantly [8–12]. As per some of the recent observations, studies had concluded the reduction in the characteristic of the shrinkage of the concrete samples by the employment of the steel fibres [13, 14]. The notable improved characteristics of the steel fibre based concrete composites makes them available for their application in the domain of pavement construction, tunnel linings, repairing of the structures including bridge decks [15–18]. The characteristics of aspect ratio (l/d) and volume fraction (Vf) of the type of fibres employed in the concrete composite affects the property of the workability of concrete significantly. The orientation of the fibres depends on different factors including fluidity of the matrix composite, duration involved in the casting of concrete samples and the geometry of the moulds [19–21]. The bridging characteristics of the fibre induced concrete composites depends upon the properties of the fibres such as their length and ITZ of the fibre – matrix composite [22].

In the present scenario, studies had widely been conducted focusing on replacing the prime components of the concrete mix with the availability of the industrial waste generated elements in lieu of lesser utilisation of its basic components for effective utilisation of the generated waste. Such an approach of utilisation of waste generated from different means of industrial processes into concrete with environmental – friendly benefits is summarized under sustainable concrete. The application and effectiveness of the sustainable concrete relies on its constituents. For example, the application of recycled aggregates from previously generated concrete are widely adopted in the generation of the sustainable concrete composite, making the product with greater durability and lesser maintenance. The employment of beneficial industrial waste into concrete composite can be done by either partially replacing or full replacement of the prime components of the traditional concrete. The characteristics of the available waste products ensure their selection in the replacement or addition into the concrete composite for the ideation of enhancement of the characteristics of the generated fibre induced concrete composite. Furthermore, the employment of such waste products into concrete production tends to observe the declining rate in the production of the clinker, thereby ensuring saving of raw materials and energy emission. Some of the commonly used industrial by – waste generated products in the production of concrete includes Fly Ash [FA], sawdust, Rice Husk Ash [RHA], sugarcane bagasse and many more. The effective utilisation of such industrial by – waste products minimizes their detrimental impact on ecosystem. In this study, the strength properties of SFRC in compression and flexure are evaluated and then, an approach of development of sustainable SFRC has been proposed by incorporating Taguchi's Approach in the formulation of the trial mixes for determination of the optimum combination with industrial by – products.

2 Experimentation on Normal Cement Concrete Sample

The prime element in the generation of concrete sample is cement. In this study, OPC 43 grade is utilized and the Indian standard codal guidelines of IS 4031 is followed in determination of its characteristics [23]; summarized under Table 1:

Table 1. Experimental Observation – Cement (Source: Author's contribution)

Characteristics	Experimental values
Fineness	7%
Specific Gravity	3.13
Standard Consistency	37 Percent
Setting Time (Initial)	180 min
Setting Time (Final)	390 min
Compressive Strength of Concrete (Average at 28th Day)	51.11 N/mm2

Followed by cement, the next component used in the preparation of concrete is aggregates, which is categorized under two domains of Fine Aggregates & Coarse Aggregates, based on the size distribution. Indian standard codal guidelines of IS 2386 is followed in determination of the characteristics of aggregates used in the study [24]; summarized under Table 2.

For preparation of concrete sample, a specified mechanism is followed, known as concrete mix designing. The mechanism of concrete mix designing is employed to determine the quantity of the prime components required in the production of concrete. Indian standard codal guidelines of IS 10262: 2019 is followed in determination of concrete mix proportioning of M25 concrete grade used in the study [25]; summarized under Table 3.

Table 2. Experimental Observation – Aggregates (Source: Author's contribution)

Characteristics	Fine Aggregates	Coarse Aggregates
FM	2.09	7.49
Absorption of water	1.15 percent	3.39 percent
Specific Gravity	2.49	2.29
Bulking	21 percent	40 percent
Elongation Index	-	7.8 percent
Flakiness Index	-	34 percent

Table 3. Experimental Observation – M 25 (Source: Author's contribution)

State of Property	Experimentation	Values
Green State	Workability (Slump, in mm)	160
	Workability (Compaction Factor)	0.90
Hardened State	7th Day Compressive Strength (MPa)	18.89
	7th Day Split Tensile Strength (MPa)	1.90
	7th Day Flexural Strength (MPa)	3.10
	28th Day Compressive Strength (MPa)	27.98
	28th Day Split Tensile Strength (MPa)	2.51
	28th Day Flexural Strength (MPa)	3.68

3 Experimentation on Steel Fibre Reinforced Concrete Samples

The study inculcated the addition of three different types of steel fibres as: 1. Two – Dimensional; 2. Undulated and 3. Flat Steel and focusing on compressive and flexural strength with addition of 1% of fibres by weight of concrete. The results obtained for various experimentation conducted onto the samples are listed in Tables 4, 5 and 6.

Table 4. Properties of Steel Fibre – Reinforced Concrete (Source: Author's contribution)

State of Property	Experimentation	Average Values
Two – Dimensional Steel Fibres	28th Day Compressive Strength (MPa)	44.01
	28th Day Flexure Strength (MPa)	4.78
Undulated Steel Fibres	28th Day Compressive Strength (MPa)	32.25
	28th Day Flexure Strength (MPa)	4.40
Flat Sheet Steel Fibre	28th Day Compressive Strength (MPa)	36.00
	28th Day Flexure Strength (MPa)	4.49

Table 5. Between Normal Cement Concrete and Fibre Reinforced Concrete (Compressive Strength) (Source: Author's contribution)

SFRC	Percentage Increase on 7th Day	Percentage Increase on 28th Day
Two-Dimensional Steel Fibre	55.95	56.10
Undulated Steel Fibre	18.97	15.15
Flat Sheet Steel Fibre	20.80	28.25

Table 6. Between Normal Cement Concrete and Fibre Reinforced Concrete (Flexure Strength) (Source: Author's contribution)

SFRC	Percentage Increase on 7th Day	Percentage Increase on 28th Day
Two-Dimensional Steel Fibre	28.95	23.25
Undulated Steel Fibre	2.0	11.85
Flat Sheet Steel Fibre	21.15	16.10

4 Taguchi's Approach

Taguchi's approach is considered for employment of the sustainability practice in the concrete composite mix by replacement of its prime components with industrial by – waste elements. This study observes the application of FA, Ground Granulated Blast Furnace Slag [GGBS] and recycled aggregates [RCA] as a partial replacement, as summarized under Table 7.

Table 7. Specimen formation based on Taguchi's Approach (Source: Author's contribution)

Sample Representation	Partially Replacement	1st Replacement (%)	2nd Replacement (%)	3rd Replacement (%)
A	Fly Ash	10	20	30
B	GGBS	10	20	30
C	RCA	10	20	30

Based on the experiments conducted, the concrete sample with Two – Dimensional Steel Fibre is considered for adopting the Sustainability approach in the concrete sample preparations with the considerations of Taguchi's Approach. The results obtained after performing the strength – based laboratory experiments are summarized under Table 8 (Fig. 1).

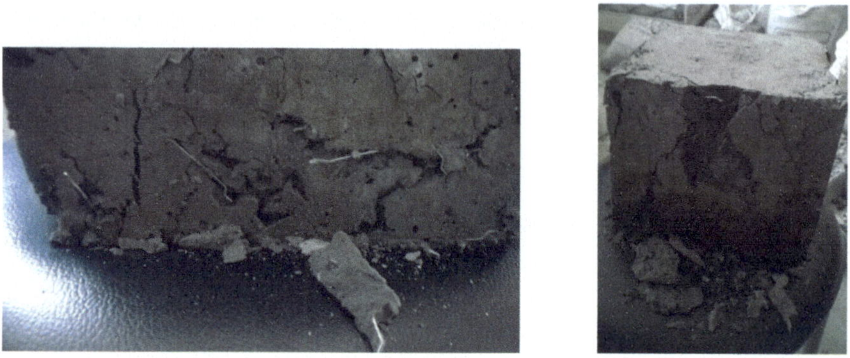

Fig. 1. Samples of Two – Dimensional Steel Fibres (Source: Author's contribution)

Table 8. Experimental Observations – Sustainable Steel Fibre Reinforced Concrete (Source: Author's contribution)

Specimen	Trial Mix	7th Day Compressive Strength (MPa)	28th Days Compressive Strength (MPa)	7th Flexure Strength (MPa)	28th Day Flexure Strength (MPa)
S0	-	30.31	44.01	3.13	4.78
S1	A1-B1-C1	33.44	43.91	3.29	5.15
S2	A1-B2-C2	35.28	50.95	3.38	5.55
S3	A1-B3-C3	41.45	59.25	3.66	5.09
S4	A2-B1-C2	32.44	50.17	3.24	5.08
S5	A2-B2-C3	34.87	49.76	3.36	5.08
S6	A2-B3-C1	51.47	68.87	4.09	5.99
S7	A3-B1-C1	45.69	23.48	3.85	5.67
S8	A3-B2-C1	51.71	67.78	4.10	5.93
S9	A3-B3-C2	53.21	70.72	4.15	6.11

5 Conclusion

The study incorporates the utilization of three different types of steel fibres: 1. Two – Dimensional; 2. Undulated and 3. Flat Steel, with consideration of addition of 2 percent by weight of the concrete, attained by concrete mix proportioning of M 25. Since the addition of Two – Dimensional Steel Fibres into concrete composite yields maximum magnitude in strength, the same sample had been considered for generation of a sustainable steel fibre reinforced concrete composite. With the application of Taguchi's Approach, it had been noted that the magnitude of strength has been increased by nearly 60 percent in compression and 30 percent in flexure after the prime components of the concrete are replaced partially in fraction of 30, 30 and 20 of FA, GGBS and RCA with cement, fine aggregates and coarse aggregates respectively. The future scope of the study is the assessment of the durability aspects of the sustainable steel fibre reinforced concrete composite.

Acknowledgments. No financial constraint has been considered in the study.

Conflict of Interest. The authors have no conflicts of interest to declare.

References

1. Xu, Y., Xu, T., Li, G., Xie, T.: Modeling smeared fiber–matrix pullout strength of steel fiber-reinforced concrete with various types of steel fiber and its tensile and compressive strength. Case Stud. Constr. Mat. **22**, e04242 (2025). https://doi.org/10.1016/j.cscm.2025.e04242
2. Yang, M., Huang, C., Wang, J.: Characteristics of stress–strain curve of high-strength steel fiber reinforced concrete under uniaxial tension. J. Wuhan Univ. Technol.-Mat. Sci. Edit. **21**(3), 132–136 (2006)
3. Zhang, Z., Xu, T., Castel, A.: Damage of non-steam-cured UHPC under axial compression with and without short-term sustained loading history. Structures **38**, 1066–1078 (2022)
4. Rambo, D.A.S., de Andrade Silva, F., Toledo Filho, R.D.: Mechanical behavior of hybrid steel-fiber self-consolidating concrete: Materials and structural aspects. Mater. Des. **54**, 32–42 (2014)
5. Bastien-Masse, M., Denarié, E., Brühwiler, E.: Effect of fiber orientation on the in-plane tensile response of UHPFRC reinforcement layers. Cement Concr. Compos. **67**, 111–125 (2016)
6. Gwon, S., Shin, M.: Direct-tensile and flexural strength and toughness of high-strength fiber-reinforced cement composites with different steel fibers. J. Asian Concrete Federat. **2**(1), 67–80 (2016)
7. Le Hoang, A., Fehling, E.: Influence of steel fiber content and aspect ratio on the uniaxial tensile and compressive behavior of ultra-high performance concrete. Constr. Build. Mater. **153**, 790–806 (2017)
8. Feo, L., Ascione, F., Penna, R., Lau, D., Lamberti, M.: An experimental investigation on freezing and thawing durability of high-performance fiber reinforced concrete. Compos. Struct. **234**, 111673 (2020)
9. Papachristoforou, M., Anastasiou, E., Papayianni, I.: Durability of steel fiber reinforced concrete with coarse steel slag aggregates including performance at elevated temperatures. Constr. Build. Mater. **262**, 120569 (2020)
10. Gao, D., Zhang, L., Zhao, J., You, P.: Durability of steel fibre-reinforced recycled coarse aggregate concrete. Constr. Build. Mater. **232**, 117119 (2020)
11. Moghadam, M.A., Izadifard, R.A.: Effects of steel and glass fibers on mechanical and durability properties of concrete exposed to high temperatures. Fire Saf. J. **113**, 102978 (2020)
12. Koushkbaghi, M., Kazemi, M.J., Mosavi, H., Mohseni, E.: Acid resistance and durability properties of steel fiber-reinforced concrete incorporating rice husk ash and recycled aggregate. Constr. Build. Mater. **202**, 266–275 (2019)
13. Wu, Z., Shi, C., Khayat, K.H.: Investigation of mechanical properties and shrinkage of ultra-high performance concrete: Influence of steel fiber content and shape. Compos. B Eng. **174**, 107021 (2019)

14. Meng, W., Khayat, K.H.: Effect of hybrid fibers on fresh properties, mechanical properties, and autogenous shrinkage of cost-effective UHPC. J. Mater. Civ. Eng. **30**(4), 04018030 (2018)
15. Xu, T., Yang, J., Wang, C., Guo, T., Deng, K., Xie, T.: Comparative sustainability and seismic performance analysis of reinforced conventional concrete and UHPC bridge piers. J. Clean. Prod. **467**, 142959 (2024)
16. Xu, T., Zhang, Z., Liu, Z., Bian, X., Zhou, Y., Deng, K.: Linear and nonlinear creep of UHPC under compression: Experiments, modeling, and verification. J. Build. Eng. **72**, 106566 (2023)
17. Zhang, Z., Leng, J., Bian, X., Xie, T., Xu, T.: A refined model for the flexural behaviour of reinforced UHPC members under sustained loading. J. Build. Eng. 109853 (2024)
18. Abbas, M.Y., Khan, M.I.: Fiber–matrix interfacial behavior of hooked-end steel fiber-reinforced concrete. J. Mater. Civ. Eng. **28**(11), 04016115 (2016)
19. Li, V.C.: A simplified micromechanical model of compressive strength of fiber-reinforced cementitious composites. Cement Concr. Compos. **14**(2), 131–141 (1992)
20. Hooda, Y., Bansal, S., Gupta, A.: Parametric strength of sustainable concrete using fly ash, GGBS and recycled aggregates as per Taguchi's approach. In: Advances in Construction Materials and Sustainable Environment: Select Proceedings of ICCME 2020, pp. 491–503. Springer Singapore, Singapore (2021)
21. Bureau of Indian Standards: IS 4031: Methods of physical tests for hydraulic cement. BIS, New Delhi, India (1996)
22. Bureau of Indian Standards: IS 2386: Methods of test for aggregates for concrete. BIS, New Delhi, India (1963)
23. Bureau of Indian Standards: IS 10262: Concrete mix proportioning—Guidelines. BIS, New Delhi, India (2019)

Review of Rubberized Concrete from Waste Tires for Sustainable Construction and Smart Infrastructure

N. R. Asha Rani(✉), Sheetal Thapa, G. V. Akhil, Tanmay Yedellu, and M. Akalesh

Department of Civil Engineering Alliance College of Engineering and Design, Alliance University, Anekal Bangalore, Karnataka 562106, India
asharaninr@gmail.com

Abstract. An innovative and sustainability method of reducing the environmental impact of concrete as it uses recycled material to replace a fine aggregate, waste tire rubber, without compromising on construction. In this review, the mechanical and durability performance of rubberized concrete is ascertained under compressive, tensile, flexural strength, and modulus of elasticity. Results are always found to be consistent showing that the increase of rubber replacement levels causes a decrease in strength because of lesser interfacial bonding tension, lesser stiffness, and porosity. However, the ductility, crack arrest, and energy absorption may be beneficial as the elastic nature of the rubber particles enhance ductility, crack arresting and energy absorption. Contributions of the durability area indicate higher values of water absorption and a sorptivity value with increased quantities of rubber, although shrinkage strains and freeze-thaw durability can be improved through the flexible and impervious properties of rubber. There are workability problems, great strength losses at high rubber levels and no standardized solutions to design and quality control. Future studies must concentrate in the fields of hybrid reinforcement systems, mechanical processing of rubber particles, thorough long-term field tests and life cycle analysis to properly measure sustainability advantages. In general, this review reveals the prospective of rubberized concrete as eco-efficient building material whereas, outlines the research and standardization gaps, which need to be filled up to implement this material practically.

Keywords: Rubberized concrete · waste tires · sustainable construction · durability · smart cities and infrastructure

1 Introduction

Concrete is the most common construction resource in the world, but its manufacturing cause significant environmental expenses because of excessive use of natural resources, especially aggregates. Since urbanization is on the rise, it is causing havoc on the ecological balance through loss of habitats, erosions along rivers, and source depletion due to the extraction of natural fine aggregates in riverbeds and quarries. At the same time,

L. Choudhary et al. (Eds.): TIMES 2025, SCI, pp. 25–38, 2027.
https://doi.org/10.1007/978-3-032-34636-0_3

rubber waste tires have become a huge environmental issue as the number of scrap tires continue to grow every year being a source of fire risks and space issues and of the chances of spreading a vector-born disease, when the waste tires are placed in landfills. The use of recycled waste rubber in concrete is a good approach in encouraging conservation of waste material diverting them out of landfills as well as decreasing the usage of regular aggregates(Gueboudji, 2024).

However, use of rubber particles as a replacement of fine aggregates in concrete or in other words crumb rubber or ground rubber creates the possibility of producing materials that are more environmentally conscious and resource wise (less material is produced through recycling of rubber in this way). This plan would not only prefer the principles of circular economy but also allow reducing the ecological impact of building concrete. Nevertheless, when fine aggregates are substituted with rubber, complex interactions are encountered due to possible influences in compressive, tensile, flexural properties of concrete. Adequate comprehension of the effects is important so that there are safe and feasible uses of rubberized concrete in both structural and non-structural components(Mostafaei et al., 2025).

This review attempts to study the current research situation on the mechanical characteristic of concrete with rubber instead of fine aggregate in a systematic way. The issues on the kinds of rubber to be employed, replacement percentages, consideration on the mixture design, and subsequent mechanical performance of rubberized concrete were discussed in this paper. The review will outline a plan of future research and practical application of rubberized concrete as a green building material by suggesting the knowledge gap and challenges, and opportunities of this strategy(Gomez et al., 2025).

1.1 Background of Sustainable Concrete and Waste Rubber Valorization

The global consumption of resources and environmental implication of concrete production are very high especially to its consumption of non-renewable aggregates. Meanwhile the concern of the disposal of wastes rubber, particularly the end-of-life tires, has turned out to be a vital environmental and human health issue as well as a million tons of used tires are produced every year. Such waste tires usually end up in landfills where they constitute a fire hazard, a breeding ground to disease vectors and a loss of precious land resources. The exploitation of rubber waste as a fine aggregate replacement in concrete appears as an attractive alternative to this twin-problem by promoting the doctrine of circular economy and environment-friendly constructional processes through the upcycling of harmful waste products to be used as a recycled secondary commodity(Petrescu et al., 2024). The incorporation of rubber pieces in the concrete works with the targets of the United Nations concerning waste outflows and sustainable and environmental-friendly construction materials(Gueboudji, 2024). Figure 1(a) shows the top 10 of the world rubber-producing countries in the year 2025. It is evident that the concentration of world production is led by Thailand (35%), Indonesia (23%), and Vietnam (10%) that contribute to about 68 percent. Other secondary producers are Ivory Coast (9%), India (6%), China (6%) and some localities such as Philippines, Cambodia, Malaysia and Myanmar contributing 2–3%. In this distribution, the extreme concentration of natural rubber in Southeast Asia is quite interesting since it underlines the opportunities and risks of regional dependence on supply, and Fig. 1(b), shows the top 10 of the world exporters of rubber waste in the year 2023. The United States contributes 25 percent of the total exports in the world with United Kingdom standing at 20 percent. France and Italy have a contribution of 10% and the rest is 9 percent in Germany. Other exporters, notable are Canada (8%) and the Netherlands (6%), and together Poland, Belgium and Spain make up less. Such statistics demonstrate the strong presence of developed economies in international trade in rubber wastes, North America and Europe have more than 80 percent of the rubber wastes exports. This trend suggests that the management of waste rubber is highly skewed in favor of highly industrialized countries implying that recycling and valorization of waste rubber is difficult to take place equitably around the globe.

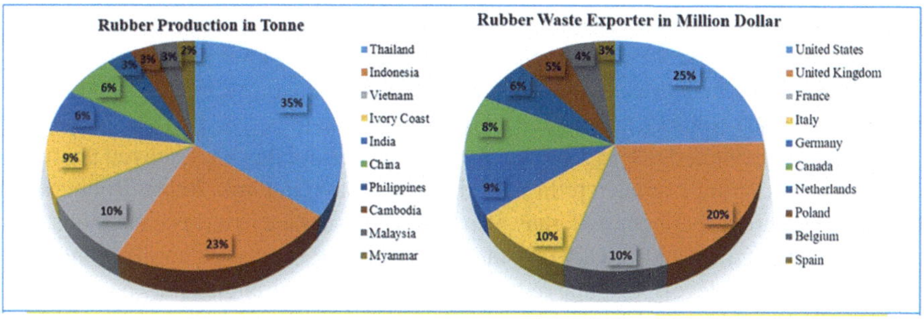

Fig. 1. (a) Top worldwide rubber manufacturers in 2025 (metric tonnes), and 1(b) Top largest rubber waste exporters of the world in 2023 (in million USD) (Source: Author's contribution)

1.2 Stress on Reducing the Usage of Natural Fine Aggregate

The natural fine aggregates, which are generally obtained after excavating riverbeds and in quarries are getting highly pressurized because of the increasing construction rates. Excessive mining of sand among other fine aggregates is undesirable, as it leads to environmental problems such as erosion of the riverbanks, loss of habitat, and depletion of the underground water. This environmental destruction is putting a threat to biodiversity and makes the surrounding people dependent on these ecosystems easy prey. So, it is now of great importance to find alternative materials capable of partly or wholly substituting the natural fine aggregates. The substitution of some fine aggregates by waste rubber serves two roles-One is that it will also help in preservation of natural resources whereas the second role is that it will also help in diverting the non-biodegradable waste rubber away to the landfills. These measures are a step towards a more balanced construction towards sustainable construction as it will ensure we do not over-exploit finite natural aggregate resources.

1.3 Purpose and objectives of the Review

The scope of the review aims to thoroughly investigate the mechanical performance of rubber as a substitute of fine aggregates in concrete. The review has summarized the available literature on application of rubber particles in concrete based on important performance weights like compressive strength, tensile strength, flexural strength and modulus of elasticity. The paper shall dwell into summarizing the type of rubber used in mixes, percentage of such rubber when mixed with cementitious matrices, and mechanical performance that will be evident in both laboratory tests and field tests. The review also outlines the challenges, knowledge gaps and research gaps that can be incorporated in further advancement of rubberized concrete to make it an environmentally friendly material of building. The paper will capture the evidence that surrounds this topic area, hence, will be utilized to inform research by scholars, engineers and policymakers interested in leading green-concrete technologies (Naresh et al., 2025).

2 Replacement of Finer Aggregates in Concrete Using Rubber

Given that recycled rubber may be a potential or complete replacement of fine aggregate in concrete, the recycled rubber has been attributed with a heavy level of concern attached to the prospects of solving two important questions of sustainability, namely excess supply of natural sand resources and the management of waste tires. The particle dimension of the rubber is acclaimed due to its elasticity and low density that has changed the fresh and hardened structure of concrete in a fundamental way(Olatunji, 2024). Rubberized concrete, when properly integrated, may also have an advantage in terms energy absorption, increased ductility and impact resistance, appropriate to specific applications. But at lower replacement levels, average reduction in mechanical strengths may be observed because rubbers are less stiff and has a weaker bond with cement paste(Eisa et al., 2025). It has been demonstrated that the physical and chemical traits of rubber, such as the size and shape of the particles, the texture of the surface and degree of cleanliness have a determinant characteristic on the way it behaves when in the concrete structure. Consequently, study on the interaction between the rubber particles and cementitious constituents is a prerequisite to optimal mix formulations. Also, the surface modification process, chemical pretreatment or Auxiliary cementitious additives have been researched to enhance the interfacial bonding and recover the loss of performance(Valliappan et al., 2025). If appropriately engineered, the use of rubber as a replacement of fine aggregates should make the concrete industry more sustainable since it would encourage the valorization of waste and minimize the use of natural aggregates(D'Alessandro, 2024).

2.1 Types of Rubber

Recycled rubber that has received the most exploration is Crumb rubber used as a substitute to fine aggregates in a concrete mix. Crumb rubber is commonly made by grinding waste tires into tiny granules of sizes less than 1 mm to 5 mm, which gives the crumb rubber a similar gradation to that of the natural sand resulting in the relative simplicity of its mix with concrete. The mechanical properties of the hardened concrete are, however, affected by the fact that the crumb rubber is elastic, hence causing low density and less stiffness compared to mineral aggregates. Besides crumb rubber, recycled tire rubber powder is also gaining more delving as an even smaller substitute substance. Tire rubber powder could be obtained by further grinding of the crumb rubber to a fine powder even in the micro scale to increase the dispersion and possibly binding between the cement paste. Studies reveal that the rubber-powdered surface area is low, thus the risk factor of weak interfacial zones becomes minimal though there is a possibility of adverse influence of rubber powder on workability and a higher water consumption(Luhar and Luhar, 2025). Another alternative, which is used in concrete is rubber in the form of shreds, which are produced by cutting scrap tires into pieces of irregular shapes by using a mechanical cutter. Its bigger particle size and rougher surface texture tends to enhance the mechanical interlock of the cementitious matrix though the hard edges means that it can cause non-uniform dispersion unless thoroughly mixed (Roy et al., 2025). In addition, the shredded rubber can incorporate increasing porosity and air entrapment due to which the mechanical behavior of concrete would also be altered. All in all, these alternative types of rubber give flexibility to the researchers and engineers to customize mixes according to the performance or sustainability goals.

2.2 Mix Designs and Proportions

The percentage of rubber as a substitute natural fine aggregate of a concrete has major influence on the fresh and hardened characteristics of a rubberized concrete. Replacement levels of up to and including 50 percent by the volume of the fine aggregate is customary to the best of literature. The lower replacement rates, on the one hand, like 5 percent or 10 percent will not make much impact on compressive strength, but higher rates will result in larger decline in the mechanical properties because of the low modulus of elasticity and hydrophobicity of rubber(Furtado et al., 2024). Technologists have investigated optimization of these levels of replacement by options between sustainability advantages and permissible losses to strength to render the structure and non-structural applications to become financially viable (Dheyaaldin et al., 2024). Mix design and dispersion methods also become an important technique for homogeneous performance. Both wet pre-mixing and dry mixing have been attempted by scientists to improve the dispersion of rubber particles in the cement matrix. They are usually dry mixed (where the rubber particles are mixed with the aggregates before adding the water and cement to the aggregate) or wet pre-mixed; the latter could involve soaking or washing of the rubber, to remove any surface contaminants to ensure better bonding to the cement paste. Also, superplasticizers and other chemical admixtures can be added to overcome the restricted work properties of the addition of the rubber; and supplementary cementitious materials can be added to improve the interfacial transition zone between the particles of rubber. Such mix design innovations play an important role in ensuring usable workability in addition to performance and durability of rubberized concrete.

3 Rubberized Concrete- Mechanical Properties

3.1 Compressive Strength

Compressive strength of concrete is one of the main factors which are used to illustrate the performance of concrete, and the rubberized concrete has a tendency of reducing in compressive strength when compared with the conventional mixes. The presence of rubber powder whose stiffness and deformability are lower compared to mineral aggregates contributes to increased porosity and reduced interfacial bonding in cementitious matrix(Sangeetha et al., 2024). This continuously grows with selected rubber filler levels, and the strength losses contributed to be among 10 to 60 percent on the level of replacement, the size of the particles and the mix design. The results have always been that composites are less in compressive strength the higher the percentage of the rubber content, but at a point of replacement hugely low (5–10 percent), these drops can be minimal and can be acceptable to be applied in non-structural areas. Rubberized concrete mixes usually have adequate compressive strength to be used in secondary structural components or paving purposes, as compared to concrete mixes without rubber (heat-cured control mixes) with the provision that ratios of replacement are carefully regulated(D'Alessandro and Ubertini, 2024).

3.2 Tensile Strength

The addition of rubber aggregates also influences tensile strength especially on splitting tensile strength. The lesser adhesion of the cement paste and the hydrophobic rubber decreases the tensile stress transfer ability of the matrix to ensure decline in tensile strength. The size and distribution of rubber particles, however, determine the extent of such reduction. Studies have indicated that small rubber particles have a superior uniform dispersion as well as having fewer stress concentrations that enhance the tensile characteristics as compared to the coarse rubber additions. The rubberized concrete also experiences additional strain capacity, as well as resistance to tensile loading cracking, despite the reduction in values of absolute tensile strength than in control mixes, which is useful in structures prone to impact(Battista et al., 2021).

3.3 Flexural Strength

Rubberized concrete also exhibits flexural strength like its collapse wherein its flexural strength decreases relative to control mixes, mostly due to the decrease in stiffness and the bad bonding relationship between the rubber and cement. Concrete higher in rubber contents under bending loads also shows increased ductility with the ability to deform before collapse, a desirable trait under energy absorption requirements. Some lower values of flexural strength are lost as a percentage of up to 50 depending on the content of the rubber, particle size and preparation of the mixture. Rubberized concrete has nevertheless displayed promise in non-load bearing components in which the flexural ductility and resistance to impact is a higher priority than ultimate strength(Indraratna et al., 2023).

3.4 Modulus of Elasticity

The rubberized concrete therefore has much less modulus of elasticity than the ordinary concrete and this demonstrates the great elastic property of the rubber grains. Such stiffness decreases to the fact that rubberized concrete is more compressible to loads that could restrict its usage in components that experience high loads. Modulus of elasticity has been found, generally by investigators, to be inversely proportional to the quantity of rubber added to the mixes with the loss as high as 60 percent observed in high-rubber-content mixes. Nevertheless, in cases where damp/absorption of the vibration is high such as street guard wall or shock-resistant roads, this low modulus can be useful. It will be necessary that there are clear relationships between rubber content, modulus, and overall performance so that safe design practices can be supported on using rubberized concrete(Indraratna et al., 2023; Manggapis et al., 2023). Figure 2 shows the change in strength retention of rubberized concrete at 0-30replacement levels. A noticeable downwards trend can be seen in compressive strength, tensile strength, flexural strength and modulus elasticity of the specimens as the number of rubber replacement increases. An example would be where the strength losses at 10% replacement are not very large (10–15%) whereas those experienced at 30% replacement can exceed 40–60% depending on the property in question. These results show that the Rubber particles, due to its reduced rigidity, and a break at low values, and weak interfacial bonding are liable to reduce load bearing capacity but at the same time provide better ductility and energy absorbing capabilities.

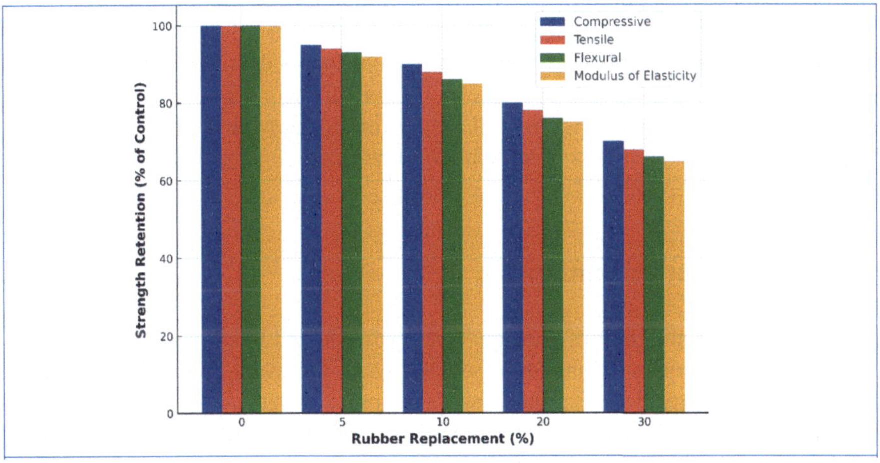

Fig. 2. Retention of the strength of the rubberized concrete containing varied percentage replacement of the fine aggregate by waste rubber. (Source: Author's contribution)

4 Failure mechanism and Microstructure

The microstructure of rubberized concrete is unique as compared to that of conventional concrete because of existence of rubber particles, softer and less dense phase within the cementitious continuum. The bonding of the rubber cement interface is one of the most important critical parameters that attract such a large part of control on the mechanical performance of rubberized concrete. As opposed to natural aggregates, rubber particles are hydrophobic, and they are also described as having low surface energy that means that they will not be adhesive to the cement paste around. This faulty interfacial transition zone (ITZ) forms concentration points of stress and it may permit introductions of microcracks within a load. It has been identified that there are micro-gaps existing at the interface of rubber cement which is normally filled with trapped air or even unhydrated cement hence contributing more strength degrading factors to the composite. Blocked outcomes of scientists or investigators who have experimented with different methods, such as chemical process or washing to alterability of the rubber surface to become more wettable and compatible in terms of bonding also come as an encouragement. One of the greatest challenges that is, however, experienced in ensuring the high mechanical performance of the rubberized concrete is the interface. Also, important roles in the porosity and crack propagation patterns are the presence of rubber. It is likely that the porousness of the concrete will be more pronounced by the fact that rubber has less density and elastic modulus as compared to the conventional aggregates because the rubber particles can retain air and bind during the casting process. Additional porosity can also reduce strength and increase permeability and result in a system experiencing increased durability issues over time, e.g. carbonation or chloride ingress. Considering the crack propagation, the soft rubber particles are considered as the crack stopper and are useful in increasing the tortuous path of cracks and preventing formation of major cracks. The special crack-bridging mechanism as well as the augmented injury absorption and post peak deformability, will come into play in the impact-sensitive or vibration damping. However, it has such tortuous microcracks; hence, it is possible to talk about rubberized concrete as being acted more like a ductile composite instead of brittle, and the failure mode is entirely different than under conventional mixes. These micro structural relationships will play a role to know and maximize them in subsequent development of the structural use of rubberized concrete in more demanding uses (Singaravel et al., 2024).

5 Durability Considerations

Durability is an essential part of concrete's performance and longevity, and the cost of structural rehabilitation is a function of durability. In the event of the application of rubber, as a fine aggregate substitute, water absorption effects are great. The additive particles hydrophobic properties may at first seem beneficial to minimize water uptake. However, the hydrophobic rubber particles are elastic and have a tendency of trapping air while mixing, thus increasing total porosity within the solidified matrix. Consequently, high water absorption of rubberized concrete has been recorded in most studies surpassing that of control mixes, which may have an influence on durability under a moist environment. Partial replacement, better mixing techniques and addition of pozzolanic materials have been found to mitigate these effects through bringing water absorption levels to acceptable levels(Kumari et al., 2024). Figure 3 shows a comparison of the water content absorbed and sorptivity of the concrete with the various proportion of rubber in it. It indicates that the two properties are likely to rise as an increase in the quantity of rubber is used in it, thereby signifying high porosity and permeability.

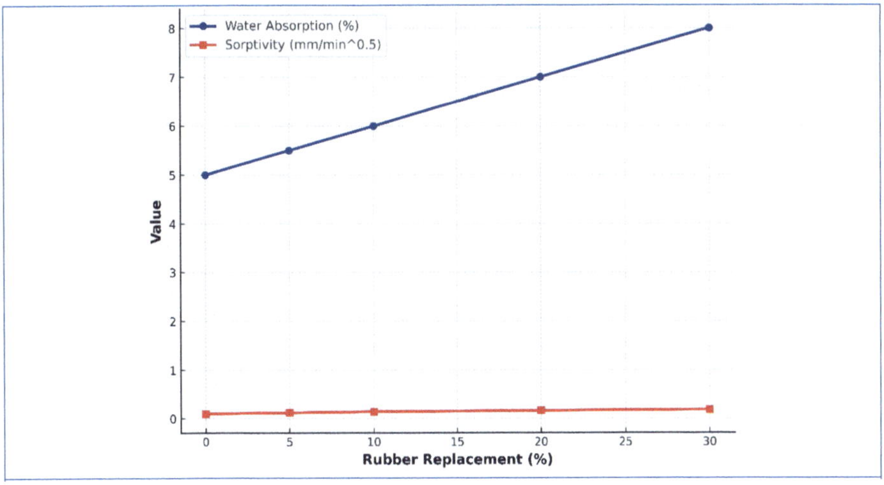

Fig. 3. Water Absorption and Sorptivity versus rubber Replacement. (Source: Author's contribution)

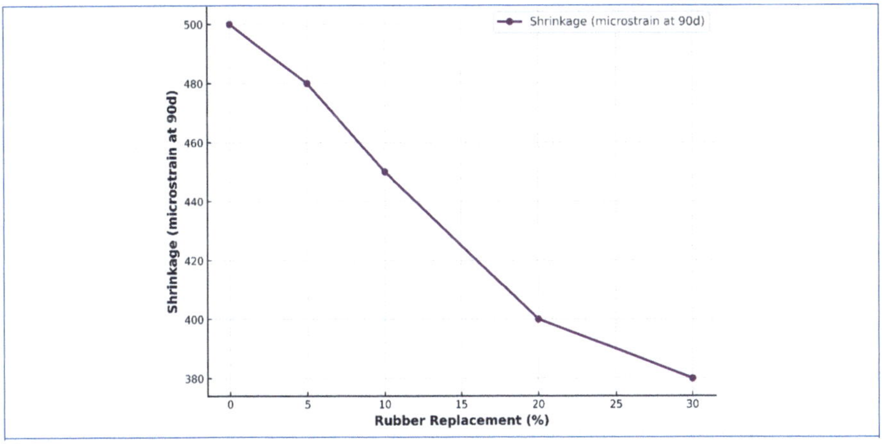

Fig. 4. Shrinkage versus Rubber replacement. (Source: Author's contribution)

Other parameters that are significant for durability and depend on the content of rubber are shrinkage. Concrete usually suffers from drying shrinkage, which causes loss of water after pouring and results in cracking and low performance. Rubberized concrete tends to have reduced shrinkage strains as opposed to standard concrete. This is credited to the rubber particles which behave in the form of internal stress relievers and can absorb shrinkage induced stresses in the cementitious matrix. Thus, inclusion of rubber may reduce the risk of shrinkage-induced cracking, but lower tensile strength of rubberized concrete is still capable of necessitating respective assessment of the structure(Manggapis et al., 2023; Singh et al., 2023). Figure 4 shows drying shrinkage for different amounts of rubber replacements at a range of 90 days. The evidence indicates that the inclusion of rubber can potentially allow abating strains related to shrinkage, proving its contribution to the minimization of the cracking possibilities.

Freeze-thaw performance of concrete with rubber aggregates is also one which has been tested in literature. Rubber particles enhance flexibility of concrete which enables it to obtain better resistance to internal stresses due to freeze-thaw cycling. The rubber additions can also act like air-entrainment but the air voids created are able to relieve the pressure in case of ice formation, and hence structural failure by cracking internally, can be drastically reduced. But this advantage is conditional to the attainment of a dense matrix around the rubber to prevent the channels of aggressive agents such as chlorides or sulfates that may enhance other facets of degradation. In general, although freeze-thaw signs are encouraging among rubberized concrete, further field testing of its long-term behavior under various exposure circumstances is nevertheless in order(Radwan & AbdelRaheem, 2025). Figure 5 indicates the link between the percentage of rubber change and mass loss regulated by the freeze-thaw cycles. The mass loss rate of rubberized concrete is slightly high, indicating that there are moderate durability issues in cyclic freezing weather.

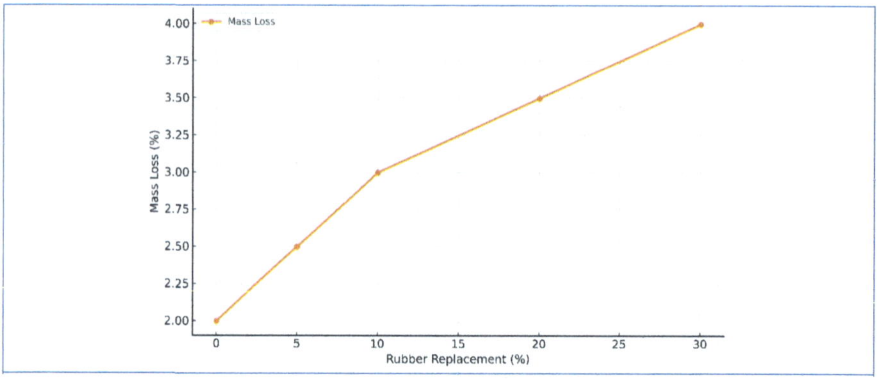

Fig. 5. Mass Freeze-Thaw Loss vs. Rubber Replacement. (Source: Author's contribution)

6 Challenges and Limitation

6.1 Problem with Workability

The main disadvantages related to the usage of rubber as fine aggregate substitute include the adverse effect of rubber on the properties of fresh concrete. The rubber particles tend to be lightweight, have irregular shapes and hydrophobicity that interferes with the cohesive flow of the cement paste and unsolicits the need for more mixing water. This has the tendency of causing segregation or the uneven distribution of rubber in the mix resulting in uneven areas or zones that defeat performance. To enhance satisfactory flowability and compaction, large concentrations of super plasticizers or viscosity modifying admixtures may be necessary to achieve this, this increases the cost and makes mix design processes difficult. Additionally, the effects of rubber inclinations to trap air also decrease the workability and require advanced mixing methods and, in other situations, rubber particles being pre-treated to make them suited to cement paste(Ince et al., 2022).

6.2 The Reduction of Strength at High Levels of Replacement

Loss of strength is another widely recorded disadvantage of rubberized concrete, especially in high replacement levels. Compressive strength, tensile and flexural strength have the propensity of reducing with increase in rubber content in that they depend on the low modulus of elasticity and low bond strength in the rubber cement interface. Very large percentage of rubbers or when they exceed 20 percent by volume of fine aggregate, they may cause serious losses of compressive strength, on some instances more than 50 percent, and thus can only be used in non-structural or secondary construction. Weak interfacial transition zones, large amounts of air void, and porosity by the elastic and hydrophobic nature of rubber particles have been attributed to lack of the reduction in stiffness and load bearing capacity. Other strategies like using rubber together with additional cementitious materials or using fiber reinforcements have yielded encouraging results toward full compensation for the loss of strength at high rates of rubber replacement(Amran, 2025).

7 Gaps in Standardization

The main constraint to the proliferation of rubberized concrete is that there are currently no common testing procedures as well as design specifications particularly regarding these mixes. Majority of the existing concrete codes and standards depend on the classical mineral aggregates that fail to consider unusual performance of rubber particles in respect to bonding, durability, and deformation properties. Consequently, engineers and practitioners have become confused on how they can design and specify rubberized concrete reliably in different applications. There is also the problem of lack of standardization that makes certification and approval unfriendly to the application of rubberized concrete in the industry although it has great potential in sustainability. Establishing overall standards and best practices that facilitate addressing material selection, mix design, product performance assessment, and quality control will be vital in establishing rubberized concrete into mainstream in the construction sector(Indraratna et al., 2023; Nasser Eddine et al., 2023; Radwan and AbdelRaheem, 2025).

8 Future research Directions

The most important direction of rubberized concrete research in the future is the investigation of hybrid resisting mixes, i.e., the use of rubber mixes with additional reinforcements in terms of steel or polypropylene fibers. Although rubber on its own has a disadvantage of mechanical strength, the combination of fibers can offset that disadvantage through the tensile strength, flexural capacity and crack control. The hybrid mixes may provide a perfect mix of ductility, toughness and durability with a maximum reuse of waste rubber in a sustainable fashion. Exploration of the synergetic effect of the combination of various waste-based additives to form well-engineered hybrid systems is one of the avenues of pursuit of high-performance eco-efficient concrete that can be applied on a larger scale on structure. The other important research priority is enhancement of the surface properties of rubber particles in terms of either making chemical- or mechanical treatment. Poor bonding between the rubber and the cement can be enhanced by pre-treatment that will help include washing, grinding, the use of plasma, or coating with pozzolanic materials. Improved interfacial transition zone can be obtained by better bonding which will result in better compressive, tensile and flexural behavior. The formation of cost effective, scalable and environmentally friendly processes of surface treatment will also be necessary in ensuring the practical implementation of rubberized concrete in the construction sector. Besides, it is essential to carry out long-term studies on the performance of such concrete as rubberized concrete to gain confidence in using rubberized concrete. Most of the existing studies are conducted based on the laboratory testing of a limited time-range prior to the controlled conditions. Nevertheless, to confirm the performance forecast, field tests that last longer and those that are carried out under a different weather, load, and exposure conditions are needed. This ought to be conducted to determine long-term durability, shrinkage, resistance to freeze-thaw strength, penetration of chloride and possibility of microstructural degradation giving a deeper insight on how rubberized concrete behaves in a real-life setting. Lastly, the incorporation of the life cycle assessment (LCA) in the research on rubberized concrete will provide an account of the environmental advantages and trade-offs of the product. Although it is obvious that the process of rubber substitution saves wastes and natural resources, energy, emission, and sustainability elements of rubber mining, rubber processing, transportation, and concrete manufacturing require further investigation. The use of LCA tools will enable researchers, policymakers and practitioners to determine the extent to which rubberized concrete can be said to achieve green constructive targets along its entire life cycle, including in the construction (production) and when a recycling or a disposal option is involved. This will facilitate the process of evidence-based decisions and standardization of the material to be accepted by the wider community.

9 Conclusion

This review has discussed the mechanical and durability responses of the concrete which has been prepared by using waste tire rubber as a partial replacement of fine aggregate. Varied amounts of compressive strength loss have been experienced in the studies reviewed with 10 to 60 percent decrease in compressive strength being noted depending on the mixing design, particle size and level of replacements. Likewise, tensile and flexural strengths tended to decline by 20350%, and modulus of elasticity declined by as much as 60% in high-rubber contents mixes. Such losses were attributed to the weak interfacial bonding, higher porosity and low stiffness of rubber and mineral aggregates.

The rubberized concrete has a few merits despite such loss of strength. Increased ductility, energy absorption and crack propagation resistance were always reported to make it applicable in areas that require impact and vibration resistance. Regarding the durability, the water absorption and sorptivity were also higher when higher rubber was used due to increased porosity. Shrinkage strains were decreased by about 15–20%, freeze thaw performance enhanced up to a 20% replacement level due to elastic and pressure relieving characteristics of rubber.

Actual areas of challenge persist, worse workability, loss of strength at more than 20% replacement and the lack of standardized design codes. The solutions to these challenges will involve research in hybrid reinforcement techniques (such as rubber/fibers), modification of rubber particle surfaces to enhance bonding, and exposure to scientific trials under a variety of conditions. Also, life cycle assessment (LCA) is necessary to quantify the sustainability benefits and to inform them at the policy level whether they should be adopted. The study in summary shows that although rubberized concrete can never supersede conventional concrete in major load bearers, it is of great potential as a sustainable and eco-efficient material in non-structural and secondary structures purposes, thus serving the activities of the circular economy and climate-resilience infrastructure.

References

Amran, M.: Innovative technology for converting automobile tire waste bead wires into recycled steel fibers for sustainable concrete composites: insights for the Al-Kharj Governorate construction industry. Int. J. Build. Pathol. Adap. **43**(4), 665–692 (2025). https://doi.org/10.1108/IJBPA-03-2024-0067

Battista, M., Gobetti, A., Agnelli, S., Ramorino, G.: Post-consumer tires as a valuable resource: review of different types of material recovery. Environ. Technol. Rev. **10**(1), 1–25 (2021). https://doi.org/10.1080/21622515.2020.1861109

D'Alessandro, A.: Smart sustainable concrete materials and structures. In: Sustainable Concrete Materials and Structures, pp. 601–624. Elsevier (2024). https://doi.org/10.1016/B978-0-443-15672-4.00019-X

D'Alessandro, A., Ubertini, F.: Innovative and sustainable concrete materials. In: Advance Upcycling of By-Products in Binder and Binder-Based Materials, pp. 373–396. Elsevier (2024). https://doi.org/10.1016/B978-0-323-90791-0.00027-5

Dheyaaldin, M.H., Bayraktar, O.Y., Öz, A., Kaplan, G.: Sustainable Use of Waste Tire Rubbers in Eco-Friendly and Lightweight Alkali-Activated Slag–Silica Fume Mortars. J. Mat. Civil Eng. **36**(12) (2024). https://doi.org/10.1061/JMCEE7.MTENG-18152

Eisa, A.M., Tahwia, A.M., Osman, Y.A., Elemam, W.E.: Characteristics of bacteria based self healing rubberized concrete for sustainable and durable construction. Sci. Rep. **15**(1), 14758 (2025). https://doi.org/10.1038/s41598-025-97174-1

Furtado, L.S., Bessa, I.S., Soares, J.B.: Integrating Smart City Technologies for Sustainable Pavement Infrastructure. In: 14th International Conference on Asphalt Pavements ISAP2024 Montreal, pp. 273–277. Springer Nature Switzerland (2024). https://doi.org/10.1007/978-3-031-67252-1_46

Gomez, C. Y., Ronald, L. G., Andry, P.-O., Jonathan, R.F.: Literature review: rubber waste in tire manufacturing. J. Phys: Conf. Ser. **3028**(1), 012036 (2025). https://doi.org/10.1088/1742-6596/3028/1/012036

Gueboudji, Z.: From Trash to Treasure: Transforming Landfill Tyres, Plastics, and Glass into Sustainable Construction Resources, pp. 295–313 (2024). https://doi.org/10.1007/978-3-031-55665-4_12

Ince, C., Shehata, B.M.H., Derogar, S., Ball, R.J.: Towards the development of sustainable concrete incorporating waste tyre rubbers: A long-term study of physical, mechanical & durability properties and environmental impact. J. Clean. Prod. **334**, 130223 (2022). https://doi.org/10.1016/j.jclepro.2021.130223

Indraratna, B., et al.: Smart Solutions in a Circular Economy for Advancing Railroad Design and Construction Using Recycled Materials, pp. 1–32 (2023). https://doi.org/10.1007/978-981-99-1121-9_1

Kumari, N., Pandey, S., Kumar, G.: Sand Mining: A Silent Threat to the River Ecosystem. In: Rivers of India, pp. 109–132. Springer International Publishing (2024). https://doi.org/10.1007/978-3-031-49163-4_6

Luhar, S., Luhar, I.: Synergistic integration of recycled materials for enhanced self-sensing concretes: a scientific exploration towards intelligent and sustainable infrastructure. In: Binding Materials for Sustainable Construction, pp. 497–528. Elsevier (2025). https://doi.org/10.1016/B978-0-443-26566-2.00014-3

Manggapis, F.F., Kumar, S.D.A., Lucena, J.R.P.G., Carabbacan, A.P.I., Cruz, O.G. Dela: Advancements in Concrete Incorporation: Harnessing the Potential of Crumb Rubber Tires as Sustainable Alternatives to Fine Aggregates, pp. 195–205 (2023). https://doi.org/10.1007/978-3-031-42588-2_16

Mostafaei, H., Kelishadi, M., Bahmani, H., Wu, C., Ghiassi, B.: Development of sustainable HPC using rubber powder and waste wire: carbon footprint analysis, mechanical and microstructural properties. Eur. J. Environ. Civ. Eng. **29**(2), 399–420 (2025). https://doi.org/10.1080/19648189.2024.2400485

Naresh, T., Murthy, N.R.D., Mallik, M., Singh, V.: Performance assessment of concrete incorporating waste rubber powder as partial cement replacement. J. Build. Pathol. Rehabilit. **10**(2), 138 (2025). https://doi.org/10.1007/s41024-025-00646-0

Nasser Eddine, Z., Barraj, F., Khatib, J., Elkordi, A.: From waste to resource: utilizing municipal solid waste incineration bottom ash and recycled rubber in pervious concrete pavement. Innov. Infrastr. Solut. **8**(12), 319 (2023). https://doi.org/10.1007/s41062-023-01289-1

Olatunji, O.: Plastics in construction: toward green buildings and climate-resilient cities. In: Re-envisioning Plastics Role in the Global Society, pp. 219–239. Springer Nature Switzerland (2024). https://doi.org/10.1007/978-3-031-48945-7_12

Petrescu, M.G., Tănase, M., Dumitru, T.: Sustainable Technologies for Road Maintenance and Rehabilitation, pp. 3–28 (2024). https://doi.org/10.1108/S2043-052320240000023001

Radwan, I.M., AbdelRaheem, M.: Construction of electrically heated pavement using recycled waste rubber tires. J. Transport. Res. Board **2679**(2), 1862–1878 (2025). https://doi.org/10.1177/03611981241273118

Roy, R., Dutta Roy, B., Haldar, P., Modak, N.: Assessment of waste rubber- epoxy composites: A sustainable solution. Proc. Inst. Mech. Eng. C J. Mech. Eng. Sci. (2025). https://doi.org/10.1177/09544062251358072

Sangeetha, S.P., Nair, V.M., Suriya, Pa., Divahar, R., Aravind Raj, P.S.: Flexural Behaviour of Concrete Beams Embedded with PVC Pipe Sandwiched with Waste Crumbed Rubber, pp. 97–107 (2024). https://doi.org/10.1007/978-3-031-50024-4_10

Singaravel, D.A., Veerapandian, P., Rajendran, S., Dhairiyasamy, R.: Enhancing high-performance concrete sustainability: integration of waste tire rubber for innovation. Sci. Rep. **14**(1), 4635 (2024). https://doi.org/10.1038/s41598-024-55485-9

Singh, P., Goli, V.S.N.S., Singh, D.N.: Current practices and the way forward in waste rubber utilization for infrastructural and geoenvironmental engineering applications. Environ. Technol. Rev. **12**(1) (2023). https://doi.org/10.1080/21622515.2023.2252164

Valliappan, V., Vijayasimhan, S., Sumesh, M., Gautam, Ramaraju, H.K.: Evolvement and future direction of research on use of waste tires in geo-engineering practice: a systematic literature review. Asian J. Civil Eng. (2025). https://doi.org/10.1007/s42107-025-01441-5

Comparative Analysis of Biological Self-Healing Mechanisms in Concrete: A Review of Existing Methods and Their Effectiveness

T. Senthil Vadivel(✉), Abha Maheshwari, and Sayandip Ganguly

Department of Civil Engineering, Manav Rachna International Institute of Research and Studies, Sector 43, Faridabad 121004, India
tsnsenthu100@gmail.com, {abhamaheshwari.sod, sayandipganguly.set}@mriu.edu.in

Abstract. Deterioration of any concrete structure generally initiates with the formation of microcracks. To increase the longevity of the structure, it is therefore necessary to cover such types of cracks at the initial stage of their growth. Self-healing concrete offers a possible solution in this direction. Among other self-healing methods, biological self-healing mechanism-based concrete has been extensively investigated in recent years. Where traditional maintenance either seems to be costly or requires timely evaluation of the structure, biological self-healing concrete proves to be a better alternative. Bacteria, fungi, algae, etc., are used in the manufacturing of biological self-healing concrete. In the presence of air in microcracks, such biological elements become active from their dormant state. It develops some residue of reaction and fills the gap of microcracks. The chemical deposit over the surface of microcracks thus resists further deterioration of concrete. However, the practical application and suitability of different bacteria depend on many factors, which are still a concern of the researcher. To advance and accelerate the progress of self-healing concrete in the construction industry, the present work performs a comparative review between different existing biological self-healing mechanisms. It is observed that in terms of improving concrete qualities and closing cracks, the use of bacteria is superior. Fungal and algal methods, on the other hand, have shown promising results. However, it needs further optimization. The review also identifies the key areas of research for future study.

Keywords: Biological Self-Healing Concrete · Sustainability · Durability · Microbial Concrete Repair

1 Introduction

Concrete is one of the most widely used building materials in the world. Considering its brittle behavior, several measures are taken to impart ductility to structural members. Use of steel is one such measure. However, the occurrence of microcracks in members poses a threat to both concrete and embedded steel. Treatment of such surfaces is therefore necessary to restrict the deterioration of the member's structural load-carrying capacity.

L. Choudhary et al. (Eds.): TIMES 2025, SCI, pp. 39–50, 2027.
https://doi.org/10.1007/978-3-032-34636-0_4

Ingress of water or air may cause reinforcing steel to corrode and shorten the structure's lifespan (Jonkers et al., 2010). Epoxy injections and surface treatments are two examples of traditional restoration procedures; nevertheless, they may be time-consuming and expensive and may not return the concrete to its former qualities. A sustainable and efficient alternative to traditional repair processes, self-healing concrete technologies have therefore garnered increasing attention (Nair et al., 2022).

Major economic powers like China and the US face the formidable problem of an aging infrastructure that needs substantial funding for rehabilitation and repair. Nearly half of America's bridges are 50 years old or older, and fixing and rehabilitating them will cost more than $125 billion right now (Wiktor and Jonkers, 2011). So, in the last few decades, there have been major repair techniques implemented, such as using alkali-activated materials as a binding agent, adhesives, polymeric resins (e.g., epoxy, polyester, and vinyl ester), and autogenous healing of unreacted cementitious particles (e.g., in ultra high-performance concrete structures (Granger et al., 2007). However, there are many cases where choosing the best repair method is difficult due to factors such as ineffectiveness in terms of cost (Li et al., 2019), negative effects on the environment (Chahal et al., 2012), the absence of a consistent and uniform interfacial transition zone (ITZ), problems with delamination (particularly in polymer-based materials) (Wei et al., 2010), and difficulties in actual on-site applications (e.g., dealing with liquid activator for alkali-activated materials (Vijay et al., 2017).

Coating concrete surfaces with a protective layer increases their longevity, and new studies on the use of microbiological calcification for self-healing have showed encouraging results. When bacteria are introduced to concrete, they break down urea and calcium into calcium carbonate ($CaCO_3$). This process is four times more successful than standard autogenous healing procedures in sealing fractures up to 0.46 mm wide. In addition to reducing porosity and repairing micro-cracks, the mechanical characteristics of concrete are greatly improved by bacterial self-healing. In general the pH of concrete remains very high, which makes it ideal for the use of alkaliphilic bacteria. It is generally applied to concrete through the encapsulation method or by the method of spray, injection, or direct addition. The present comparative review is prepared after exploring almost 40 research publications. The aim of this study is to determine the challenges in the application of biological self-healing methods, their life cycle cost, and their effectiveness in real-scale applications. Previous research has individually contributed for the development of different self-healing methods. On the other hand, the present work focuses on summarizing the effect of different healing mechanisms on the mechanical property, durability and sustainability of the concrete.

Objective of this study: The primary objective of this study is to identify the factors that influence the selection of biological self-health methods. Along with that, it also investigates the efficiency of different methods, constraints to application, and future scope of research. In this way the review showcases the advancement in concrete technology as well as opens up the path towards sustainability.

Significance Of the Study: This extensive review can significantly contribute to advancing the construction industry towards sustainable development. This study indicates the suitability of using different microorganisms as self-healing mechanism activators, which in turn facilitates increasing the life span of the structure and durability performance even in adverse environmental conditions. Life cycle cost analysis is another approach that optimizes the selection of different repairing and retrofitting techniques. As self-healing concrete is reviewed as one of the alternatives to frequent evaluation for decisions regarding repairing, life cycle cost analysis of such concrete will provide a robust framework for the decision-making process.

2 Literature Review

2.1 Historical Background

Self-healing mechanism was introduced in the concrete in the early 1900s to naturally cover the cracks in the concrete surface. Since then, several techniques have been developed to efficiently apply self-healing methodology in the concrete. Autogenous healing was one of the initial methods derived for this purpose (Li et al., 2023). In this method, the natural characteristics of cementitious material were considered to heal small fractures in the concrete surface. A biological self-healing method is then proposed with additional advantages over the autogenous healing method. Its successful implementation attracts researchers to investigate the efficiency of different microorganisms for this purpose. In the late 20th and early 21st centuries, the research gained momentum with the motive of shifting the trend towards the development of sustainable construction. With an understanding of biological self-healing processes to seal cracks with calcium carbonate deposit (Rui and Qian, 2022), it has been possible to extend the life span or durability of the concrete structure. It inspired the researcher to add bacteria or other microorganisms (Javeed et al., 2024) into concrete mixes for improved properties both in terms of mechanical properties and environmental sustainability.

2.2 Current State of Research

Over the last few years, the biological self-healing concrete has drawn significant attention from the researchers. Several methodologies have been proposed and investigated both numerically and experimentally. The process of introducing microorganisms such as Bacillus and Sporosarcina species has also been successfully experimented with in filling up microcracks as small as 0.5 mm (Jogi and Lakshmi, 2021). Encapsulation of bacteria into concrete is gradually improved with more focus, as it allows them to survive and activate in a variety of environments (Pupa et al., 2021). In addition, an experiment is also conducted combining biological self-healing concrete with green concrete or recycled aggregate. It showed an improvement of structural performance and greater sustainability than conventional concrete (Nilimaa, 2023). The present review thus highlights the potential of biological self-healing concrete in shaping future research with its greater applicability and in improving the structural performance in natural processes.

2.3 Thematic Analysis

An analysis on different themes that emerged from the previous studies on biological self-healing concrete is also reviewed. Incorporation of different living entities such as bacteria, fungi, and algae in concrete is one of the primary themes. The analysis is performed to understand the effectiveness of each strain and the mechanism they follow to heal the cracked surface or to form the calcium carbonate deposit (Pandit et al., 2022). Selection of suitable methods to introduce the microorganisms into the concrete is considered as another theme. Methods like encapsulation, spraying, injection, and direct addition are studied in different environments. It reveals their ideal environment and other factors for use in concrete as a self-healing element (Vemmer and Patel, 2013). The effect of self-healing concrete on durability, efficiency in healing cracks, and environmental factors are also separately examined (Mohammed et al., 2024). On the other hand, the combination of biological self-healing concrete with existing sustainable practices such as green building or recycled aggregates is also studied (Peng et al., 2023). All such themes indicate the progress, unsolved objectives, and pathway for future research.

2.4 Bacteria Based Method

Cracks in bacterial self-healing concrete may be filled by bacteria that make calcium carbonate ($CaCO_3$). Bacterial techniques have been shown to increase compressive strength by as much as 30% and successfully seal fractures as broad as 0.5 mm (Elmenshawy et al., 2024). Different bacterial strains' efficacies are summarized in Table 1. Type of bacteria, manner of integration, and concrete mix all have a role in how well bacterial self-healing treatments work. A study conducted by (Etesami et al., 2023) found that bacterial survival and activation were both increased with the use of encapsulation strategies. The effectiveness, however, might be impacted by variables including the quality of the bacterial strains and the ambient conditions inside the concrete matrix. Many construction and infrastructure projects have found success using bacterial self-healing concrete. In settings where cracking and degradation are common, it shines. (Prasad et al., 2023) found that commercial applications have shown its usefulness in real-world circumstances.

Utilizing a biotechnological approach predicated on calcite precipitation considerably improves structural concrete's compressive strength. For higher-grade concrete (e.g., 50 MPa), adding bacteria such as Bacillus megaterium and Sporosarcina pasteurii may boost the compressive strength by as much as 24%. There was a 20% increase in compressive strength in fly ash and bacterial concrete, and there was an additional 20% increase in strength in silica fume concrete as a result of calcium carbonate precipitation. The porosity and permeability of cement mortar are decreased by bacterial calcite precipitation, which increases its compressive strength and durability. By limiting water absorption and chloride ion penetration, bacteria fill the pores with calcium carbonate, helping the concrete to last longer. Scanning electron microscopy (SEM), energy-dispersive X-ray spectroscopy (EDS), and X-ray diffraction (XRD) tests all agreed that bacterial additions to concrete improved its microstructure by filling holes with calcite crystals. Because of this improvement in microstructure, the concrete is stronger and lasts longer.

Table 1. Efficiency of Different Bacterial Types (Source: Author's contribution)

Bacterial Strain	Maximum Crack Width Sealed	Improvement in Compressive Strength (%)	Reference
Bacillus spp.	0.5 mm	30	(Etesami et al., 2023), (Etesami et al., 2023), (Prasad et al., 2023), (Elmenshawy et al., 2024)
Sporosarcina spp.	0.46 mm	42	(Seo et al., 2022), (Khaudiyal et al., 2022)
Bacillus subtilis	0.4 mm	35	(Karimi and Mostofinejad, 2020), (Zamba and Mohammed, 2024), (Jena et al., 2020)

2.5 Fungi Based Method

Fungal self-healing approaches include the use of enzymatic mechanisms by fungus to create calcium carbonate. (Madusanka et al., 2024) found that fungi such as Trametes versicolor may increase flexural strength by 25% and close fractures as broad as 0.3 mm. The efficacy of fungal self-healing is summarized in Table 2. The self-healing mechanisms of fungi are not as well-documented as those of bacteria. Their effectiveness is influenced by how well they are mixed with concrete and the circumstances in which fungus flourish. In order to make fungal self-healing strategies more effective overall, researchers are now working on ways to improve these parameters (Luo et al., 2018). Fungal self-healing concrete has only recently emerged from its experimental stage and has yet to find widespread use. Nonetheless, it may be useful in certain settings where fungi may persist for an extended period of time. Its potential building applications might be explored more in the future (Jones et al., 2020).

Table 2. Efficiency of Different Fungal Types (Source: Author's contribution)

Fungal Species	Maximum Crack Width Sealed	Improvement in Compressive Strength (%)	Reference
Trametes versicolor	0.3 mm	25	(Tišma et al., 2021), (Hu et al., 2020)
Aspergillus nige	0.25 mm	20	(Chaudhuri et al., 2020)

(*continued*)

Table 2. (*continued*)

Fungal Species	Maximum Crack Width Sealed	Improvement in Compressive Strength (%)	Reference
Penicillium spp.	0.2 mm	18	(Altaf et al., 2018)

2.6 Algae Based Method

Using algae to produce calcium carbonate is one way that algae may repair themselves. (Nur and Dewi, 2024) found that algae like Spirulina platensis may decrease porosity by as much as 50% and fill fractures as broad as 0.2 mm. The efficacy of algal self-healing techniques is summarized in Table 3. Current approaches for algae self-healing are promising, but they still need to be fine-tuned. The success of these techniques is conditional on the algae species and growing environment. To improve efficiency, a deeper knowledge of how algae might be integrated into concrete is important. There are currently few real-world uses for algal self-healing concrete, which is still in its early phases of development. Additional study is necessary to properly explore the possibilities of incorporating algae-based technologies into sustainable building processes.

Table 3. Efficiency of Different Algae Types (Source: Author's contribution)

Algal Species	Maximum Crack Width Sealed	Improvement in Compressive Strength (%)	Reference
Spirulina platensis	0.2 mm	05	(Wang et al., 2023)
Chlorella vulgaris	0.15 mm	45	(Tamtürk et al., 2023), (Lian et al., 2024)
Nannochloropsis *spp.*	0.1 mm	40	(Vadiveloo et al., 2016)

3 Methodology

This paper's approach is based on a systematic literature analysis that compares and contrasts three biological self-healing processes in concrete: one that utilizes fungus, one that uses bacteria, and one that uses algae. The goals of the study are to find the best way for fixing concrete that is both efficient and effective. Literature review, data extraction and synthesis, and comparison analysis are the three main components of the process.

3.1 Finding and Choosing Relevant Literature

Academic databases such as Google Scholar, Scopus, Web of Science, and ScienceDirect were used to do a thorough literature search. The keywords searched for were "bacteria-based self-healing concrete," "fungi-based self-healing concrete," "algae-based self-healing concrete," "biological self-healing mechanisms," "microstructure of self-healing concrete," along with "strength of self-healing concrete." Articles published in peer-reviewed journals, conference proceedings, and patents between the years 2000 to 2023 were the only ones considered.

- At first, more than two hundred items were found. Studies involving the use of biological processes in concrete (e.g., bacteria, fungus, algae) were not included.
- Research that quantifies the approaches' efficacy in terms of microstructure, strength, and fracture healing. The complete text of articles published in English is accessible.
- A total of 40 articles were chosen for a comprehensive evaluation after these criteria were applied. A high-level perspective of the selection procedure is shown in Fig. 1.

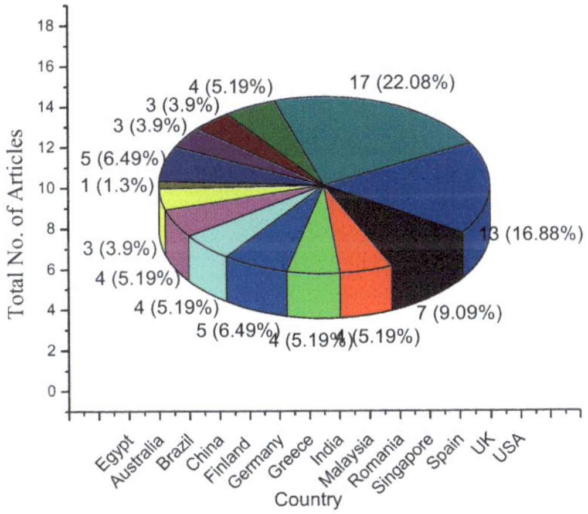

Fig. 1. Total Number of Articles Published in Different Countries (Source: Author's contribution)

3.2 Extracting and Synthesizing Data

- In order to make comparisons, data were taken and arranged into tables from the chosen research. Among the data points retrieved were:
- Strength Measurements: The research found an increase in both compressive and flexural strength.
- Alterations to the Microstructure: SEM-typically detected signs of fracture healing, reduced porosity, and mineral precipitation. The efficiency and effectiveness of the biological agents under various environmental circumstances, the time it takes for cracks to mend, and the crack width at which they heal.
- Real-World Use: The methodology' possible scalability, limits, and practical uses. Important information gleaned from the research that were evaluated is summarized in Table 4.

Table 4. Biological Self-Healing Mechanisms in Concrete: A Summary (Source: Author's contribution)

Method	Strength Improvement (%)	Crack Width Healed (mm)	Notable Microstructural Changes
Bacteria Based	Up to 25%	0.5 mm	Reduced porosity, $CaCO_3$ precipitation
Fungi Based	Up to 15%	0.3 mm	Mineral precipitation, pore filling
Algae Based	Up to 10%	0.2 mm	Surface layer mineralization

4 Discussion

There are clear benefits and drawbacks to using techniques based on bacteria, fungus, or algae for biological self-healing in concrete, according to a comparative examination of these approaches. There is a great deal of variation among the methods with regard to efficiency, practicality, and scalability, but all of them show distinct strengths in improving the self-healing, durability, and strength of concrete.

4.1 Approach Based on Bacteria

Among three different approaches for biological self-healing concrete, the bacteria-based approach has been proved to be the most efficient. In the presence of a crack, bacteria infused in concrete become active from its dormant state and metabolize in the presence of air to form calcium carbonate ($CaCO_3$). The resultant calcium carbonate deposits over the open crack area on the exposed surface of the concrete structure and thereby heal the crack. Previous studies show that using bacteria in concrete cracks up to 0.5 mm wide can be healed by the self-healing mechanism. An increase in compressive strength by 25% is also achieved in bacteria-based self-healing concrete. Its higher metabolism rate in the presence of air and resilience to acidic environments make it a primary choice as a self-healing microorganism. Yet, its high cost of cultures and requirement of enough moisture to initiate metabolism in the presence of air restrict its wide applications. The efficiency of bacteria-based self-healing concrete in different environmental conditions should be explored with more investigations in the future.

4.2 Approach Based on Fungi

Fungi is considered as one of the promising alternatives when other bacterial techniques fail to show significant improvement over conventional concrete. As an additional advantage, fungi can penetrate the concrete matrix, change the microstructure of concrete, and consequently control precipitation of minerals. Cracks up to 0.3 mm in width is proved to be effectively covered up by the use of fungi. Additionally, it could increase the compressive strength of concrete by approximately 15%. However, the possibility of potential alterations of fungi in the presence of alkaline environments poses a serious challenge to their use in such environments. Moreover, the rate of deposit on cracked surfaces in the presence of fungi is comparatively slower than that of bacterial microorganisms. The chances of biodegradation or consequent reduction in structural performance of concrete by the fungal growth need to be investigated in the future.

4.3 Approach Utilizing Algae

The most promising of the three approaches, self-healing solutions based on algae have the most untapped potential because of the algae's long lifespan and little environmental impact. In order to seal micro-cracks and decrease porosity, algae may aid in the creation of a mineralized layer on top of concrete. Tests have shown that the technique may mend fissures as broad as 0.2 mm and increase strength by as much as 10%. Because algae need little resources to grow and have little effect on the environment, solutions based on them hold great promise for the future. It is yet unclear, however, how well algae can combine with the concrete matrix and how efficiently they can produce large quantities of therapeutic chemicals. More study is needed to evaluate long-term performance and scalability, and there are few practical applications due to most research being conducted in laboratory settings.

5 Conclusion

In this study, we have looked at three different self-healing processes in concrete: one based on bacteria, one on fungus, and one on algae. We have compared and contrasted their advantages and disadvantages as well as their potential uses. In terms of study depth and practical application, technologies based on bacteria now lead the way. These approaches provide considerable advances in the strength and durability of concrete. But there are problems with cost and long-term viability that they must overcome. Despite their relative infancy, fungi-based approaches provide an alternative strategy for improving the microstructure of concrete and have shown promise in several contexts. Additional study is needed to solve the present constraints of algae-based technologies, which are promising from a sustainability standpoint but have practical and efficiency limits.

Addressing the drawbacks of each approach, especially in terms of efficiency, longevity, and scalability, should be the primary goal of future studies. There may be new opportunities for the building of robust and environmentally friendly infrastructure if these biological technologies are integrated with current construction procedures. If we want to see concrete technology advance and the demand for sustainable building materials satisfy, we need to keep digging into these biological self-healing capabilities.

Credit Authorship Contribution Statement. T. Senthil Vadivel: Conceptualization, Methodology, Reviewing, Writing, Investigation, Visualization, Editing and Supervision. Abha Maheshwari: Reviewing, Writing, Investigation and Visualization. Sayandip Ganguly: Reviewing, Writing, and Editing.

Declaration of Competing Interest. The study's authors so acknowledge that no third-party money was received from any organization in order to conduct the research stated in this publication. The writers further declare that none of them has any financial or non-financial interests that might jeopardize the accuracy of the work presented in this publication. No conflicts of interest are disclosed.

References

Jonkers, H.M., Thijssen, A., Muyzer, G., Copuroglu, O., Schlangen, E.: Application of bacteria as self-healing agent for the development of sustainable concrete. Ecol. Eng. **36**(2), 230–235 (2010)

Nair, P.S., Gupta, R., Agrawal, V.: Self-healing concrete: A promising innovation for sustainability—A review. Materials Today: Proceedings **65**(2), 1410–1417 (2022)

Wiktor, V., Jonkers, H.M.: Quantification of crack-healing in novel bacteria-based self-healing concrete. Cement Concr. Compos. **33**(7), 763–770 (2011)

Granger, S., Loukili, A., Cabot, G.P., Chanvillard, G.: Experimental characterization of the self-healing of cracks in an ultra-high performance cementitious material: Mechanical tests and acoustic emission analysis. Cem. Concr. Res. **37**(4), 519–527 (2007)

Li, L., Zheng, Q., Li, Z., Ashour, A., Han, B.: Bacterial technology-enabled cementitious composites: A review. Compos. Struct. **225**, 111170 (2019)

Chahal, N., Siddique, R., Rajor, A.: Influence of bacteria on the compressive strength, water absorption and rapid chloride permeability of fly ash concrete. Constr. Build. Mater. **28**(1), 351–356 (2012)

Wei, S., Sanchez, M., Trejo, D., Gillis, C.: Microbial mediated deterioration of reinforced concrete structures. Int. Biodeterior. Biodegradation **64**(8), 748–754 (2010)

Vijay, K., Murmu, M., Deo, S.V.: Bacteria-based self-healing concrete—A review. Constr. Build. Mater. **152**, 1008–1014 (2017)

Li, D., Zheng, H., Gu, K., Lang, L., Shi, S., Chen, B.: Autogenous healing mechanism of cement-based materials. Front. Struct. Civ. Eng. **17**, 948–963 (2023)

Rui, Y., Qian, C.: The influence of bacteria on biologically induced calcium carbonate and its evolution process. J. Cryst. Growth **581**, 126515 (2022)

Javeed, Y., Goh, Y., Mo, K.H., Yap, S.P., Leo, B.F.: Microbial self-healing in concrete: A comprehensive exploration of bacterial viability, implementation techniques, and mechanical properties. J. Mater. Res. Technol. **29**, 2376–2395 (2024)

Jogi, P.K., Lakshmi, T.V.: Self-healing concrete based on different bacteria: A review. Materials Today: Proceedings **43**(2), 1246–1252 (2021)

Pupa, P., et al.: The efficacy of three double-microencapsulation methods for preservation of probiotic bacteria. Sci. Rep. **11**, 13753 (2021)

Nilimaa, J.: Smart materials and technologies for sustainable concrete construction. Developments in the Built Environment **15**, 100177 (2023)

Pandit, C., et al.: Biological agents for synthesis of nanoparticles and their applications. Journal of King Saud University – Science **34**(3), 101869 (2022)

Vemmer, M., Patel, A.V.: Review of encapsulation methods suitable for microbial biological control agents. Biol. Control **67**(3), 380–389 (2013)

Mohammed, T.A., Kasie, Y.M., Assefa, E., Getu, Y.M., Tufa, D.H.: Enhancing structural resilience: Microbial-based self-healing in high-strength concrete. Int. J. Concrete Struct. Mat. **18**, 22 (2024)

Peng, Y., Liu, H., Peng, H., Zhang, J.: Biological self-healing strategies from mechanically robust heterophasic liquid metals. Matter **6**(1), 226–238 (2023)

Elmenshawy, Y., Elmahdy, M.A., Moawad, M., Elshami, A.A., Ahmad, S.S., Nagai, K.: Investigating the bacterial sustainable self-healing capabilities of cracks in structural concrete at different temperatures. Case Stud. Constr. Mat. **20**, e03188 (2024)

Etesami, H., Jeong, B. R., Glick, B.R.: Potential use of *Bacillus* spp. as an effective biostimulant against abiotic stresses in crops—A review. Current Research in Biotechnology **5**, 100128 (2023)

Prasad, B., Sharma, D., Kumar, P.K., Dubey, R.C.: Biocontrol potential of *Bacillus* spp. for resilient and sustainable agricultural systems. Physiol. Molecul. Plant Pathol. **128**, 102173 (2023)

Etesami, H., Jeong, B.R., Glick, B.R.: Biocontrol of plant diseases by *Bacillus* spp. Physiol. Mol. Plant Pathol. **126**, 102048 (2023)

Seo, J., et al.: *Sporosarcina pasteurii*-induced hydration and shrinkage properties of Portland cement. Constr. Build. Mater. **356**, 129213 (2022)

Khaudiyal, S., Rawat, A., Das, S.K., Garg, N.: Bacterial concrete: A review on self-healing properties in the light of sustainability. Materials Today: Proceedings **60**(1), 136–143 (2022)

Karimi, N., Mostofinejad, D.: *Bacillus subtilis* bacteria used in fiber reinforced concrete and their effects on concrete penetrability. Constr. Build. Mater. **230**, 117051 (2020)

Zamba, D.D., Mohammed, T.A.: Self-healing performance of normal strength concrete with *Bacillus subtilis* bacteria. J. Build. Pathol. Rehabilit. 9 (2024)

Jena, S., Basa, B., Panda, K.C., Sahoo, N.K.: Impact of *Bacillus subtilis* bacterium on the properties of concrete. Materials Today: Proceedings **32**(4), 651–656 (2020)

Madusanka, C., et al.: A review of recent advances in fungal mycelium-based composites. Discover Materials **4**(13) (2024)

Luo, J., et al.: Interactions of fungi with concrete: Significant importance for bio-based self-healing concrete. Constr. Build. Mater. **164**, 275–285 (2018)

Jones, M., Mautner, A., Luenco, S., Bismarck, A., John, S.: Engineered mycelium composite construction materials from fungal biorefineries: A critical review. Mater. Des. **187**, 108397 (2020)

Tišma, M., et al.: *Trametes versicolor* in lignocellulose-based bioeconomy: State of the art, challenges and opportunities. Biores. Technol. **330**, 124997 (2021)

Hu, K., et al.: Exploring the degradation capability of *Trametes versicolor* on selected hydrophobic pesticides through simultaneous evaluation of culture broth and biological matrix. Chemosphere **250**, 126293 (2020)

Chaudhuri, A., Bhattacharyya, S., Chaudhuri, P., Sudarshan, M., Mukherjee, S.: In vitro deterioration study of concrete and marble by *Aspergillus tamarii*. Journal of Building Engineering **32**, 101774 (2020)

Altaf, M.M., Imran, M., Abulreesh, H.H., Khan, M.S., Ahmad, I.: Diversity and applications of *Penicillium* spp. in plant-growth promotion. In: New and Future Developments in Microbial Biotechnology and Bioengineering, pp. 261–276. Elsevier (2018)

Nur, M.M.A., Dewi, R.N.: Opportunities and challenges of microalgae in biocement production and self-repair mechanisms. Biocatal. Agric. Biotechnol. **56**, 103048 (2024)

Wang, Y., et al.: Surface patterns of mortar plates influence *Spirulina platensis* biofilm attached cultivation: Experiment and modeling. Algal Res. **71**, 103079 (2023)

Tamtürk, F., et al.: Optimization of *Chlorella vulgaris* spray drying using various innovative wall materials. Algal Res. **72**, 103115 (2023)

Lian, F., et al.: Enhancing *Chlorella vulgaris* growth with Fe-loaded sulphoaluminate cement: A promising phosphorus adsorbent reutilization strategy. Resour. Conserv. Recycl. **209**, 107732 (2024)

Vadiveloo, A., Moheimani, N.R., Kosterink, N.R., Cosgrove, J.J., Garcia, C.G., Lubián, L.M.: Photosynthetic performance of two *Nannochloropsis* spp. under different filtered light spectra. Algal Res. **19**, 168–177 (2016)

Sustainable Utilization of Municipal Solid Waste for Paver Block Production: A Literature Review

Sonal Saluja[1](✉), Prakash Somani[2], Arun Gaur[3], and Sadiqa Abbas[1]

[1] Department of Civil Engineering, Manav Rachna International Institute of Research and Studies, Faridabad, Haryana 121004, India
salujasonal02@gmail.com

[2] Department of Civil Engineering, National Institute of Technology, Srinagar, Jammu and Kashmir 190006, India

[3] Department of Civil Engineering, Malaviya National Institute of Technology, Jaipur, Rajasthan 302017, India

Abstract. Municipal Solid Waste (MSW) has emerged as a significant environmental problem, particularly in the developing cities like India where too few spaces for safe disposal of such waste are available. Ineffective waste management practices result in ecological and health risks while also leading to the depletion of resources. In recent years, sustainable application of MSW in construction materials has gained attention as a viable solution for addressing these concerns. This review deals with the current status of municipal solid waste management in India, specifically regarding it characteristics and composition and challenges associated with its disposal. Particular emphasis is placed on the integration of diverse solid waste materials- including plastics, fly ash, glass, metals, and agricultural residues into the paver blocks construction. The purpose of the paper is to consolidate findings from multiple research investigations in order to evaluate the mechanical, physical and durability characteristics of paver blocks produced by using MSW-based materials. The review highlights the dual benefits of this approach: effective waste minimization and the development of cost-effective, sustainable construction alternatives.

Keywords: paver block · municipal solid waste · sustainable construction · composite waste characteristics

1 Introduction

Municipal Solid Wastes (MSW) includes organic material like kitchen waste, paper and cardboard, yard trimmings, among others, and inorganic material such as glass, plastics, metal, wood, leather and textiles. Solid waste is generated from a variety of sources, including residential, commercial, agricultural, and institutional activities. However, hazardous waste such as that produced by hospitals and certain industries is excluded from this category. Improper waste management practices have led to the excessive accumulation of solid waste in open areas and along streets, creating serious environmental

L. Choudhary et al. (Eds.): TIMES 2025, SCI, pp. 51–61, 2027.
https://doi.org/10.1007/978-3-032-34636-0_5

and public health concerns. In recent years, the management of municipal solid waste (MSW) has emerged as a critical global challenge for government authorities, necessitating the development and implementation of effective strategies focused on reduction, reuse, and recycling. Therefore, this paper reviewed the MSWM system, composition and characteristics of MSW, and various uses of solid waste in roads, i.e., paver blocks and as a substitute material for pavement layers.

1.1 Municipal Solid Waste Handling and Management Practices in India

The country at present produces approximately 145 million tonnes of waste, mentioned in a report by the Ministry of New and Renewable Energy (MNRE). This is estimated to rise to around 260 to 300 million tonnes per day by the year 2047. The total quantity of waste generated state-wise for the years 2015–16 to 2000–2020 and the percent change concerning the year 2015–16 is presented in Table 1 (CPCB, India, 2021). According to the CPCB report, about 132,498 metric tonnes were collected, and around 64,386 metric tonnes were only treated (CPCB, 2019). It is estimated that about 80% - 90% of the total waste produced in India is dumped in landfills, as shown in Fig. 1 (Joshi and Ahmed, 2016; Ahluwalia and Patel, 2018). This has a detrimental effect on public health and the cleanliness of the city.

Table 1. Generation of municipal solid waste over years (Source: Author's contribution)

Waste Generation							
State	Population 2011	Metric tonnes per day (MTPD)					
		2015–2016	2016–2017	2017–2018	2018–2019	2019–2020	% Change w.r.t (2015–2016)
Andhra Pradesh	49577	5842.271	5869.487	5850.25	5842.271	6138.014	5.06
Haryana	25351	4388.371	4095.033	4116.24	4205.519	4746.301	8.16
Karnataka	61095	8021.33	10147.77	10459.63	10848.12	11120.46	38.64
Madhya Pradesh	72627	6058.181	6144.364	6542.618	7257.48	7239.336	19.49
Maharashtra	112374	1983.351	21273.18	21457.34	21631.42	20815.59	949.52
Punjab	27743	4042.598	4122.566	4185.54	4204.331	4061.959	0.4789
Tamil Nadu	72147	208.6526	13297.97	13766.18	12671.56	12907.43	6086.09
Telangana	35004	6012.822	6547.24	7079.672	7708.351	8423.213	40.09
Uttarakhand	10086	831.8886	1070.478	996.9963	1385.687	1461.422	75.68
Chandigarh	1055450	335.6585	408.2333	453.5925	426.377	408.2333	21.62
Delhi	16787941	8727.12	9158.15	9465.27	9813.02	9498.744	8.84
Total	18309395	46452.24	82134.47	84373.33	85994.14	86820.7	86.90

The MSW so collected is heterogeneous, and the percentage of waste collection is found to be lowest in Delhi, which stands at only 39% (Ahluwalia and Patel, 2018). Furthermore, this collected waste is not subjected to the basic process of segregation in most states and union territories. This segregation significantly reduces the burden on municipal organizations responsible for efficiently treating solid waste.

Fig. 1. Waste dumped at the landfill site (Source: Author's contribution)

From Table 1, it was observed that the quantity of waste generated increased with time. A massive upsurge of waste generated was observed in Maharashtra and Tamil Nadu between the years 2015–16 to 2019–2020, as it recorded an increment of 949.52% and 6086.09%. The quantity of solid waste produced between 2015 and 2020 consistently increased and registered an uptake of approximately 86.90%. This trend is alarming, indicating that the rate of waste generation will continue to rise in the coming years. The current waste management system of India is inadequate for dealing with this huge expected influx of waste and accumulated waste.

2 Composition and Characteristics of Municipal Solid Waste

It is essential first to understand the characteristics of the waste stream to evaluate the alternative systems, equipment needs, and management techniques, especially for the efficient implementation of various disposal options, energy recovery, and resources (Tchobanoglous et al., 1993). An elaborate and detailed analysis of the characteristics and composition are discussed below.

2.1 Composition of Waste

As per Singh (2020) observation, the percentage of biodegradable waste is the highest, followed by inert and non-biodegradable waste and then recyclable waste. Also, it has been noticed that the quantity of biodegradable waste is higher in developing countries compared to western countries due to the major variations in the lifestyle of the people. The breakdown of the composition of waste and its quantity generated in Indian cities is shown in Fig. 2.

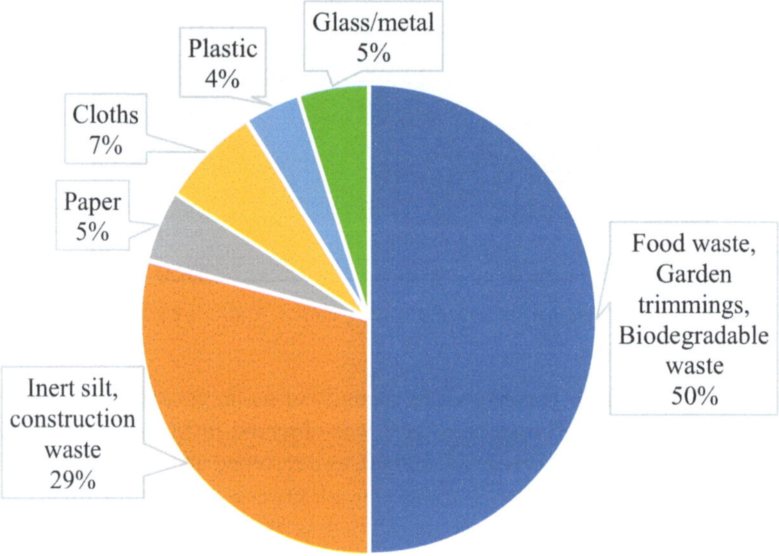

Fig. 2. Composition of various types of wastes (Source: Author's contribution)

2.2 Characteristics of Composite Waste

Several methods for characterizing municipal solid waste for reuse have been proposed since the 1990s. Waste's properties may be roughly divided into two categories: physical and chemical.

2.2.1 Physical Characteristics

As mentioned earlier, understanding the physical properties of solid waste are essential for the operation and selection of systems, management programs, analysis, and design of disposal facilities.

2.2.2 Chemical Characteristics

The information on chemical characteristics and their classification is necessary to identify waste behavior during the various treatment processes. Environmental Protection Agency (EPA) has specified the general ranges for all the parameters of proximate and ultimate analysis of MSW are presented in Table 2 (Tchobanoglous et al., 1993) below.

Table 2. Proximate and ultimate analysis of MSW (Source: Author's contribution)

Proximate analysis (% by mass)			
Parameter	Upper Limit	Typical	Lower Limit
Volatile matter	60	53	40
Moisture	40	20	15
Noncombustible (glass, metal, ash)	30	20	15
Fixed carbon	5	7	12
Ultimate analysis (% by mass)			
Carbon	60	47	40
Hydrogen	8	6	4
Oxygen	50	40	30
Nitrogen	1	0.8	0.2
Sulfur	0.3	0.2	0.05
Ash	10	6	1

NEERI prepared the population range-wise chemical characteristics of Indian MSW, and it was found that it changes with population density are presented in Table 3 (Sharholy et al., 2008; NEERI, 1995) below. This change in the composition was due to urbanization, changes in people's living standards and socioeconomic growth in various parts of the country can be easily seen in the data summarized in Fig. 3 (Joshi and Ahmed, 2016; Planning commission report, 2014) below.

Table 3. Change in chemical characteristics of MSW population-wise (Source: Author's contribution)

Year	Paper	Metal	Rags	Biodegradables	Plastic/rubber	Others	Glass	Inert
1996	3.63	0.49	-	42.21	0.60	-	0.60	45.13
2005	8.13	0.50	4.49	47.43	9.22	4.02	1.01	25.16
2011	9.63	0.63	-	42.51	10.11	-	0.96	17.00

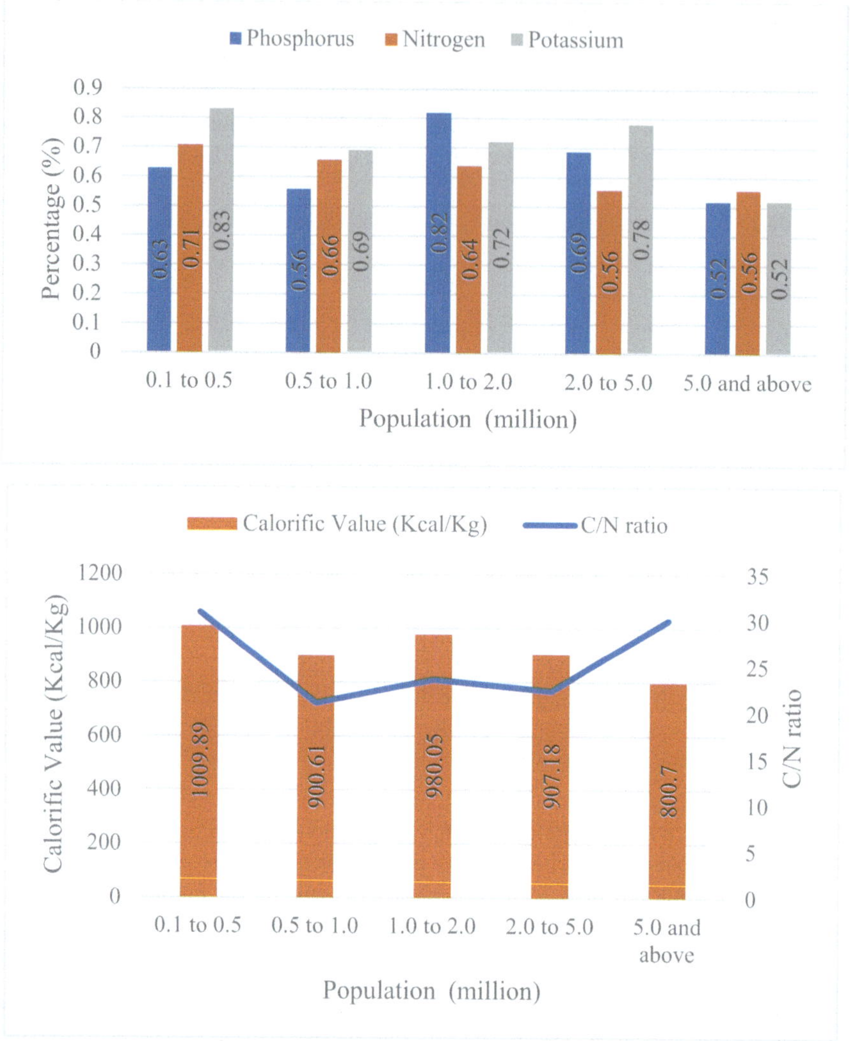

Fig. 3. Compositional changes in MSW with respect to time (%) (Source: Author's contribution)

3 Sustainable Utilization of Municipal Solid Waste in Paver Block Development

Different solid waste materials have been utilized in the manufacturing of concrete paver blocks, serving as partial replacements for cement, aggregates, fillers, or fibers. A summary of the same is provided below:

3.1 Pozzolanic Materials

Pozzolanic materials are siliceous or siliceous and aluminous substances that, in their initial state, lack any binding properties. However, when water is added, it chemically combines with the calcium hydroxide in the material to form cementitious compounds (Shetty and Jain, 2019; Choudhary et al., 2023).

3.1.1 Waste Basalt Powder

Basalt powder is a finely ground residue formed during the preparation of aggregate used in the production of asphalt mixture. As we know, the cement manufacturing industry contributes about 7% of the total CO_2 generated in the world, causing the degradation of the environment. In Tataranni (2019) studies, attempts have been made to replace cement completely with waster basalt powder. The result of the tests showed that the alkali-activated waste basalt powder could be an alternative solution for paver block construction.

3.1.2 Red Mud and Fly Ash

Red mud, also called bauxite residue, is a type of industrial waste manufactured by Bayer's process in which bauxite is processed into alumina. Approximately 175 million tonnes of red mud were generated from over 133 million tonnes of alumina production in 2020. Also, fly ash is obtained by coal combustion as a finely-grained material. It is estimated that thermal power stations across the country generate more than 75 million tonnes of fly ash annually in India (Shetty and Jain, 2019). Kumar and Kumar (2013) studied the red mud's influence upon addition (0–40%) to the fly ash for making geopolymer blocks on its reaction and structure. Various properties of fly ash geopolymer were also tested and the results of the tests revealed a remarkable improvement in the reaction intensity in all the samples.

3.1.3 Processed Waste Tea Ash

The processed waste tea ash (PWTA) in concrete paver blocks as a cement replacement used by Djamaluddin et al. (2020). The testing result revealed that the replacement of processed waste tea ash degrades the quality of the paver block due to the formation of pores in the structure reflected in the SEM analysis results.

3.1.4 Cement Kiln Dust

Sadek et al. (2017) investigated the use of CKD as a partial replacement for cement and found that it negatively impacted both durability and mechanical properties compared to mixes containing only cement.

3.1.5 PET Flakes and Fly-Ash

Bijivemula and Noolu (2021) used different classes of fly ash and PET flakes of various sizes to make sustainable concrete blocks. Alkali activators such as Na2, SiO3, and NaOH were utilized to effectively bond the particles. The test results exhibited that at higher curing temperatures and alkali solution molarity strength increases.

3.1.6 Minerals

Chkheiwer (2017) investigated the properties of paver blocks containing three varied minerals, such as fly ash, silica fume, and metakaolin, with replacement of 0% to 20% of cement. It was found that metakaolin-contained concrete had more suitable strength and workability than fly ash and silica fume concrete.

3.2 Inert Materials

Various types of inert materials are used in concrete paver block production. The commonly used inert materials are recycled aggregate, recycled plastic, waste marble, construction and demolition waste, etc.

3.2.1 Recycled Aggregates

Improper management of recycled concrete, crushed glass, construction, and demolition waste, ceramic tile waste, etc., poses a severe environmental challenge. It also increases the space requirement of a landfill to accumulate this waste. Due to road development projects in India, there is a huge quantity of recycled asphalt pavement (RAP) made by the crushing and excavating of existing flexible pavements.

Hossiney et al. (2020) attempted to evaluate the possibility of alkali-activated concrete paver blocks in which RAP aggregates are used. They reported that these paver block's compressive strength and abrasion resistance value got reduced. However, still, it was recommended to use on pedestrian paths and other than motorized roads.

Wang et al. (2019) developed sustainable concrete paver blocks by incorporating five different types of construction waste materials and test results indicated that the use of recycled aggregates led to reduced strength and increased water absorption.

Rodríguez et al. (2017) tried to understand the possibility of using recycled aggregate at an industrial level. They found that it satisfied the necessities and requirements laid under the Spanish standards for these elements. Gayarre et al. (2013) developed concrete kerbs and floor blocks from C&D recycled aggregates. Test results indicated that the physical and mechanical performance of the recycled aggregates has little influence.

The utilization of recovered aggregates from demolished concrete structures in the manufacturing of paver blocks was examined by Matar and El Dalati (2011). Their research showed that paver blocks with different percentages of recycled aggregates as partial substitutes can be used successfully in building applications.

3.2.2 Crushed Brick

Reddy et al. (2021) used brick waste aggregates and was found that brick masonry waste aggregates can be effectively used in recycled aggregate paver blocks (RAPB) up to 25% and 50% replacement without negotiating of desired compressive strength value. Jankovic et al. (2012) studied crushed bricks as the aggregate for concrete tiles. It was observed that the CSB can be used as aggregate lower in strength, density, and graduated absorption of the concrete component.

3.2.3 Other Materials

Recycled plastic

The potential use of plastic waste as a binding material to produce paver blocks was investigated by Agyeman et al. (2019). The study observed that the samples with high plastic have low water absorption and extra compressive strength than the low plastic and control samples.

Recycled tire rubber

The use of recycled tire rubber, or crumb rubber, as an aggregate was investigated by da Silva et al. (2015). The results of the tests revealed that although the compressive strength decreases as the amount of crumb rubber increases, it reaches a maximum value of 40 MPa when 50% of crumb rubber is added.

Waste marble

Gencel et al. (2012) used waste marble to replace fine aggregate and was observed that mechanical strength decreases while the abrasive wear and freeze-thaw durability resistance increases with the waste marble percentage increase.

4 Conclusion

Incorporating various waste materials into the production of paver blocks presents a sustainable and advanced key to the challenges of managing municipal solid waste while meeting the increasing demand for environmentally friendly construction materials. This review synthesizes studies on a wide spectrum of materials, including pozzolanic wastes such as waste basalt powder, red mud, fly ash, processed waste tea ash, cement kiln dust, coal waste, and PET flakes blended with fly ash. Furthermore, a critical analysis has been conducted on the potential of mineral and inert wastes, including recycled aggregates, crushed bricks, and other materials such as waste marble, recycled plastics and waste glass. The various authors results suggest that these materials can partially or fully replace conventional constituents, that can improve the mechanical, durability, and economic performance of paver blocks, while also reducing environmental burdens.

To fully utilize the potential of paver blocks made from MSW, future research should concentrate on creating optimized mix designs, conducting thorough life cycle assessments, validating findings at a field scale, and establishing relevant codal provisions. Collaboration among researchers, policymakers, and industries will be vital to scale up this approach. The responsible use of these diverse waste materials can greatly enhance circular economy efforts, promote resource conservation, and support the overall goals of sustainable urban infrastructure.

Conflict of Interest. The authors declare that they have no known financial or personal relationships that could have appeared to influence the work reported in this paper. All authors confirm that there are no competing interests related to the research, its analysis, or its publication. Any potential conflicts have been fully disclosed, ensuring transparency and adherence to ethical standards.

References

Agyeman, S., Obeng-Ahenkora, N.K., Assiamah, S., Twumasi, G.: Exploiting recycled plastic waste as an alternative binder for paving blocks production. Case Stud. Constr. Mat. **11**, e00246 (2019). https://doi.org/10.1016/j.cscm.2019.e00246

Ahluwalia, I.J., Patel, U.: Solid waste management in india: an assessment of resource recovery and environmental impact. Indian Council for Research on International Economic Relations (ICRIER), New Delhi (2018). https://hdl.handle.net/10419/203690

Bijivemula, S.K.R., Noolu, V.: Analysis of mechanical and durability properties of alkali activated blocks using PET flakes and Fly-ash. Materials Today: Proceedings **43**, 1093–1097 (2021). https://doi.org/10.1016/j.matpr.2020.08.326

Central Pollution Control Board (CPCB): Consolidated annual review report on implementation of Municipal Solid Wastes (Management and Handling) Rules, 2000 (2019)

Central Pollution Control Board (CPCB): Status of STPs: national inventory of sewage treatment plants as on 08 March 2021 (2021). https://cpcb.nic.in/status-of-stps

Chkheiwer, A.H.: Improvement of concrete paving blocks properties by mineral additions. Journal of Babylon University/Engineering Sciences **25**(1), 157–164 (2017)

da Silva, F.M., Barbosa, L.A.G., Lintz, R.C.C., Jacintho, A.E.P.: Investigation on the properties of concrete tactile paving blocks made with recycled tire rubber. Constr. Build. Mater. **91**, 71–79 (2015). https://doi.org/10.1016/j.conbuildmat.2015.05.027

Djamaluddin, A.R., Caronge, M.A., Tjaronge, M.W., Lando, A.T., Irmawaty, R.: Evaluation of sustainable concrete paving blocks incorporating processed waste tea ash. Case Studies in Construction Materials **12**, e00325 (2020). https://doi.org/10.1016/j.cscm.2019.e00325

Gayarre, F.L., López-Colina, C., Serrano, M.A., López-Martínez, A.: Manufacture of concrete kerbs and floor blocks with recycled aggregate from C&DW. Constr. Build. Mater. **40**, 1193–1199 (2013). https://doi.org/10.1016/j.conbuildmat.2011.11.040

Gencel, O., Ozel, C., Koksal, F., Erdogmus, E., Martínez-Barrera, G., Brostow, W.: Properties of concrete paving blocks made with waste marble. J. Clean. Prod. **21**(1), 62–70 (2012). https://doi.org/10.1016/j.jclepro.2011.08.023

Hossiney, N., Sepuri, H.K., Mohan, M.K., Arjun, H.R., Govindaraju, S., Chyne, J.: Alkali-activated concrete paver blocks made with recycled asphalt pavement (RAP) aggregates. Case Studies in Construction Materials **12**, e00322 (2020). https://doi.org/10.1016/j.cscm.2019.e00322

Jankovic, K., Nikolic, D., Bojovic, D.: Concrete paving blocks and flags made with crushed brick as aggregate. Constr. Build. Mater. **28**(1), 659–663 (2012). https://doi.org/10.1016/j.conbuildmat.2011.10.036

Joshi, R., Ahmed, S.: Status and challenges of municipal solid waste management in India: A review. Cogent Environm. Sci. **2**(1), 1139434 (2016). https://doi.org/10.1080/23311843.2016.1139434

Kumar, A., Kumar, S.: Development of paving blocks from synergistic use of red mud and fly ash using geopolymerization. Constr. Build. Mater. **38**, 865–871 (2013). https://doi.org/10.1016/j.conbuildmat.2012.09.013

Matar, P., El Dalati, R.: Strength of masonry blocks made with recycled concrete aggregates. Phys. Procedia **21**, 180–186 (2011). https://doi.org/10.1016/j.phpro.2011.10.027

Protection, E.: Environmental protection agency (EPA). Semiannual Regulatory Agenda, 68 (2003)

Reddy, S.V., Tarun, N., Basutkar, S.M., Wali, R.S., Renukadevi, M.V.: Influence of brick waste and brick waste fines as fine aggregate on the properties of paver blocks–Preliminary investigation. Materials Today: Proceedings **43**, 1496–1502 (2021). https://doi.org/10.1016/j.matpr.2020.09.312

Rodríguez, C., Miñano, I., Aguilar, M.Á., Ortega, J.M., Parra, C., Sánchez, I.: Properties of concrete paving blocks and hollow tiles with recycled aggregate from construction and demolition wastes. Materials **10**(12), 1374 (2017). https://doi.org/10.3390/ma10121374

Sadek, D.M., El-Attar, M.M., Ali, A.M.: Physico-mechanical and durability characteristics of concrete paving blocks incorporating cement kiln dust. Constr. Build. Mater. **157**, 300–312 (2017). https://doi.org/10.1016/j.conbuildmat.2017.09.107

Choudhary, L., Sahu, V., Dongre, A., Tonk, A.: Macro-and microstructural durability investigations of sustainable ternary geopolymer concrete paver blocks. European Chemical Bulletin **12**(6), 5474–5494 (2023)

Sharholy, M., Ahmad, K., Mahmood, G., Trivedi, R.C.: Municipal solid waste management in Indian cities–A review. Waste Manage. **28**(2), 459–467 (2008). https://doi.org/10.1016/j.wasman.2007.02.008

Shetty, M.S., Jain, A.K.: Concrete Technology (Theory and Practice), 8e. S. Chand Publishing (2019)

Singh, S.: Solid waste management in urban india: imperatives for improvement (2020)

Tataranni, P.: Recycled waste powders for alkali-activated paving blocks for urban pavements: A full laboratory characterization. Infrastructures **4**(4), 73 (2019). https://doi.org/10.3390/infrastructures4040073

Tchobanoglous, G., Theisen, H., Vigil, S. (eds.): Integrated solid waste management: engineering principles and management issues, pp. xxi+-978. McGraw Hill Inc., New York (1993)

Wang, X., Chin, C.S., Xia, J.: Material characterization for sustainable concrete paving blocks. Appl. Sci. **9**(6), 1197 (2019). https://doi.org/10.3390/app9061197

Advancing Sustainable Infrastructure: A Technical Review of Steel Fibre-Recycled Aggregate Synergy in Self-Compacting Concrete

A. Nisha Rani[1(✉)], Vaishali Sahu[1], and Lokesh Choudhary[2]

[1] Department of Multidisciplinary Engineering, The NorthCap University, Gurugram, Haryana 122017, India
{23cvd002,vaishalisahu}@ncuindia.edu

[2] Department of Civil Engineering, Manav Rachna International Institute of Research and Studies, Faridabad, Haryana 121004, India
lokeshchoudhary.set@mriu.edu.in

Abstract. In recent years, Self-compacting concrete (SCC) has become a major trend in the construction industry. SCC is used in the heavily reinforced structure, as it can fill formwork without vibration and less wear and tear. SCC consumes natural aggregates heavily, causing brittle failure and raising sustainability concerns. Recycled aggregates (RA) from construction and demolition waste can be incorporated in SCC, promoting sustainability and reducing landfill space. The tensile capacity of SCC increased with the addition of steel fibres (SF). This review mainly focuses on the fresh and hardened properties of the steel fibre-reinforced SCC (SFR-SCC) with and without recycled aggregates (RA). Fresh and hardened properties of the concrete mix, such as slump flow, T-500, V-funnel tests, compressive strength, split tensile strength, and flexural strength, are assessed. This review showed that workability steadily decreases as the content of steel fibres and recycled aggregate increases. This was due to the increased water absorption, interparticle friction and greater viscosity. RA reduces strength, and this reduction is compensated for by the SF. SF shows improved compressive, tensile, and flexural behaviour due to its crack- bridging characteristic, which enhances ductility and shifts the failure mode from brittle to ductile. The interfacial transition zone is further improved by the combined effect of SF and other supplementary cementitious materials, such as fly ash, silica fume, slag and treated RA, allowing SFR-SCC with RA contents to perform mechanically on par with, or better than, conventional SCC, in a number of mixes. Overall, the combined effect of recycled aggregates and steel fibres has significant potential for resource conservation, production of structure-grade sustainable concretes, waste valorisation, and adherence to global sustainable goals.

Keywords: Self-compacting concrete · fibre reinforcement · recycled concrete aggregates · fresh and hardened properties · comparative review

L. Choudhary et al. (Eds.): TIMES 2025, SCI, pp. 62–73, 2027.
https://doi.org/10.1007/978-3-032-34636-0_6

1 Introduction

SCC is a modified form of high-performance concrete. (Najim et al., 2018) It was mainly intended for civil engineering, infrastructure, and housing applications. (Kamal et al., 2015) SCC has been widely used in building construction and structural applications in countries such as the United States of America, Japan, and Canada. (Mohammad et al., 2023). SCC can be pumped under low pressure with minimal wear and tear. (Ghosh et al., 2024; Youvaraj et al., 2023) Compared to vibrated concrete, SCC reduces construction and labour costs. (Wang et al., 2020). However, the production and use of SCC may result in increased water consumption, waste generation, and increased emission of pollutants. (Anurag & Singh, 2021; Khatib et al., 2025).

Infrastructure development has increased due to global population growth, leading to increased consumption of construction materials. (Anurag & Singh, 2022) Concrete usage remains high and is projected to continue at significant levels in the foreseeable future. (Al-Rekabi & Abo Dhaheer, 2024; Kumar et al., 2023) At the global level, carbon footprints increase due to the production of construction and demolition waste and the consumption of natural resources, posing significant environmental concerns. (Gao et al., 2023).

Natural aggregates can be replaced with waste from construction and demolition, resulting in energy savings, reduced carbon dioxide emissions, and reduced landfill space. (Maameri et al., 2023). Incorporating recyclable and renewable resources into concrete is part of the Sustainable Development Goals (SDGs). (Al-Rekabi & Abo Dhaheer, 2024) Natural resources can be conserved, and the cost of new concrete can also be reduced by using recycled materials (Simalti & Singh, 2021).

Recycled aggregates obtained from demolished buildings have the capability to meet the demand for alternative development at a low cost. (Aslani et al., 2019) Also, these aggregates enhance the flowability of SCC without much altering its basic properties. (Abdulhaleem et al., 2023) However, the adhered old mortar attached to the surface of recycled aggregate has a negative effect on fresh and hardened properties of concrete. (Maameri et al., 2023) Compared to natural aggregates, recycled aggregates possess lower density, stiffness, rougher surface, and larger porosity, limiting their application in civil engineering. (Ji et al., 2023; Kiruthiga et al., 2025).

Fibre-reinforced concrete has garnered more attention in recent years. (Wang et al., 2020) Fibre-reinforced self-compacting concrete exhibits enhanced mechanical properties, including tensile strength, toughness, durability, fatigue life, and impact resistance. (Aslani et al., 2019; Shukla et al., 2023) But SCC has certain drawbacks, such as lower tensile and flexural strength, poor impact resistance, and the possibility of cracks occurring due to shrinkage, which are improved by incorporating steel fibres (Cuong, 2024).

The type of fibre, fibre volume fraction, aspect ratio of fibres, fibre geometry, orientation, and fibre distribution all affect the performance of steel fibres in concrete. (Koushkbaghi et al., 2019; Bansal et al., 2023) Fibres, such as propylene, basalt, and steel, are known to enhance the mechanical properties of concrete, with steel fibre being the most commonly used. (Kalra et al., 2018; Liu et al., 2024). The addition of steel fibres improves tensile and flexural strength, energy absorption, and ductility. (Ghasemi et al., 2018; Mohammed et al., 2023) Due to the rising cost and high demand for steel fibres, alternative sources, including those derived from waste tires are being investigated (Hamza et al., 2018).

In this literature, fresh and hardened characteristics of steel fibre-reinforced SCC and steel fibre-reinforced SCC with recycled aggregates already carried out by previous researchers are compiled and compared.

2 Fresh Properties of Self-Compacting Concrete

2.1 Slump Value

Slump test, is one of the most popular techniques for determining SCC's free horizontal flow in the absence of impediments. (Ashteyata et al., 2024; Roy et al., 2018).

Khaloo et al. (2014) investigated the influence of steel fibres with percentages of 0.5%, 1%, 1.5%, 2% and silica fume on SCC. When a 2% steel fibre volume fraction was added, the workability was below the minimum limit specified by EFNARC. (Khaloo et al., 2014). Steel fibres with an aspect ratio of 80 and a percentage of up to 1, along with fly ash, resulted in a slump value ranging from 684 to 651 mm. (Nalanth et al., 2014) Mahmod et al. (2018) investigated the influence of 0.75% volume fraction of waved or crimped steel fibres with an aspect ratio of 80 on the SCC of strength M20, M50 and M60, along with limestone powder. The values were within the limits suggested by EFNARC. (Mahmod et al., 2018) The steel fibre of aspect ratio 50, with a percentage up to 0.5% and limestone powder had the slump value between 715 mm and 590 mm. (Ghasemi et al., 2018). The addition of high content of steel fibres reduced the fluidity of SCC and caused congestion. (Hamza et al., 2018) Simalti & Singh (2021) Experimented on both manufactured steel fibre and recycled steel fibre, including supplementary materials such as fly ash and silica fume and observed that manufactured steel fibre was superior. The dense network structure formed by the steel fibres, prevented the flow of cement paste. (Ashteyata et al., 2024). The larger surface area of the fibres required a higher volume of mortar or fluid paste to surround and lubricate them. (Kannam et al., 2018) According to Cuong, 2024, High-fly ash content, along with viscosity-modifying admixtures, enhances the slump value, contributing to the stability of the concrete mixture. (Cuong, 2024).

SCC with recycled aggregates are brittle in nature, and the steel fibres are added to increase the ductility.As the steel fibre and recycled aggregate content increased, the slump value decreased. (Nalanth et al., 2014; Kamal et al., 2015; Malathy et al., 2022) In high-strength concrete M70, the slump value is almost the same for natural aggregates with steel fibres as for recycled concrete aggregate with steel fibres, along with cementitious materials such as fly ash and silica fume. (Kannam et al., 2018) It was observed that there was a marginal difference in the slump value between M20 and M80 mixes with recycled concrete aggregate content of 25% and steel fibres of different aspect ratios, along with supplementary materials, fly ash, and micro-silica. (Nitesh et al., 2019) Rubber recycled aggregates along with steel fibres adversely affected the Slump value. (Wang et al., 2020; Eren et al., 2021) The hydrophobic nature of rubber, generated more friction than normal fine sand and steel fibres obstructed the flow due to high aspect ratios and mechanical interactions. (Wang et al., 2020) The workability was much affected due to the combined effect of Recycled Coarse Aggregate (RCA) and Recycled Fine Aggregate (RFA) when compared to the individual effect of RCA. (Ji et al., 2023) The percentage variation of recycled concrete aggregate from 20% to 100% with 0.55% steel fibre by volume caused only a slight difference in the slump value. (Mardani et al., 2025) The slump flow of steel fibre-reinforced SCC improved due to the inclusion of supplementary materials. (Liu et al., 2024).

The reduction in coherence between the granules of coarse aggregates and the heterogeneity contribution of steel fibres to the mixture decreased the strength. (Abdulhaleem et al., 2023; Mohammed et al., 2023) Workability of the mixtures would be reduced due to greater water absorption by steel fibres, congestion due to the recycled steel fibres and recycled aggregates, increased interparticle and internal friction. (Mohseni et al., 2017; Ghashouieh et al., 2024; Mardani et al., 2025) However, the occurrence of bleeding and segregation would be minimized due to the uniform distribution of fibres in the fresh concrete mixture. (Mohseni et al., 2017) The, inter-particle friction and interlocking between fibres and aggregates, and between fibres, reduced the workability. (Kannam et al., 2018) The irregular shapes of the increased recycled aggregates increase interparticle friction, and the excess water cannot offset the friction between the rough recycled aggregates and fibres. (Gao et al., 2023) Recycled aggregate properties can be enhanced by impregnating them in a cement–silica fume slurry with a 75:25 cement: silica fume ratio. (Al-Rekabi & Abo Dhaheer, 2024) Replacing industrial steel fibres with recycled steel fibres had little effect on slump values. (Hichem et al., 2024).

2.2 T-500

The time taken for the concrete to reach a circle with a 500mm diameter is called the T-500 time. (Mahmod et al., 2018) T-500 indicates the concrete's viscosity and stability. A shorter time indicates greater fluidity or lesser workability loss. (Ahmed et al., 2023) The addition of supplementary cementitious materials, such as silica fume, along with 0.75% recycled steel fibres, increased the T-500 time. (Mastali & Dalvand, 2016) T-500 time was estimated at 81.39% as the fibre addition rate increased from 0% to 1%. Friction between the fibres increased the flow time. (Hamza et al., 2018) Simalti & Singh (2021) experimented with both manufactured and recycled steel fibres in SCC. T-500 values were higher in the manufactured steel fibre mix than in the recycled steel fibre mix. (Simalti & Singh, 2021).

Kamal et al. (2015) conducted experiment both on crushed red brick with steel fibre volume and crushed ceramic with steel fibre volume. For the same steel fibre volume fraction, crushed ceramic recycled aggregate showed a lower T-500 time than red brick recycled aggregate. (Kamal et al., 2015) T-500 value increased from 2.18 to 3.85 s for steel fibre replacement from 0% to 1.25% and crushed red brick replacement from 0% to 50%. (Kotwal et al., 2022) T-500 time was higher in mixes with 100% recycled aggregates and steel fibres than in mixes with 100% recycled aggregates and no steel fibres. (Kannam et al., 2018) When steel fibres of different aspect ratios are combined with 25% recycled concrete aggregate, along with fly ash and micro silica, a marginal decrease in T-500 values was observed. (Nitesh et al., 2019).

2.3 V-funnel

V-funnel indicates the filling ability of SCC, provided that blocking and/or segregation do not take place. (EFNARC 2005) The V-funnel test's flow time is to some degree related to plastic viscosity. (Nalanth et al., 2014) The addition of supplementary cementitious materials, such as silica fume, along with 0.75% recycled steel fibres, resulted in a maximum increase in V-funnel time. (Mastali & Dalvand, 2016) V-funnel time increased from 2.2 to 9 s when the recycled fibre content increased from 0% to 1.5%. The friction between the fibres increases the flow time. (Hamza et al., 2018).

Kamal et al. (2015) Conducted an experiment on crushed red brick with steel fibre volume and crushed ceramic with steel fibre volume. V-funnel time was within the EFNARC limits for concrete mixes with recycled coarse aggregates and hooked-end steel fibres with an aspect ratio of 80. (Ghasemi et al., 2018) V-funnel time was higher in mixes with 100% recycled aggregates and steel fibres than in mixtures with 100% recycled aggregates and no steel fibres. (Kannam et al., 2018) The V-funnel value ranged from 6 to 11 s when steel fibres of different aspect ratios were combined with 25% recycled concrete aggregate, fly ash, and micro-silica. (Nitesh et al., 2019) Mohseni et al. (2017) observed that increased plastic viscosity due to the combined effect of recycled coarse aggregates and steel fibre increased the V-funnel time for the mixtures.

3 Mechanical Properties of Hardened Self-Compacting Concrete

3.1 Compressive Strength

Compressive strength is one of the most important properties of concrete (Aslani et al., 2019). Compared with other fibres, such as polypropylene, Steel fibres showed a much more pronounced effect on compressive strength. (Bansal et al., 2023; Roy et al., 2018) The greater macro-crack bridging efficiency was attributed to the higher elastic modulus and strength of steel fibres. (Mohseni et al., 2017). The compressive strength of the SCC mix increased with a high steel fibre content, irrespective of the steel fibre aspect ratio. (Nitesh et al., 2019) The compressive strength of the SCC mix reinforced with recycled steel fibres increased with the addition of supplementary cementitious materials such as silica fume. (Mastali & Dalvand, 2016). Rubber particles attached to therecycled steel fibres formed a weak zone in concrete, causing a decrease in strength, and hence, recycled steel fibres extracted from rubber tires exhibited low compressive strength. (Hamza et al., 2018) Simalti & Singh (2021) compared recycled steel fibres with manufactured steel fibres in SCC and found that recycled steel fibres were more effective at arresting cracks. This improvement was due to better micro-crack-arresting properties and good bonding between the matrix and the fibre. (Simalti & Singh, 2021) The compressive strength of the high-strength self-compacting concrete increased by 10% with the incorporation of discarded hybrid steel fibres, due to the strong bond between the matrix and the fibres, which limited the crack propagation. (Mezzal et al., 2021) The increase in strength with the addition of steel fibres, was due to the bridging effect of steel fibres which arrested the further extension of cracks. (Shukla et al., 2023) Recycled steel fibres extracted from elevator cables significantly increased the bonding strength at the interface between the steel fibres and the matrix, leading to increased compressive strength. (Cuong, 2024)

The addition of steel fibres to the SCC changed the failure mode from brittle to ductile. (Mahmod et al., 2018).

Researchers observed that the compressive strength of steel fibre-reinforced SCC containing recycled concrete aggregate was lower than that of steel fibre-reinforced SCC containing normal coarse aggregate, regardless of the steel fibre aspect ratio. (Nitesh et al., 2019) Steel fibres reduce crack width, and adequate curing reduces shrinkage-related cracking. (Nalanth et al., 2014) SCC mixtures containing fibres along with recycled coarse aggregate had lower compressive strength than SCC with the same number of fibres alone. The weaker bond between recycled coarse aggregate and the cement matrix reduced the concrete's strength. (Mohseni et al., 2017) The reduction in strength of the mix containing recycled concrete aggregate was due to microcracks in the old mortar of the recycled concrete aggregate. (Koushkbaghi et al., 2019) When recycled tire rubber is used as a partial replacement for natural aggregates along with steel fibres, the compressive strength decreased with an increase in recycled tire rubber, and it was independent of the rubber size. (Eren et al., 2021) SCC mixed with 15% crushed clay brick waste aggregates and scrap tires recycled steel fibres of 40 and 65 as aspect ratio had 23% increased compressive strength when compared to the SCC containing only 15% waste aggregates and no steel fibres. This was due to the prevention of cracks and better bridging effect from short steel fibres, as well as better bonding between the matrix and the steel fibres. (Salam & Khalil, 2022) Recycled aggregates, along with steel fibres, improved the compressive strength of the M30 SCC. (Kotwal et al., 2022) The addition of steel fibres to the SCC mix, with the replacement of natural coarse aggregates with recycled coarse aggregates by more than 50%, decreased the compressive strength. (Abdulhaleem et al., 2023). The anhydrous cement in the recycled fine aggregate compensated for the inferior influence on the compressive strength, causing a lower decrease in strength by the recycled fine aggregates when compared to the recycled coarse aggregate. (Gao et al., 2023). The weaker interfacial transition zone between the old mortar and the recycled aggregate, compared to that new interfacial transition zone between the new and old mortar, led to a reduction in the compressive strength of the SCC mix. (Maameri et al., 2023).

3.2 Split Tensile Strength

Splitting tensile strength affects the length and size of cracks in concrete. (Ashteyata et al., 2024). Adding fibres to self-compacting concrete increases its tensile strength, mainly because fibre bridge cracks and prevents fractures at the fibre-matrix interface. (Mohseni et al., 2017). The bridging action of steel fibres arrests the cracks. (Mastali & Dalvand, 2016) Splitting tensile strength increased due to the addition of fibres to the self-compacting concrete. (Mahmod et al., 2018) Simalti & Singh (2021) observed that SCC with 1.5% manufactured steel fibres had the highest tensile strength of 4.45 MPa, whereas the maximum split tensile strength of SCC mix containing 1.5% recycled steel fibres, was 3.47MPa. When compared to recycled steel fibres,the high stiffness and the long, uniform, undulated length of manufactured steel fibres resulted in greater bridging effect and increased crack resistance. (Mezzal et al., 2021) SCC mix containing 1% and 1.5% steel fibres had higher split tensile strength than the reference mix without fibres. Fibres act as connecting bridges in the cracked areas. (Abdulhaleem et al., 2023)

The increase in tensile strength of the SCC was due to the cumulative effect of the bond between the concrete and individual steel fibres, which resisted the tensile forces. (Cuong, 2024).

The concrete's strength reduces with the addition of recycled aggregates due to the weaker bond between recycled coarse aggregate and the cement matrix. (Mohseni et al., 2017) As the recycled concrete aggregate increased, the split tensile strength of the SCC containing steel fibres and recycled concrete aggregate decreased and this reduction in strength was due to the porous surface mortar and the high water absorption of the recycled concrete aggregate. (Koushkbaghi et al., 2019) Split tensile strength of the SCC mix reinforced with steel fibres of different aspect ratios increased regardless of SCC mix contained recycled concrete aggregate or normal coarse aggregate. (Nitesh et al., 2019) Steel fibre reinforced SCC containing recycled tire rubber as partial substitution of natural aggregates had higher split tensile strength than the SCC mixtures containing recycled tire rubber as partial substitution of natural aggregates with no steel fibre reinforcement. (Eren et al., 2021) SCC mixes with 15% crushed clay brick waste aggregates and scrap tires recycled steel fibres of 40 and 65 as aspect ratio had 38.90% increased split tensile strength when compared to the green SCC containing only 15% waste aggregates and no steel fibres. Long fibres bridged macro cracks while short fibres bridged micro cracks. (Salam & Khalil, 2022) Splitting tensile strength for the M30 SCC mix was higher at all stages when compared to the reference mix with no recycled aggregates and steel fibres. (Kotwal et al., 2022) The addition of steel fibres to SCC containing brick demolishing waste improved the split tensile strength. (Youvaraj et al., 2023). The adhesion between the recycled coarse aggregate and the mortar, and the ability of steel fibres to bridge cracks or to prevent their development, led to an increase in strength. (Abdulhaleem et al., 2023) The increment in strength in hybrid fibre reinforced with recycled fine aggregate was due to the larger amount of anhydrate cement on the surface of recycled fine aggregate, which compensated for the adverse effect of recycled fine aggregate on the splitting tensile strength of SCC. (Gao et al., 2023) SCC mix reinforced with 2% of steel fibres and containing granite industry by-product as 100% recycled fine aggregate had increased tensile strength of the SCC mix by 106% compared to the reference mix with only recycled fine aggregates and no fibres. The fibres delay the crack propagation and improve the capacity and behavior of concrete. (Zamani et al., 2023). Tensile strength is reduced due to weak cohesiveness between recycled coarse aggregate and the paste of cement, resulting in decreased strength of the interfacial transition zone. (Shawais et al., 2024).

3.3 Flexural Strength

Flexural strength, also referred to as modulus of rupture, is an indirect way to measure tensile strength. (Ashteyata et al., 2024). Maximum flexural strength was obtained when the content of silica fume was 14%, and steel fibres was 0.75%. (Mastali & Dalvand, 2016) The flexural strength of the SCC samples increased with the addition of steel fibres. (Mahmod et al., 2018) Recycled steel fibres from rubber tires increased the flexural strength, and this was due to the random distribution of recycled steel fibres in concrete, which led to the transfer of the fracture loads from one point to another, thus controlling the cracks. (Hamza et al., 2018) Manufactured steel fibre produced higher strength than recycled steel fibres due to its uniform length, improving the bridging effect. (Simalti & Singh, 2021).

Addition of steel fibres to the SCC mix containing recycled aggregates increased the flexural strength and this was due to the random distribution of steel fibres in the SCC mix which helped in transferring of load from point to point. (Maameri et al., 2023) Addition of steel fibres changed the failure pattern of concrete from brittle to ductile. (Jahandari et al., 2021) In both M30 and M70 mixes, flexural strength was higher with steel fibres, irrespective of the mixture containing natural aggregates or recycled concrete aggregates. (Kannam et al., 2018).. The tensile crack formation is restrained due to the high modulus of steel fibres, improving the mechanical strengths and also due to the hydrophilic nature of steel fibres. (Eren et al., 2021) SCC mixes with 15% crushed clay brick waste aggregates and scrap tires recycled steel fibres of 40 and 65 as aspect ratios had the highest flexural strength. Also, long fibres bridged the macro-cracks, while short fibre bridged the micro-cracks. (Salam & Khalil, 2022) Steel fibres resist the growth of micro-cracks, improving flexural strength. (El Ouni et al., 2022) SCC mix containing 2% steel fibres and granite industry by-product as 100% recycled fine aggregate increased the flexural strength of the SCC mix by 100% compared to the reference mix with only recycled fine aggregates and no fibres. Steel fibres high tensile strength, elasticity and efficient anchoring capabilities, inhibiting micro-crack growth, contributed to the improvement of flexural strength. (Zamani et al., 2023).A higher volume fraction of recycled steel fibres decreased the average fibre spacing, resulting in a greater number of fibres being engaged in bearing thestress and leading to multiple cracks. (Liu et al., 2024).

4 Conclusions

The impacts of steel fibres and their combined effect with recycled aggregates on SCC have been reviewed.

- Both steel fibres and recycled aggregates had a negative impact on the fresh characteristics such as slump flow, T-500 and V-funnel. The steel fibres decreased workability due to congestion and increased surface area, requiring a greater amount of paste. Furthermore, the combined effect of steel fibres and recycled aggregates reduced the workability. This reduction in strength was due to the angularity, rough texture, and increased internal and interparticle friction. Workability values were below EFNARC limits when recycled steel fibre content exceeded 1.5% without supplementary materials.
- Steel fibres improved the hardened properties of SCC. It improved the tensile and flexural strength compared to compressive strength. This was due to effective crack bridging, stress redistribution, improved post-cracking behaviour, and enhanced ductility of steel fibres. Also, the steel fibres transformed the SCC failure mode from brittle to ductile by delaying crack initiation and propagation. Manufactured steel fibres exhibited better performance than recycled steel fibres.
- Recycled aggregates in SCC, reduced compressive strength, tensile strength and flexural strength due to high porosity, lower density, water absorption due to the old mortar

attached to them and a weak interfacial transition zone between the mortar and the concrete. However, the incorporation of steel fibres improved tensile and flexural strength through bridging action, enhanced crack resistance, and improved bending capacity and ductility.

- The combined application of steel fibres and recycled aggregates, together with supplementary cementitious materials such as fly ash, silica fume, slag, or limestone, enhances the mechanical performance of Self-compacting concrete (SCC).

Conflict of Interest. The authors declare that they have no conflict of interest.

References

Abdulhaleem, K.N., Van Mohammed, R., Hamada, H. M., Humada, A.M., Majdi, A.: Effect of recycled aggregate concrete and steel fibers on the strength of self-compacting concrete. E3S Web Conf. **427**, 02012 (2023). https://doi.org/10.1051/e3sconf/202342702012

Ahmed, H.U., Mohammed, A.A., Mohammed, A.S.: Effectiveness of nano-SiO2 on the mechanical, durability, and microstructural behavior of geopolymer concrete at different curing ages. Arch. Civ. Mech. Eng. **23**(2), 129 (2023). https://doi.org/10.1007/s43452-023-00668-w

Al-Rekabi, A.H., Abo Dhaheer, M.S.: Flexural performance of sustainable hybrid fibre-reinforced SCC beams made of treated recycled aggregates. J. Build. Pathol. Rehabil. **9**(1), 33 (2024). https://doi.org/10.1007/s41024-023-00382-3

Anurag, & Singh, S.: Impact of using recycled demolition waste as aggregates in steel fiber reinforced self-compacting concrete on its sulphate resistance. IOP Conf. Ser.: Earth Environ. Sci. **889**(1), 012064 (2021). https://doi.org/10.1088/1755-1315/889/1/012064

Anurag, Singh, S.K.: Estimation of the impacts of adding recycled demolition waste and steel fibers on different mechanical properties of self-compacting concrete. Mater. Today: Proc. **48**, 1723–1730 (2022). https://doi.org/10.1016/j.matpr.2021.10.031

Ashteyata, A., Obaidat, A.T., Qerba'a, R.: The rheological, mechanical, and durability behavior of Self-Compacted Concrete (SCC) mixed with hybrid fibers after exposure to high temperatures and cycles of freezing and Thawing. Eng. Sci. **27**, 1042 (2024). https://doi.org/10.30919/es1042

Aslani, F., Hou, L., Nejadi, S., Sun, J., Abbasi, S.: Experimental analysis of fiber-reinforced recycled aggregate self-compacting concrete using waste recycled concrete aggregates, polypropylene, and steel fibers. Struct. Concr. **20**(5), 1670–1683 (2019). https://doi.org/10.1002/suco.201800336

Bansal, S., Choudhary, L., Kalra, M., Dave, N., Sharma, A.K.: Mechanical and microstructural characteristics of structural concrete containing RCA treated with sodium metasilicate. J. Eng. Des. Technol. (2023). https://doi.org/10.1108/JEDT-03-2023-0109

Cuong, N.H.: Influence of recycled steel fiber on the properties of self-compacting concrete with high fly ash content. J. Sci. Technol. Civ. Eng. (JSTCE)-HUCE, **18**(1), 55–67 (2024). https://doi.org/10.31814/stce.huce2024-18(1)-05

El Ouni, M.H., et al.: Mechanical performance, water and chloride permeability of hybrid steel-polypropylene fiber-reinforced recycled aggregate concrete. Case Stud. Constr. Mater. **16**, e00831 (2022). https://doi.org/10.1016/j.cscm.2021.e00831

Eren, N.A., Alzeebaree, R., Çevik, A., Niş, A., Mohammedameen, A., Gülşan, M.E.: The effects of recycled tire rubbers and steel fibers on the performance of self-compacting alkali activated concrete. Periodica Polytech. Civ. Eng. **65**(3) (2021). https://doi.org/10.3311/PPci.17601

Gao, D., Ji, D., Gu, Z., Yan, H., Zhang, Y.: Workability and mechanical properties analysis of hybrid fibers reinforced self-compacting concrete incorporating recycled aggregates based on acoustic emission technique. Structures **51**, 1722–1741 (2023). https://doi.org/10.1016/j.istruc.2023.03.139

Ghasemi, M., Ghasemi, M.R., Mousavi, S.R.: Investigating the effects of maximum aggregate size on self-compacting steel fiber reinforced concrete fracture parameters. Constr. Build. Mater. **162**, 674–682 (2018). https://doi.org/10.1016/j.conbuildmat.2017.11.141

Ghosh, S., Bansal, S., Choudhary, L., Sahu, V.: Effects of coconut fiber on mechanical properties of pervious concrete incorporating recycled concrete aggregates. Indian Conc. J. **98**(3), 43–52 (2024).

Hamza, B., Said, K., Belkacem, M.: The influence of recycled steel fibers on self-compacting concrete performance. IOP Conf. Ser.: Mater. Sci. Eng. **431**, 102008 (2018). https://doi.org/10.1088/1757-899X/431/10/102008

Hichem, M., Kada, A., Karim, B., Azzaz, M., Mohammed, E., Leklou, N.: Performance of self-compacting concrete containing recycled aggregates and recycled steel fibers. Int. J. Eng. Res. Afr. **68**, 51–66 (2024). https://doi.org/10.4028/p-GamA0s

Jahandari, S., et al.: Mechanical properties of recycled aggregate concretes containing silica fume and steel fibres. Materials **14**(22), 7065 (2021). https://doi.org/10.3390/ma14227065

Ji, D., Wu, Z., Gao, D., Gu, Z., Fang, D.: Analysis on rheological and workability properties of steel-polypropylene fibers reinforced self-compacting concrete with recycled coarse and fine aggregates. J. Build. Eng. **79**, 107931 (2023). https://doi.org/10.1016/j.jobe.2023.107931

Kalra, M., Kumar, G., Choudhary, L.: Seismic response of RCC framed structure with floating columns. Int. J. Sustain. Building Technol. Urban Dev. **9**(1), 18–30 (2018). https://doi.org/https://doi.org/10.22712/susb.20180003

Kamal, M.M., Safan, M.A., Etman, Z.A., Abd-elbaki, M.A.: Effect of steel fibers on the properties of recycled self-compacting concrete in fresh and hardened state. Int. J. Civ. Eng. **13**(4), 400–410 (2015)

Kannam, P., Rao, S.V, Kumar, P.R.: Influence of recycled aggregate on shear behavior of steel fibrous SCC. J. Mater. Eng. Struct. **5**(2), 185–205 (2018). https://asjp.cerist.dz/en/article/117812

Khaloo, A., Molaei Raisi, E., Hosseini, P., Tahsiri, H.: Mechanical performance of self-compacting concrete reinforced with steel fibers. Constr. Build. Mater. **51**, 179–186 (2014). https://doi.org/10.1016/j.conbuildmat.2013.10.054

Khatib, M., Saleh, Z., Jaber, W., Al-Amine, D., Choudhary, L., Mortada, H.: Effects of electromagnetic waves on bees' lives: implications for SDG 15 (Life on Land) and SDG 13 (Climate Action). In: Innovations in the Development of Sustainable Infrastructure (pp. 207-224) (2025). https://doi.org/10.1007/978-3-031-83750-0_14

Kiruthiga, P., Dave, N., Guduru, R. K., Choudhary, L.: Investigation of coal bottom ash as pozzolan in binder and mortar along with strength prediction using machine learning model. Innovative Infrastruct. Solutions. **10**(5), 184 (2025). https://doi.org/10.1007/s41062-025-01992-1

Kotwal, S., Singh, H., Kumar, R.: Experimental investigation of Steel Fibre reinforced Self Compacting Concrete (SCC) using recycled aggregates as partial replacement of coarse aggregates. Mater. Today: Proc. **48**, 1032–1037 (2022). https://doi.org/10.1016/j.matpr.2021.06.436

Koushkbaghi, M., Kazemi, M.J., Mosavi, H., Mohseni, E.: Acid resistance and durability properties of steel fiber-reinforced concrete incorporating rice husk ash and recycled aggregate. Constr. Build. Mater. **202**, 266–275 (2019). https://doi.org/10.1016/j.conbuildmat.2018.12.224

Kumar, N., Kumar, D., Swami: Mechanical properties of fiber reinforced geo-polymer concrete made with three mineral admixtures and steel fibre at different ambient curing periods. Adv. Sustain. Constr. Mater. 060021 (2023). https://doi.org/10.1063/5.0144501

Liu, Y., et al.: Fresh and hardened properties of sustainable RSF-Reinforced recycled aggregate SCC. J. Mater. Civ. Eng. **36**(9) (2024). https://doi.org/10.1061/JMCEE7.MTENG-17621

Maameri, N.E., Bensaci, H., Kenai, S.: The effect of treated recycled steel fibers and slag on the properties of recycled self-compacting concrete. MRS Adv. **8**(10), 613–617 (2023). https://doi.org/10.1557/s43580-023-00616-z

Mahmod, M., Hanoon, A.N., Abed, H.J.: Flexural behavior of self-compacting concrete beams strengthened with steel fiber reinforcement. J. Build. Eng. **16**, 228–237 (2018). https://doi.org/10.1016/j.jobe.2018.01.006

Malathy, R., et al.: Use of industrial silica sand as a fine aggregate in concrete—An explorative study. Buildings **12**(8), 1273 (2022). https://doi.org/10.3390/buildings12081273

Mardani, A., Hatungimana, D., Mardani, N., Assaad, J., El-Hassan, H.: Feasibility of steel fiber-reinforced self-compacting concrete containing recycled aggregates–compliance with EFNARC guidelines. Int. J. Sustain. Eng. **18**(1) (2025). https://doi.org/10.1080/19397038.2025.2538858

Mastali, M., Dalvand, A.: Use of silica fume and recycled steel fibers in self-compacting concrete (SCC). Constr. Build. Mater. **125**, 196–209 (2016). https://doi.org/10.1016/j.conbuildmat.2016.08.046

Mezzal, S.K., Al-Azzawi, Z., Najim, K.B.: Effect of discarded steel fibers on impact resistance, flexural toughness and fracture energy of high-strength self-compacting concrete exposed to elevated temperatures. Fire Saf. J. **121**, 103271 (2021). https://doi.org/10.1016/j.firesaf.2020.103271

Mohammed, V.R., Abdulhaleem, K.N., Hamada, H.M., Humada, A.M., Majdi, A.: Effect of recycled aggregate concrete and steel fibers on the fresh properties of self-compacting concrete. E3S Web Conf. **427**, 02013 (2023). https://doi.org/10.1051/e3sconf/202342702013

Mohseni, E., Saadati, R., Kordbacheh, N., Parpinchi, Z.S., Tang, W.: Engineering and microstructural assessment of fibre-reinforced self-compacting concrete containing recycled coarse aggregate. J. Clean. Prod. **168**, 605–613 (2017). https://doi.org/10.1016/j.jclepro.2017.09.070

Najim, K.B., Saeb, A., Al-Azzawi, Z.: Structural behaviour and fracture energy of recycled steel fibre self-compacting reinforced concrete beams. J. Build. Eng. **17**, 174–182 (2018). https://doi.org/10.1016/j.jobe.2018.02.014

Nalanth, N., Venkatesan, P.V., Ravikumar, M.S.: Evaluation of the fresh and hardened properties of steel fibre reinforced self-compacting concrete using recycled aggregates as a replacement material. Adv. Civ. Eng. **2014**, 1–6 (2014). https://doi.org/10.1155/2014/671547

Ghashouieh, M.N., Malekinejad, M., Amiri, M.: Microstructural analysis of the effect of using nano-silica on the mechanical properties of cement-sand mortar under the effect of heat. Int. J. Concr. Struct. Mater. **18**(1), 74 (2024). https://doi.org/10.1186/s40069-024-00715-7

Nitesh, K.J.N.S., Rao, S.V., Kumar, P.R.: An experimental investigation on torsional behaviour of recycled aggregate based steel fiber reinforced self compacting concrete. J. Build. Eng. **22**, 242–251 (2019). https://doi.org/10.1016/j.jobe.2018.12.011

Roy, D., Choudhary, L., Sharma, N., Sharma, N.: Study on physical properties of quaternary cement concrete with novocon XR steel fibers. Int. J. Sustain. Building Technol. Urban Dev. **9**(4), 197–208 (2018). https://doi.org/https://doi.org/10.22712/susb.20180020

Salam, N., Khalil, W.I.: The effect of scrap tires recycled steel fibers on some hardened properties of green SCC. Eng. Technol. J. **40**(5), 1–9 (2022). https://doi.org/10.30684/etj.v40i5.1133

Shawais, Z.A., Abdulhaleem, K.N., Ahmed, S.H., Hamada, H.M., Mohammed, V.R.: Influence of recycled coarse aggregate and steel fiber on the workability and strength of self-compacting concrete. IOP Conf. Ser.: Earth Environ. Sci. **1374**(1), 012084 (2024). https://doi.org/10.1088/1755-1315/1374/1/012084

Shukla, N.K., et al.: Predicting the compressive strength of SCC containing nano silica using surrogate machine learning algorithms. Comput. Concr. **32**(4), 373–381 (2023). https://doi.org/10.12989/cac.2023.32.4.373

Simalti, A., Singh, A.: Comparative study on performance of manufactured steel fiber and shredded tire recycled steel fiber reinforced self-consolidating concrete. Constr. Build. Mater. **266**, 121102 (2021). https://doi.org/10.1016/j.conbuildmat.2020.121102

Wang, J., Dai, Q., Si, R., Ma, Y., Guo, S.: Fresh and mechanical performance and freeze-thaw durability of steel fiber-reinforced rubber self-compacting concrete (SRSCC). J. Clean. Prod. **277**, 123180 (2020). https://doi.org/10.1016/j.jclepro.2020.123180

Wang, R., Yu, N., Li, Y.: Methods for improving the microstructure of recycled concrete aggregate: a review. Constr. Build. Mater. **242**, 118164 (2020). https://doi.org/10.1016/j.conbuildmat.2020.118164

Youvaraj, K., Prasanath, R.R., Manishankar, S., Dhamodhran, P., Mounika, M.: Study on recycled aggregate and steel fibers in self compacted concrete. AIP Conf. Proc. (2023). https://doi.org/10.1063/5.0155234

Zamani, A.A., Ahmadi, M., Dalvand, A., Aslani, F.: Effect of single and hybrid fibers on mechanical properties of high-strength self-compacting concrete incorporating 100% waste aggregate. J. Mater. Civ. Eng. **35**(1) (2023). https://doi.org/10.1061/(ASCE)MT.1943-5533.0004528

Section 2: Energy-Efficient Buildings, Net-Zero Infrastructure, Policy and Regulatory Frameworks

Decarbonizing Civil Infrastructure: Innovative Materials, Methods, and Metrics

Milad S. Khatib[1,2,3](✉) and Lokesh Choudhary[4]

[1] School of Engineering, Lebanese International University, Mouseitbeh, Beirut, Lebanon
milad.khatib@liu.edu.lb

[2] IFFT, University of Balamand, Koura, Lebanon

[3] Engineering Department, ISSEA-Cnam, City Sport, Beirut, Lebanon

[4] School of Engineering and Technology, Manav Rachna International Institute of Research and Studies, Faridabad 121004, India
lokeshchoudhary.set@mriu.edu.in

Abstract. Decarbonizing the civil infrastructure is not only about cutting emissions; it is the very essence of sustainable cities and industries. It is also the direct link to the UN's Sustainable Development Goals 9, 11, and 13. The construction sector contributes 40% of the global CO_2 emissions, largely due to the high energy intensity of the production of construction materials, the construction industry itself, and the high CO_2 footprint of the life cycle of buildings. In this article, the possibilities to cut the footprint will be explored along three avenues: the use of new, low-carbon materials; the use of forward-thinking construction techniques; and the use of standardized assessment techniques. New materials, such as geopolymers, recycled aggregates, bio-based binders, and CO_2-absorbing concretes, offer great potential for the reduction of the footprint. The use of forward-thinking techniques, such as digital optimization, modular construction, and circular economy principles, also offers great opportunities. The use of standardized techniques, especially life cycle analysis and carbon footprint analysis, offers the opportunity to set standards and align the way forward. The global case studies show that the reduction of the footprint is possible by as much as 80%, and as little as 10%. Unlike prior reviews that focus primarily on either material innovations or policy frameworks, this paper fills a research gap by synthesizing technical, methodological, and governance dimensions into a unified roadmap for decarbonizing civil infrastructure. The plan outlines a course of action that marries technology, policy, and governance to achieve decarbonization. The plan outlines quick wins for the present, such as CO_2-mineralized concretes and recycled aggregates, followed by mid-term strategies such as geopolymers and modular construction, and, finally, bold steps into the future with fossil-free steel and circular demolition. The sum of these strategies promises up to 90% reductions in embodied emissions by 2050, which will set the stage for a climate-resilient, low-carbon infrastructure future.

Keywords: decarbonization · civil infrastructure · low-carbon materials · life-cycle assessment · embodied carbon · sustainable construction

L. Choudhary et al. (Eds.): TIMES 2025, SCI, pp. 77–96, 2027.
https://doi.org/10.1007/978-3-032-34636-0_7

1 Introduction

One of the most resource-intensive sectors across the globe, civil infrastructures are among the biggest guzzlers of cement, steel, and other building materials while at the same time spewing out considerable greenhouse gases into the atmosphere. About 40% of the yearly CO_2 emissions are caused by constructions and buildings, according to UNEP 2020 statistics. As cities are expanding at a much faster rate than ever before, civil infrastructures are expected to increase by 70% by 2050. Hence, reducing the carbon footprint of civil infrastructures has become a pressing task.

The process of reducing the carbon footprint of civil infrastructures encompasses three aspects: the materials, the process of building these infrastructures, and measuring the impact. Most reviews have focused on either the materials or the process of building these infrastructures without bringing the methodological aspects into a cohesive plan to achieve the task at hand. Figure 1 presents the share of global CO_2 emissions from materials and construction, and operational phase. This work attempts to fill that gap by bringing various strands into a cohesive plan to weigh the options in both high-capability and resource-limited environments.

The development of new low-carbon building materials such as geopolymer cement has shown considerable promise to reduce the carbon footprint of civil infrastructures. New techniques of building these infrastructures, such as modular buildings, along with the concept of the circular economy, have shown considerable promise to reduce waste and increase energy efficiency. An equally important aspect is the development of standardized metrics such as life cycle assessment to measure the impact.

The paper reviews the state-of-the-art in these fields and develops a comprehensive framework to achieve the target of net-zero civil infrastructures by 2050. Rather than focusing on either the technical or policy aspects alone, the paper attempts to integrate these two aspects along with industry practices at different time scales. The case studies were studied from a variety of both developed and developing countries, such as Scandinavia, North America, and Lebanon, allows them to provide a comparative analysis, which is not thoroughly discussed in previous literature. This means that the main contribution of the paper is in its integrative approach. Specifically, the originality lies in unifying materials, construction methods, assessment tools, and policy frameworks into one integrated roadmap, an approach not previously attempted in a single review.

The work is underpinned by a narrative review of sources from 2017 to 2025, including peer-reviewed publications, as well as policy reports and case studies from across the globe. A total of 150 sources were scanned from major databases such as Scopus, Web of Science, and Google Scholar. Although not a systematic search, a structured search was conducted to obtain a wide scope of recent tech, policy, and market developments in the field. The literature search was conducted with the help of Scopus, Web of Science, and Google Scholar, with keywords such as "low carbon materials," "embodied carbon," "LCA construction," and "zero emission sites." Case studies were chosen from countries with both developed and developing resources, ensuring a worldwide scope of the subject matter, although not a rigorous systematic search, a wide scope of recent developments and recent advancements have been included, and a cohesive framework was developed.

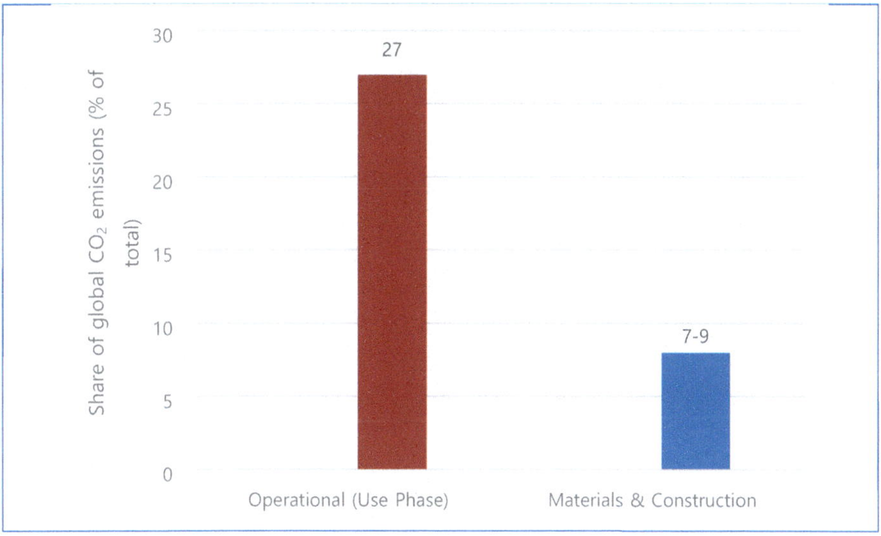

Fig. 1. Buildings-related CO_2 emissions split (2022) based on Global ABC/UNEP. (Source: Author's contribution)

To ensure openness, a systematic search of 152 peer-reviewed articles and 27 policy and industry reports was conducted, all of which were done following the established principles of PRISMA:

- Databases: Scopus, Web of Science, Google Scholar.
- Period covered: 2017–2025.
- Inclusion criteria focused on embodied carbon, life cycle assessment, low-carbon materials, innovative construction techniques, and policies related to infrastructure
- Exclusion criteria: papers focused solely on operational building energy use or non-infrastructure applications.

A literature search was conducted using a combination of keywords with Boolean logic. For example, the search terms used were "low carbon materials" AND "infrastructure" AND "life cycle assessment," "embodied carbon" AND "civil engineering," and "zero emission construction sites." The literature screening was conducted following the PRISMA method of identification, screening, eligibility, and inclusion. From the initial 450 sources, 230 duplicates were removed, giving 179 sources. The complete flowchart is shown in Fig. 2.

A systematic literature search was not conducted due to the extensive literature being scattered across various fields of engineering, policies, and industry practices. The structured narrative review was used to ensure an appropriate balance between breadth and depth of literature review. The drawback of this method is that there is a possibility of bias towards English literature and recently published literature.

Fig. 2. Screening Flow (Source: Author's contribution)

2 Decarbonization Drivers and Global Targets

The decarbonization of civil infrastructure is being driven by a convergence of international agreements, national and regional policies, market forces, technological advances, and societal expectations. With infrastructure accounting for a large share of global greenhouse gas (GHG) emissions, it has become a critical focus area in the transition to a sustainable, low-carbon future (UNEP, 2020). This section reviews the major global drivers shaping decarbonization strategies and highlights the barriers that should be addressed for successful implementation. Table 1 summarizes selected international, national, and municipal policies that are shaping infrastructure decarbonization and influencing project-level decisions.

Table 1. Selected policies shaping infrastructure decarbonization (Source: Author's contribution)

Jurisdiction	Policy	Year	Scope	Implications for Projects
European Union	Level(s) Framework	2021–2024 (updated guidance)	Common indicators for whole-life performance (incl. Embodied carbon), aligned with EN 15978	Harmonizes metrics, supports taxonomy-aligned reporting

(*continued*)

Table 1. (*continued*)

Jurisdiction	Policy	Year	Scope	Implications for Projects
California (USA)	Buy Clean California Act	2022 (EPD requirement); 2025 (updated GWP limits)	EPDs required; GWP limits for steel, glass, mineral wool; concrete guidance emerging	Drives low-carbon procurement using facility-specific EPDs
United Kingdom	UK Net Zero Carbon Buildings Standard	2024 (Pilot version)	Cross-industry limits/targets for operational & embodied carbon	Defines evidence-based net-zero alignment; informs future regulation
Oslo (Norway)	Zero Emission Construction Sites	2025	Municipal construction machinery/logistics to be zero-emission	Accelerates shift to electric equipment and clean site power

International and national policies increasingly address whole-life carbon. The EU's Level(s) framework harmonizes indicators for whole-life performance (CDGS, 2024); and the UK Net Zero Carbon Buildings Standard (CIBSE, 2024) pilots validated limits and reporting rules. Cities such as Oslo (City of Oslo, 2025) have gone further by mandating emission-free municipal construction sites, accelerating electrification of machinery and logistics.

A side-by-side comparison reveals that European systems, such as Level(s), promote the need for harmonization and benchmarking. Developing regions lack the institutional muscle to impose similar measures. Definitions of policy-driven embodied carbon vary from country to country, making it complicated to benchmark between nations. One of the main underlying causes of the lack of enforcement of embodied carbon measures is the fragmented governance framework. In many countries, national policy does not come with municipal implementation plans. Developing regions are hindered in the implementation of embodied carbon measures by institutional capacity and procurement systems. In the case of developed nations, it is not the lack of intent but the inconsistency in definitions and methods used to hinder enforcement.

2.1 International Climate Commitments

Multilateral institutions are also refining the goal of limiting global warming to well below 2 °C above pre-industrial levels, with an aspiration that could limit it to 1.5 °C. The International Energy Agency (IEA) has stated that to remain on the 1.5 °C trajectory, cement emissions should be reduced by 95% globally by 2050 compared to 2019 levels (IEA, 2024). The realization of these reductions depends on rapid innovation in low-carbon technologies.

2.2 Regional and National Initiatives

In all these places, governments are launching measures to reduce greenhouse gas emissions from the construction industry. For instance, the European Union has developed the Level(s) initiative to provide a common set of indicators to measure carbon footprint across the building life cycle. The United Kingdom has launched a Net Zero Carbon Buildings Standard that mandates government projects to adhere to embodied carbon requirements (CIBSE, 2024).

In the United States, the Infrastructure Investment and Jobs Act (IIJA) allocate considerable funding to upgrade roads, bridges, and transit systems with low-carbon materials and climate-resilient design (2021). Several states have also implemented "Buy Clean" policies that mandate structural steel, concrete, and asphalt to meet carbon disclosure and performance requirements for government projects.

Asian countries are also advancing aggressively to meet the Paris Agreement goals. For instance, Japan's Strategic Energy Plan targets low-carbon production of cement and steel through hydrogen-based production and CCUS technologies (METI, 2021).

2.3 Market and Industry Drivers

Government policies are now being supplemented by the actions of the market. Investors are demanding better reporting on environmental, social, and governance (ESG) factors, and companies with high carbon footprints are facing pressure from shareholders and creditors.

The construction industry is also moving in the right direction. The top companies are now pledging to Science-Based Targets (SBTi) to ensure that their strategies are aligned with the 1.5 °C trajectory. The industry is expanding the portfolio of its sustainable products, while digital tools such as Building Information Modeling (BIM) and Life Cycle Assessments (LCA) are entering day-to-day operations to monitor and reduce emissions (Volvo Construction Equipment, 2024).

2.4 Social and Environmental Pressures

The awareness of the effects of climate change, such as flooding, rising sea levels, heat, and infrastructure vulnerability, has accelerated the need to change. Civil society, environmental groups, and professional bodies of engineers are advocating for infrastructure that is environmentally friendly and sustainable. The people of the cities, especially those in the most exposed areas of Europe (Rider Levett Bucknall, 2025), are demanding infrastructure designs that are environmentally friendly and resilient to extreme weather conditions.

2.5 Barriers to Decarbonization

Despite the momentum, there are still significant challenges to overcome:

- Cost and economic factors: New technologies such as geopolymer cement and carbon capture concretes are generally more expensive to produce than traditional cement and concrete, which acts as a deterrent.

- Supply chain challenges: As coal-fired power plants are phased out, so too are the sources of industrial waste such as fly ash and slag used for geopolymer-based materials.
- Technical challenges: There is little understanding of the longevity of these new materials and technologies.
- Standardization challenges: In the absence of standardized carbon reporting, there are inconsistencies that make it difficult to track progress.
- Institutional challenges: Traditional construction methods are often resistant to change.

2.6 Synthesis

The pace of moving towards a low-carbon world is accelerating with increased political, market, and people support globally. However, there are still challenges related to costs, supply chains, technology, and standards that need to be overcome for a wide-scale deployment.

To meet the goals of the Paris Agreement and the United Nations Sustainable Development Goals, specifically SDG 9 Industry, Innovation and Infrastructure, SDG 11 Sustainable Cities and Communities, and SDG 13 Climate Action, a multi-faceted approach for the decarbonization of infrastructure needs to be taken into consideration, where the government needs to provide clear regulations and financial incentives, industries need to accelerate innovation and scale up their supply chains, and academia needs to accelerate their research and data to close the performance gap.

A comparative analysis of these challenges and solutions reveals that, in most cases, policy ambition lags far behind enforcement capabilities, such as the case with the EU's Level(s) framework, which provides standardized measurement, yet adoption of the framework in most EU countries is still on a voluntary basis, whereas Oslo's zero-emission sites demonstrate the power of mandatory procurement for a faster rate of adoption, where regulatory clarity and enforcement capabilities are key drivers of decarbonization, not ambition, as demonstrated in Oslo (City of Oslo, 2025).

3 Low-Carbon and Carbon-Neutral Materials

Substitutions and mix optimizations have the potential to reduce embodied carbon significantly. Some indicatives were summarized cradle-to-gate global warming potential (GWP) factors, and recent literature was consulted to assist with early decision-making. These values are location-dependent, energy mix-dependent, and supplier environmental product declarations (EPDs), and where available, project teams should prefer verified supplier EPDs.

3.1 Geopolymer and Alkali-Activated Binders

Geopolymers, derived from fly ash, slag, and other aluminosilicates, achieve reductions in CO_2 emissions of up to 80% compared to PC (Provis, 2018). Additionally, they provide improved durability properties, making them suitable for applications in bridges and marine structures. Alkali-activated slag-fly ash binders and calcined clays achieve reductions in global warming potential of 40–80% compared to OPC. This has been achieved in precast and in-situ applications. The main challenges are the scarcity of materials and the need for mixed design expertise.

3.2 Recycled Aggregates and Industrial By-Products

It reduces the demand for raw aggregates and can also help in reducing the amount of carbon in structures by up to 30% (Xiao et al., 2018). Similarly, slag and silica fume can also act as alternatives to cement. When RAC is used along with cement-efficient mix and quality control measures, it results in significant savings. Industrial waste products like GGBS/slag, fly ash, and silica fumes contribute to substitution and durability.

3.3 Bio-Based Materials

There are also renewable products like hempcrete, bamboo composites, and mycelium. Though their mechanical properties are still behind those of conventional products, hybrid products are closing the gap. Mass timber products like CLT and glulam are reducing environmental footprints and speeding up construction times. Carbon sequestered in biogenic materials should be tracked separately in the ledger, forestry practices should be sustainable, and fire/insurance issues are design related. CLT and other mass timber products are expanding their applications. The projects completed in 2022–2023 have potential for mid-to high-rise applications (Santos et al., 2025). However, barriers remain, including stringent fire safety requirements, increased insurance premiums, and regulatory uncertainty in many jurisdictions. Addressing these issues will be critical for scaling adoption.

3.4 Carbon-Sequestering Concretes

Technologies like CO_2 curing and mineral carbonation enable concrete to store carbon in the long term (Monkman & Shao, 2006). Pilot studies have shown that concrete curing can store up to 15% of injected CO_2. When mixed with concrete and exposed to injected CO_2, some of it gets fixed in the concrete in the form of stable carbonates. This reduces the amount of cement required in concrete mix by about 4–6% without reducing strength. This has shown potential in reducing global warming potential in cradle-to-gate studies. Yet the rate of scaling up is restricted to the availability of curing technologies and verification systems. Compared to geopolymers and bio-based materials, CO_2 curing of concrete is incremental in potential but more viable in the short term given the existing supply chains.

A meta-analysis of about 40 studies indicates average potential savings of about 65% in embodied carbon for geopolymers, 45% for bio-based materials, 25% for recycled aggregates, and 12% for CO_2-mineralized concretes. Regional studies also indicate larger potential savings in European studies compared to Asian studies. This is attributed to the cleaner energy mix in European nations. This indicates that material potential cannot be considered universal and should always be considered in regional contexts.

Despite the potential of these materials in reducing greenhouse gases, trade-offs are also involved. The future availability of GGBS and fly ash is uncertain in the coming decades due to the shutdown of coal plants. Bio-based materials are also considered to be more sustainable but also raise concerns about regional certification and land use. Geopolymers also indicate high potential in reducing greenhouse gases but are restricted in scaling up potential since the supply chains are not fully developed. The comparative embodied-carbon performance of conventional and alternative materials is presented in Table 2.

Table 2. Comparison of Low-Carbon Materials (Source: Author's contribution)

Material	Embodied Carbon Range (kg CO_2 e/m^3)	Notes	References
Normal-strength concrete (30 MPa, OPC-based)	250–350	Standard Portland cement mixes	(UNEP, 2020)
Concrete with SCMs (30–40% GGBS or fly ash replacement)	150–220	Lower-carbon alternative	(Provis, 2018)
High-strength concrete (60 MPa +)	400–500	Due to higher clinker and admixture demand	(UNEP, 2020)
Recycled aggregate concrete (RAC)	200–300	Depends on RCA replacement ratio	(Xiao et al., 2018)
Structural steel (Virgin production)	11,000–13,000	Very high due to energy intensity	(IEA, 2024)
Structural steel (70–90% recycled, EAF route)	500–1,200	Much lower with scrap-based production	(Structural Engineering Institute, 2024)
Cross-laminated timber (CLT)	150–200	Often carbon-negative when biogenic storage included	(Santos et al., 2025)
Bamboo composites	100–150	Highly renewable, fast-growing feedstock	(Bredenoord, 2023)
Fired clay bricks	200–250	Strong regional variation	(UNEP, 2020)
Glass (float)	1,000–1,200	High embodied energy	(SE2050, 2023)

4 Innovative Methods and Construction Practices

Emerging practices, digital design optimization, prefabrication, 3D printing, and zero-emission construction sites, reduce both material intensity and site emissions. Cities including Oslo report > 80% zero-emission machinery usage on municipal projects (City of Oslo, 2025), with a 2025 mandate for emission-free sites; similar measures are being evaluated globally. Table 3 lists selected tools and standards that are widely applied for whole-life and embodied carbon assessment in infrastructure projects.

Table 3. Selected tools and standards for whole-life/embodied carbon assessment. (Source: Author's contribution)

Tool / Standard	Purpose	Typical Use	Notes
EN 15978 (standard)	Method for assessing environmental performance of buildings	Basis for Level(s), WLC reporting	Defines system boundaries, modules A–D
EU Level(s) framework	Common indicators for sustainability incl. WLC	Design-to-operation benchmarking	Supports Taxonomy-aligned reporting
EC3 (Building Transparency)	Free database of EPDs; benchmarking materials' GWP	Procurement stage comparisons	Relies on product-specific EPDs
One Click LCA	Commercial LCA tool with databases and reporting	Whole-building LCA & credits	Integrates with BIM workflows

4.1 Digital Design and Optimization

Building Information Modeling (BIM) offers a robust digital environment that supports parametric design, enabling architects and engineers to iteratively explore design alternatives with immediate insight into performance and sustainability. Integration of digital twins and AI optimization has been increasingly applied in Europe and Asia to monitor and reduce embodied emissions in real-time (Yang et al., 2024). This integration empowers stakeholders to make data-driven choices from the outset, embedding sustainability into the project's DNA and avoiding late-stage redesigns.

4.2 Modular and Off-Site Construction

Modular and prefabricated construction methods represent a shift toward efficiency and environmental responsibility. For instance, LCA assessments of modular high-rise residential buildings in the UK found embodied carbon to be approximately 10% lower than conventional builds (Hu et al., 2025). These findings underscore off-site construction's potential for meaningful carbon reductions.

Nevertheless, scalability depends on supply chain integration, factory availability, and transport logistics, which may limit impact in regions with dispersed populations or weak industrial bases. Although modular systems demonstrate clear carbon reductions in urbanized contexts, their benefits diminish in dispersed geographies due to transport emissions and the absence of prefabrication facilities. This context dependency is rarely accounted for in current LCA studies, suggesting a research gap in quantifying logistics-related trade-offs. Despite these benefits, modular approaches also face potential rebound effects from shorter construction times accelerating urban expansion. These trade-offs highlight the need for context-sensitive deployment and careful life-cycle boundary setting in LCAs.

4.3 Renewable-Powered Construction Processes

Industry pilots are proving that construction equipment powered by renewables can actually work in the real world. The Electric Worksite research in Sweden tested battery-powered and grid-connected electric construction equipment ranging from 3.5 to 30 tons and showed they can work as well as, or better than, their diesel-powered counterparts. And without the emissions and racket. Volvo CE News (2024) writes that "electric cells are becoming increasingly viable options for decarbonizing the construction industry."

4.4 Circular Economy in Construction

The implementation of circular economy strategies is on the rise, and this is being made possible through the use of material passports, which provide digital information on the origin, properties, and recyclability of materials used within buildings, thus guiding their reuse towards sustainability. Current industry insights show that material passports improve transparency, retain material value, and promote effective reuse and reduction of waste (RLB Europe, 2025).

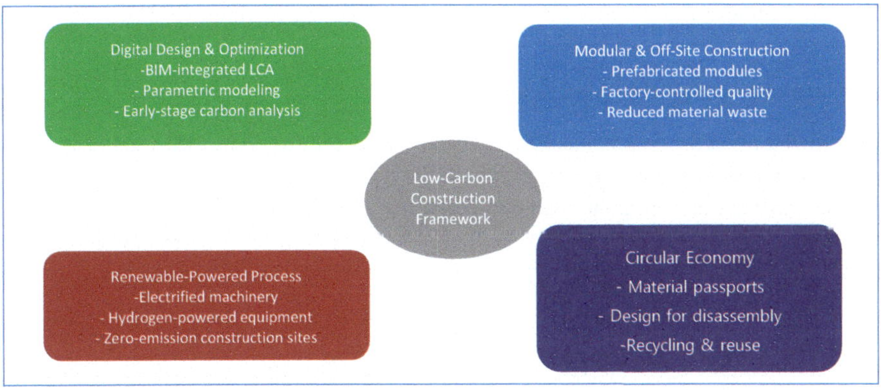

Fig. 3. Framework for Innovative Low-Carbon Construction Methods (Source: Author's contribution)

The framework, as shown in Fig. 3, arranges innovative practices in three levels of digital optimization, modularization, and renewable-powered processes. The figure helps visualize the relationships between different innovative practices at different stages of the project lifecycle, from design-stage changes to site implementation. The figure also helps to highlight relationships between different practices that may be overlooked in single-case descriptions of an innovative practice. After checking the numbers, digital optimization and modular construction showed to shave 10 to 22% off average impacts, while sites powered by renewables have been shown to have reduced emissions by as much as 80% in controlled environments. These are averages that give us a practical guideline to scale theory to practice.

5 Metrics, Assessment Tools, and Standardization

Metrics and tools are the basis upon which embodied and operational carbon are measured in a standardized manner, from project to project. More recently, the focus has been on aligning methodologies and expanding open datasets to increase comparability and transparency in the field.

5.1 Life-Cycle Assessment (LCA)

Recent studies proposed "lcpy," which is an open-source Python package that facilitates dynamic and parametric life cycle assessment and life cycle costing, including time-variant factors and uncertainty analysis.

5.2 Embodied Carbon Benchmarks

However, databases of materials are still important for benchmarking carbon embodied in buildings. There was an update to tools such as tallyLCA and OneClick LCA in 2023, which improves the representation of end-of-life scenarios (Module D) for carbon credits related to steel and wood materials (Structural Engineering Institute, 2024).

5.3 Carbon Footprinting Methodologies

Standardized procedures, such as PAS 2050, which is published by BSI (2011), govern the calculation of carbon footprint, although the distinction in defining boundaries and interpreting data does not guarantee comparability among projects.

5.4 Toward Unified Frameworks

However, unified platforms are gaining traction. Consider the Embodied Carbon in Construction Calculator (EC3), a free and open-access solution that has access to a massive database of verified Environmental Product Declarations (EPDs). This enables benchmarking, specification, and sourcing of materials with reduced environmental impact. When integrated into BIM systems, it enables designers and builders to easily identify areas to reduce embodied carbon in the early stages of design (OneClickLCA, 2025).

Despite the number of tools being developed, the industry still has problems to solve in terms of fragmentation. The allocation of recycled content in materials shows significant variability between tools, leading to inconsistent results from LCA software. Harmonization of results will not only require the usual reported results but also the underlying methodological assumptions. For example, EC3 emphasizes transparency with EPD-based benchmarking, while OneClickLCA integrates BIM workflows but relies on commercial databases (OneClickLCA, 2025). This divergence illustrates the lack of harmonization across tools, designing barriers to consistent global benchmarking.

6 Global Case Studies

Having established the key metrics, the second part turns into international case studies to show how the theory works in practice. It highlights real-life examples of low-carbon infrastructures such as bridges, pavements, and buildings in various geographical locations to showcase the innovative practices.

6.1 Bridges

For instance, the Queensferry Crossing in Scotland reduced about 7,000 tons of embodied carbon through the reworking of the Ferrytoll Viaduct. This involved trimming the concrete through smarter design and the use of materials from the site. This is an instance of how structural changes to large infrastructural projects can result in reduced emissions (Ramboll, 2025).

6.2 Pavements

Across the European continent, including the Netherlands, tests are underway to use asphalt that is mixed with recycled plastic waste. A recent study indicates that for every kilometer, as much as 3 tons of CO_2 savings could be achieved by using plastic asphalt, as opposed to burning the waste or using other materials (Highways Today, 2025). In addition to the storage of plastic waste, the approach also improves the lifespan of the road.

6.3 Buildings

The Punggol Digital District, located in Singapore, is a prime example of a successful implementation of the concept of integrated decarbonization. The district includes an AI-enabled smart grid, a solar array, and district energy systems, all designed to help cut 1,700 tons of operational carbon annually, as well as providing 3,000 MWh of free and clean energy, which is equivalent to powering 11,000 three-room HDB flats annually (Energy Transition Net, 2025). In addition to the smart grid and the solar array, the district also includes cooling systems that work at the district level, thus providing enhanced energy efficiency by as much as 30% when compared to traditional standalone systems. The district also includes an integrated procurement system that requires the use of low-carbon concrete mixes, which have Environmental Product Declarations.

6.4 Expanded Case Studies

Global decarbonization of the built environment is happening via a variety of strategies, each tailored to the unique circumstances, resources, and policy environments of the respective regions. The Lebanese efforts using recycled aggregates, combined with the EU's ambitious plan to achieve fossil-free steel and zero-emission sites. These case studies highlight the acceleration of innovation in both resource-scarce and technologically advanced economies. At the same time, the US and Canada are also focusing on procurement strategies and mass timber as new areas of focus, and the Scandinavian countries are leading the way to construction sites that use zero emissions. In all these regions, the level of embodied carbon has decreased by as much as 90% when comparing the traditional use of steel to the use of recycled aggregates. In the case of the city of Oslo, the level of emissions is 70–80% lower than traditional diesel engines. These case studies offer valuable lessons for the global community. There are two different routes that the global community is taking: systemic innovation in high-income economies, like the zero-emission sites of the city of Oslo, 2025, and incremental material substitution in resource-constrained areas, like the Lebanese case, UNDP Lebanon, 2022. While both routes are important, systemic innovation is the key to setting the global agenda. The incremental approach, on the other hand, is useful for the short term, especially when resources and capabilities are scarce.

This comparative analysis shows that there is a clear dichotomy between the strategies of richer and less resourceful areas, where richer areas seek systemic breakthroughs such as mass timber and zero emissions, while less resourceful areas prefer step-by-step material substitutions. This dichotomy shows that one size does not fit all and that strategies should be appropriate to each local context. If systemic breakthroughs, such as Sweden's push for fossil-free steel or Oslo's drive for zero emission sites, are encouraged, then there is great hope for real breakthroughs and quick change through good regulation and market engagement. However, even with less ambitious strategies, such as Lebanon's use of recycled aggregates, there is still hope for real benefits, especially when resources and institutions are limited. Table 4 summarizes regional strategies, showing CO_2 savings and their level of transferability.

Table 4. Summarizing Region, Strategy, Reported CO_2 Savings, Scalability. (Source: Author's contribution)

Region	Strategy	CO_2 Savings	Scalability/Transferability
Scotland	Structural optimization (bridge)	7,000 t saved	High, limited only by design culture
Netherlands	Plastic-modified asphalt	3 t/km saved	Medium, depends on plastic supply
Singapore	District smart grid + solar	1,700 t/year	High in dense urban contexts
Lebanon	RCA pilot projects	15–20% reduction in aggregates	Early stage, requires policy support

6.4.1 Lebanon

Despite its limited resources, Lebanon has initiated early research into low-carbon concretes and recycled aggregates, particularly through collaborating with local universities. In Beirut, early pilot projects have explored the use of recycled concrete aggregates (RCA) as partial substitutes for virgin aggregates in non-structural concrete, achieving embodied carbon reductions of approximately 15–18%. (UNDP Lebanon, 2022). Life cycle assessment information from pilot projects undertaken in Lebanon has shown that using 20% of recycled concrete aggregate (RCA) instead of virgin aggregate can lead to a reduction of 15 to 18% of embodied carbon in non-structural components of buildings, which can be cost-neutral at scale (UNDP Lebanon, 2022). However, some issues still need to be addressed, such as the quality of RCA, the lack of a standard framework for EPDs, and the scarcity of recycling facilities.

Furthermore, to ensure the adoption of low-carbon technologies in resource-scarce countries, supportive policies are necessary, such as the mandatory segregation of waste and the inclusion of quotas of RCA in public projects. Development programs have also been undertaken to promote solar-powered construction equipment for resilient rebuilding.

6.4.2 European Union

Scandinavia has become a pioneer of emission-free construction. Oslo mandated that all municipal construction sites should be zero-emission by 2025, with pilot sites already using battery-electric excavators and hydrogen-powered trucks. Meanwhile, Sweden's *HYBRIT* project demonstrates fossil-free steel production, which could cut embodied carbon in structural elements by up to 90% (HYBRIT Development, 2021).

6.4.3 United States

California's Buy Clean legislation has produced a model for embodied-carbon procurement. San Francisco International Airport's Terminal 1 redevelopment used low-carbon concrete mixes with EPD reporting, achieving significant reductions (San Francisco International Airport, 2021).

6.4.4 Africa, South Asia, and Latin America

In Africa, Rwanda's Green City Kigali pilot integrates local low-carbon building materials with renewable energy infrastructure (UN-Habitat, 2022). In India, several metro rail projects have mandated the use of fly-ash blended concretes, achieving up to 25% reductions in embodied carbon (Kumar and Verma, 2025). In Latin America, Colombia's Vivienda VIS program is piloting bamboo and recycled aggregate housing as low-cost and low-carbon alternatives (Bredenoord, 2023).

6.4.5 Mass Timber Pilots

Canada and Austria are leaders in mass timber high-rises. Vancouver's Brock Commons Tallwood House, an 18-storey student residence, cut construction time by ~30%

and embodied carbon by ~25% relative to a concrete equivalent (University of British Columbia, 2018).

6.4.6 Zero-Emission Sites

Beyond Oslo, Helsinki has piloted similar sites powered entirely by renewable electricity (Helsinki City Construction Services, 2023). These pilots provide insight into the challenges in charging networks in the real world.

7 Roadmap for Decarbonizing Civil Infrastructure

To achieve net-zero targets for civil infrastructure by 2050, momentum in four parallel streams of work should be sustained: Materials, Methods, Metrics, and Policy. Each of these streams has its own short-term, medium-term, and long-term milestones to achieve a whole-system transformation. The roadmap below illustrates the parallel evolution of these streams of work.

7.1 Materials

- Short-term (0–2 years): Increase supplementary cementitious materials, launch CO_2-mineralized concrete pilot projects, and step up recycled aggregates for non-structural purposes. Launch demonstration projects for bio-based materials such as hempcrete, bamboo, and mycelium composites.
- Medium-term (3–10 years): Develop geopolymer-based binders and calcined clays, and grow mass timber hybrid systems for mid-rise buildings, and carbon sequestering concretes in building specifications. Establish industrial symbiosis for a stable fly ash and slag feedstock.
- Long-term (10–25 years): Transition to near zero-clinker cements, fossil-free steels, and bio-based composites. In 2050, structural materials should have a negligible carbon footprint, accompanied by a circular economy for demolition and mineralization of construction waste.

7.2 Methods

- Short-term (0–2 years): Meld BIM with life cycle assessment to inform early-stage carbon decisions; test modular and prefabricated products in housing and schools; launch electrification of light construction equipment.
- Medium-term (3–10 years): Accelerate modular and off-site construction in infrastructure such as bridges and pavements; launch zero-emission fleets for urban infrastructure; scale up renewable-powered sites. Consider other ideas like the concept of material passports or digital tracking of reusable items.
- Long-term (10–25 years), digital twin technology will be fully implemented to track carbon footprint in real-time, autonomous zero-emission machines will be introduced for the construction industry, and material passports will be implemented using blockchain technology to ensure the reuse/recycling of materials.

7.3 Metrics

- Short-term (0–2 years): The focus is on specifying Environmental Product Declarations (EPDs) for the most critical materials, promoting open-access benchmarking platforms like EC3, and aligning regional databases with international specifications.
- Medium-term (3–10 years): The plan is to establish whole-life carbon (WLC) specifications by asset type, integrate carbon benchmarking into the tendering process, and enhance compatibility of LCA with BIM platforms for automatic reporting.
- Long-term (10–25 years): The vision is to achieve international standardization of carbon accounting schemes, provide real-time dynamic carbon monitoring for infrastructure assets, and integrate carbon performance into finance and insurance mechanisms.

7.4 Policy

- Short-term (0–2 years): The plan is to improve how to buy and use materials, such as by advancing strong public procurement policies such as Buy Clean and Level[s], providing tax incentives for clean materials, and mandating pilot projects that achieve net-zero emissions.
- Medium-term (3–10 years): The goal is to extend the regulatory net to cap embodied carbon, align building codes with net-zero aspirations, develop carbon pricing for construction materials, and ensure that large cities achieve emission-free municipal sites by 2030.
- Long-term (10–25 years): The aim is to achieve binding international agreements on embodied carbon, mainstream circular economy rules into construction, link carbon performance with urban planning approvals, and set global standards for net-zero certified infrastructure.

The entire plan is presented as a feedback loop, with the following elements:

- Drivers: climate, market, and societal expectations.
- Strategies: low-carbon materials, digital construction, circularity, and environmental product declarations (EPDs).
- Barriers: economics, technology, institutions, and society.
- Outcomes: reduced emissions, enhanced resilience, and better alignment with the Sustainable Development Goals.

Furthermore, the identification of the major contribution of the manuscript, namely, viewing the decarbonization problem as a system, with actions that are interconnected and interdependent, rather than a set of separate actions.

7.5 Integrated Roadmap

The four tracks support each other. For example, for short-term materials innovation such as CO_2-mineralized concrete, supportive policies such as EPD regulations and metrics definition through LCA data bases are necessary. On the other hand, for long-term innovation such as fossil-free steel and autonomous zero emission sites, supportive governance and strong finance are necessary. Finally, by 2050, these four tracks come together to provide infrastructure that is carbon neutral, resilient, and circular.

Figure 4 shows four parallel tracks, each for Materials, Methods, Metrics, and Policy, each with short-term, medium-term, and long-term milestones. When these four tracks come together, they provide the necessary foundations to reach net-zero by 2050.

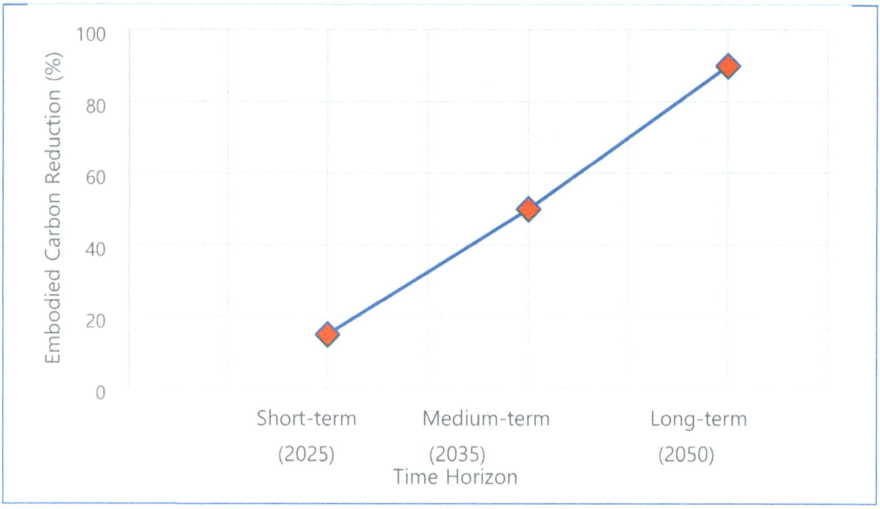

Fig. 4. Roadmap to Net-Zero Civil Infrastructure by 2050 (Source: Author's contribution)

If these milestones are rolled out in order, they could reduce embodied emissions by 40–50% by 2035 and 80–90% by 2050, in step with IPCC net zero pathways. To be specific, some of the targets are: at least 30% of Portland cement replaced with supplementary cementitious materials/geopolymers by 2035; 50% BIM-LCA integration in major infrastructure projects by 2030; zero-emission municipal construction sites across all OECD countries by 2035; and fossil-free steel covering 40% of global demand for structural steel by 2045. These values are drawn from IEA (2024) and Hybrit (2021), and where not directly referenced, they should be regarded as indicative scenario targets rather than binding forecasts. These targets provide a framework that policymakers and players in the industry can work with to monitor their progress. They drive home the fact that time is of the essence, and at the same time, communicate that the technology and processes required to meet these targets and reach these goals are possible.

There are four tracks, each with its own set of milestones, that are developed for the roadmap: materials, methods, metrics, and policy, each with their own set of short-term, mid-term, and long-term goals. This framework recognizes that, for instance, breakthroughs in nearly zero-clinker cements and fossil-free steel are linked with the development of standardized regulations for a circular economy, while early policy interventions such as mandated environmental product declarations are linked with the adoption of CO_2-mineralized concrete.

8 Conclusion

This paper seeks to underscore the pressing need to decarbonize civil infrastructure, particularly through materials, construction techniques, and carbon metrics. Although challenges persist, it is argued that an amalgamation of good policies, innovation, and industry efforts can accelerate progress towards this end. Not only is the pursuit of carbon-neutral infrastructures possible, but it is also necessary to achieve climate change goals set at the global level. What makes the current study stand out is its ability to integrate different aspects such as materials, construction techniques, measurement indicators, and policies into a single framework that is phase-based. It builds on previous studies such as the RICS Whole Life Carbon Standard (CIBSE, 2024), PAS 2080 (BSI, 2011), among other such studies, by moving forward with the latest advancements in technology and policy.

Conflict of Interest. The authors declare that there is no conflict of interest.

References

United Nations Environment Programme (UNEP). 2020 global status report for buildings and construction (2020). https://globalabc.org/sites/default/files/inline-files/2020%20Buildings%20GSR_FULL%20REPORT.pdf

California Department of General Services. Buy Clean California Act (2024). https://www.dgs.ca.gov/pd/resources/page-content/procurement-division-resources-list-folder/buy-clean-california-act

Chartered Institution of Building Services Engineers. UK net zero carbon buildings standard: pilot version (2024). https://www.cibse.org/policy-insight/key-policy-areas/uk-net-zero-carbon-buildings-standard-uknzcbs

City of Oslo, Klimaetaten. Barriers and key success factors for zero-emission construction in Oslo (2025). https://www.klimaoslo.no/rapport/barriers-success-factors-for-zero-emission-construction-in-oslo

International Energy Agency. Cement: tracking report (2024). https://www.iea.org/energy-system/industry/cement

U.S. Congress. Infrastructure Investment and Jobs Act (Public Law No. 117–58) (2021). https://www.congress.gov/bill/117th-congress/house-bill/3684

Ministry of Economy, Trade and Industry. Strategic energy plan (2021). https://www.enecho.meti.go.jp/en/category/others/basic_plan/

Provis, J.L.: Alkali-activated materials. Cem. Concr. Res. **114**, 40–48 (2018). https://doi.org/10.1016/j.cemconres.2017.02.009

Xiao, J., Li, W., Fan, Y., Huang, X.: An overview of study on recycled aggregate concrete in China (1996–2011). Constr. Build. Mater. **31**, 364–383 (2018). https://doi.org/10.1016/j.conbuildmat.2011.12.074

Santos, F.L.D.L., Santos, H.F., Costa, L.L., Lima, V.S., Cunha, R.D.A., César, S.F.: Mass timber as a tool to sustainable construction: a review. Revista de Gestão Soc. e Ambiental **19**(3), e011329 (2025). https://doi.org/10.24857/rgsa.v19n3-007

Monkman, S., Shao, Y.: Assessing the carbonation behavior of cementitious materials. J. Mater. Civ. Eng. **18**(6), 768–776 (2006). https://doi.org/10.1061/(ASCE)0899-1561(2006)18:6(768)

Yang, Z., Tang, C., Zang, T., Doan, D.T.: Digital twins in construction: Architecture, applications, trends and challenges. Buildings **14**(9), 2616 (2024). https://doi.org/10.3390/buildings14092616
Hu, Y., Luyao, X., Shang, K.: The research review on life cycle carbon emissions in the operational process of modular buildings. Buildings **15**(12), 2085 (2025). https://doi.org/10.3390/buildings15122085
Volvo Construction Equipment. Electric worksite study proves viability of urban electric construction (2024). https://www.volvoce.com/global/en/news-and-events/news-and-stories/2024/study-proves-viability-of-urban-electric-construction/
Rider Levett Bucknall. The role of material passports in advancing construction's circular economy (2025). https://www.rlb.com/europe/insight/the-role-of-material-passports-in-advancing-constructions-circular-economy/
Structural Engineering Institute. Comparison of LCA tools from ECIDs round 2 (2024). https://se2050.org/resources-overview/embodied-carbon/ecid-comparison-of-lca-tools/
British Standards Institution. PAS 2050:2011: Specification for the assessment of the life cycle greenhouse gas emissions of goods and services (2011). https://shop.bsigroup.com/ProductDetail/?pid=000000000030180242
One Click LCA. Top LCA software for 2025 (2025). https://oneclicklca.com/en/resources/articles/lca-software
Ramboll. Queensferry crossing: huge carbon saving (2025). https://www.ramboll.com/projects/transport/queensferry-crossing-huge-carbon-saving
Highways Today. Transforming asphalt in the age of climate pressure (2025). https://highways.today/
Energy Transition Net. Case study: decarbonization success in Singapore's utility sector (2025). https://www.energytransitionnet.com/2025/07/07/case-study-decarbonization-success-in-singapores-utility-sector/
United Nations Development Programme Lebanon. Sustainable construction and reconstruction initiatives in Lebanon (2022). https://www.undp.org/lebanon/
HYBRIT Development. HYBRIT projects and initiatives (2021). https://www.hybritdevelopment.com/
San Francisco International Airport. Terminal 1 redevelopment sustainability fact sheet (2021). https://www.flysfo.com/
UN-Habitat. Green city Kigali: Africa's first green city pilot (2022). https://unhabitat.org/
Kumar, S., Verma, A.: A research on the feasibility of using landfill fly ash as Portland cement substitute. Int. J. Multi. Res. **7**(3) (2025). https://doi.org/10.36948/ijfmr.2025.v07i03.48445
Bredenoord, J.: Sustainable building materials for low-cost housing and the challenges facing their technological developments: examples and lessons regarding bamboo, earth-block technologies, building blocks of recycled materials, and improved concrete panels. J. Architectural Eng. Technol. **6**(1), 104550 (2023). https://doi.org/10.4172/2168-9717.1000187
University of British Columbia. Brock commons tallwood house: a case study (2018). https://cwc.ca/wp-content/uploads/2018/07/CS-BrockCommon.Study_.23.lr_.pdf
Helsinki City Construction Services. Pilot projects on emission-free construction sites (2023). https://www.hel.fi/

A Review on Energy Efficiency Analysis & Occupant Comfort in High Rise Buildings

Rajesh Kumar Yadav(✉), Anu Tonk, and Akanksha Mathur

Department of Multidisciplinary Engineering, The Northcap University, Gurugram, Haryana 122017, India
{23med002,anutonk,akankshamathur}@ncuindia.edu

Abstract. Energy consumption in high-rise hotel buildings has emerged as a critical sustainability challenge owing to their continuous operation, high dependence on heating, ventilation, and air-conditioning (HVAC) systems, intensive lighting demand, vertical transportation requirements, and stringent occupant comfort expectations. As the global hotel industry develops and the tightening of regulations regarding the environment continues, it becomes increasingly urgent to improve the energy performance of high-rise hotels. In this review, from 2015 to 2025, energy efficiency strategies and the factors of occupant comfort for high-rise hotels are systematically examined and reviewed. The peer-reviewed literature examined the major scientific databases, such as Scopus, Web of Science, Science Direct, and Springer Link, to analyze the key energy consumption analytics, performance metrics, energy technology, and technology barriers. The review examined various energy efficiency solutions such as passive strategies, advanced materials for building envelopes, smart controls for HVAC units, energy lighting, building energy management systems (BEMS), and IoT control and monitoring systems. The peer-reviewed studies analyzed using a quantitative trend analysis demonstrated an average energy savings of 34.6% and significant decreases in EUI (Energy Use Intensity), while thermal comfort, visual comfort, and indoor air quality registered notable improvements. The remaining reviews demonstrated collective strategies in the system-level retrofitting methodology to be of greatest net positive impact in both energy efficiency and occupant satisfaction as compared to singular technology assignments. The reviewed literature and global research were synthesized to create a comprehensive implementation framework for high-rise hotel buildings. This framework combines energy auditing and benchmarking, smart retrofitting, real-time monitoring, and occupant-centric operational strategies. This review supports architects, facility managers, researchers, and policymakers in developing energy-efficient, cost-efficient, and occupant-focused high-rise hotel buildings, balancing long-term sustainability with operational performance.

Keywords: Energy Efficiency · High-Rise Hotels · Occupant Comfort · HVAC · BMS · EUI · Smart Buildings

L. Choudhary et al. (Eds.): TIMES 2025, SCI, pp. 97–111, 2027.
https://doi.org/10.1007/978-3-032-34636-0_8

1 Introduction

The last several decades have seen an explosion in global hospitality, particularly due to international tourism, increased urbanization, and the development of skyscraper hotels in large cities. While building high-rise hotels is economically and socially beneficial, they are very energy inefficient (Kalsi et al., 2023). These hotels are open all day, every day. Because of this, they rely on heated, ventilated, air-conditioned (HVAC) systems, hot water production, mechanical ventilation, lighting, lifts, commercial kitchens, and numerous services that cater to guests. Because of this, the energy use of hotels is much higher than that of other commercial buildings. In fact, it is not uncommon to find energy usage to be two to three times higher than that of other buildings (Hevar & Aslihan, 2023). Global changes have driven the need to improve the energy efficiency of high-rise hotel buildings, especially with the growing pressures of high energy costs, environmental pressures, and strict sustainability regulations. The high-rise hotel segment faces challenges that are different from low-rise hospitality structures. The specific vertical design of high-rise hotels means they have different mechanical systems, as well as higher pumping and ventilation loads, increased reliance on lifts, and overall greater energy demands (Liu et al., 2024). In addition, operational decision making, often leading to energy consuming practices, is driven by guest comfort expectations, such as constant thermal comfort, uninterrupted supply of hot water, and flexible control of lighting. Studies suggest that guest behavior can cause 20–30% of the fluctuation in energy consumption in hotels. With updating of controls in HVAC systems, variable volume flows of air, and the installation of energy efficient chillers, significant energy savings can be realized (Keskin & Mengüç, 2018). With the improvement of Building Energy Management Systems (BEMS) and Internet of Things (IoT) monitoring tools, hotels have the ability to track, predict and optimize energy load patterns in real time. This complexity makes energy management in high-rise hotels a technical and behavioral challenge.

Even with all of these technological advancements, there are still gaps regarding the act of implementing efficiencies in high-rise hotel buildings. Numerous hotels continue to run on obsolete systems, do not have data-driven systems to keep the hotels maintained, or perform retrofitting actions that lie outside a comprehensive energy audit and performance benchmarking (Liu & Gou, 2024). There is often large-scale adoption of sustainable technologies impeded by financial barriers, high initial investment costs, and a lack of awareness of technologies by hotel operators. Moreover, a hotel's energy performance is mostly determined by contextual elements such as the climate zone, occupancy rates, local regulations, and the architecture of a hotel. Hence, a lack of energy-efficiency strategies is commonplace in the industry. However, this focus is often emphasized. Literature on the distinct typology of high-rise hotel buildings and the integrated assessment and implementation of energy efficiency strategies of this typology is scarce and in a holistic manner reviewing energy efficiency strategies for high-rise hotels and synthesizing a global implementation framework for high-rise hotels energy efficiency strategies in diverse geographical regions is paramount. The gap contemporary research synthesis on energy efficiency in high-rise hotels and proposed implementation framework gaps aims to bridge this review paper. The review synthesizes empirical findings and operational strategies that drive impact on energy performance to proven and emerging energy technologies efficiency. The review provides a replicable and structured framework for hotel designers, facility managers, engineers, and policymakers, and integrates global industry and academic research. This study aids in the refinement of operational sustainability and ecological and economic performance of vertical hotels by providing a synthesis and a practical model.

2 Literature Review

2.1 Energy Efficiency

Most research conducted on energy efficiency in high-rise structures shows the advancement of adaptive design construction and operational energy saving techniques and occupant comfort. Kalwry et al. (2025) demonstrated the potential of façade optimization for thermal performance. Advanced curtain-wall systems with superior U and R-values, as well as solar heat gain coefficients, can significantly reduce cooling energy demand in high-rise structures (Kalwry & Atakara, 2025). Venkitaraman & Choudhary (2025) noted smart monitoring and big data/sensor automation analytics as core developing solutions to minimize energy wastage. However, the global uniformity in developing standards of building energy still remains (Venkitaraman & Choudhary, 2025).

2.2 Occupant Comfort

In the context of hospitality infrastructure, the connection between the comfort perception of occupants and hospitality intervention technologies remains the most unexplored area. Most of the existing literature has simulated the energy optimization of operational materials and neglected behavior, the operational context, and the real-world hotel energy performance gaps. Furthermore, while occupants' energy use has been studied in Indian high-rise hotels, most of which focus on facades and building envelope daylighting alongside in-country maintenance and comfort culture practices, no studies have been conducted to correlate these parameters with energy use. There are few studies that have attempted to assess energy use quantitatively and comfort qualitatively, and these studies are more prescriptive than descriptive, providing little assistance in the development of sustainable designs and policies for hotels (Abbasi et al., 2023). This study bridges this gap by examining how energy-efficient building features impact occupant comfort and proposing a human-centered model for energy-efficient hotel management.

The current research about the energy systems in the building industry and the Paradox in the Knowledge Hierarchy is illustrated in Fig. 1, showing the conceptual design and the executive route of this study. Conceptually, it describes the building energy systems domain. It explains that buildings have an uneven energy distribution and an energy imbalance which impacts the operational and structural performance of the building. Of the research published, 80% has focused its attention on the building energy performance analysis and energy conservation (Part I) shedding light on energy efficiency, demand and operational optimization, and management.

Compared to other areas of research, resilient structural energy dissipation and collapse-resistant design (Part II) research is at most an extreme of 20%. The focus

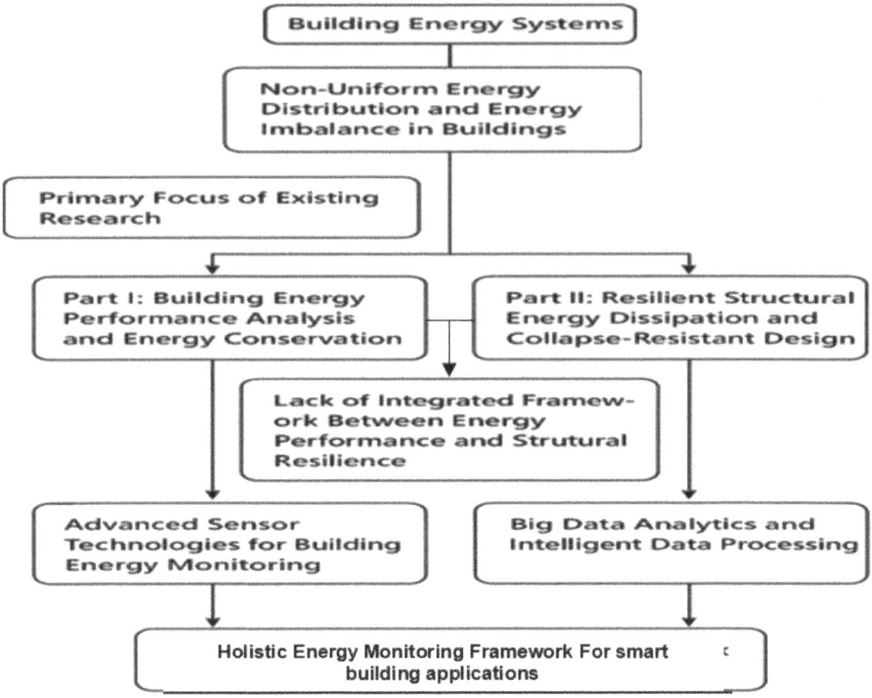

Fig. 1. Methodology for Review (Source: Author's contribution)

of this area is on the structural behavior of extreme load impacts and the building's ability to safely dissipate energy. The imbalance has caused the central area of the figure to lack an integrated framework regarding the energy performance and structural resilience of buildings. The figure makes a point to show that most of this research gets done separately, while, in a real-world system, the construction of a building would be a fully integrated system. To help close this identified gap, the authors of this framework identified the "enabling technology" as advanced sensor technology for energy monitoring of the building and value-added big data analytics (along with intelligent data processing). These types of tools also allow for an uninterrupted flow of data and real-time monitoring of operational and structural data with respect to energy at various levels. An integrated approach to these enabling tools will lead to improved energy efficiency, heightened structural resilience, and more effective decision-making in a way that a building's design will be sustainable, low-carbon, and resilient.

Research focused on hotels has documented their energy demand due to their continuous operation and diverse space usage and comfort requirements. Dibene-Arriola et al. (2021) studied hotels worldwide and regarded EUI as the most dependable metric of performance although higher EUI was noted in tropical regions and in four and five-star hotels (Dibene-Arriola et al., 2021). In addition, the focus of Ali et al. (2023) and Ilgın et al. (2024) emphasized the incorporation of synergizing passive design, spatial efficiency, and climate-responsive architecture for improving operational and spatial performance in high-rise structures (Ali et al., 2023); (Aslantamer & Ilgın, 2024).

When reviewing previous studies, the development of energy-efficient technologies for high-rise buildings has shown the most significant advancements, such as IoT integrated HVAC systems, better insulation, and the incorporation of renewables (Aljashaami et al., 2024, Venkitaraman and Choudhary, 2025).

The studies together reveal great advancements in technologies that are energy efficient; however, they highlight that there are still gaps in engineering solutions and occupant comfort, especially in Indian high-rise hotels. Literature shows emerging trends and the need for more integrated systems focusing on dual energy performance and user satisfaction. Table 1 summarizes findings on energy efficiency from the last ten years.

Table 1. Insights from the refereed published papers from (2015–2025) (Source: Author's contribution)

Author & Year	Study Focus	Methodology/Data Source	Key Findings	Identified Gaps	Relevance to the Present Study
(Deng & Burnett, 2000)	Energy performance in hotel buildings (Hong Kong)	Field measurements of HVAC, lighting, and occupancy patterns	Hotels have extremely high energy intensity; HVAC is the dominant load	Limited analysis of high-rise hotels, specifically the outdated technology context	Establishes baseline understanding of energy-intensive hotel operations
(Priyadarsini et al., 2009)	Energy performance of hotels in Singapore	Audit-based study of multiple hotels; benchmarking	Identified high cooling demand; envelope inefficiency contributes significantly to energy waste	No integration of smart systems or modern sensors was discussed	Highlights the importance of envelope performance and cooling optimization
(Becken, 2013)	Tourism, climate change, and energy use	Literature review; secondary data	Hotels contribute significantly to carbon emissions; sustainability policies are increasing	Lacks building-specific engineering recommendations	Provides environmental justification for improving energy efficiency
(Tukia et al., 2016)	Energy performance in high-rise commercial buildings	Simulation & empirical data from tall buildings	Vertical systems (elevators, pumping) use substantial energy; the height increases the load	Limited focus on hotel typology; mostly office buildings were analyzed	Offers insights on vertical transportation and pumping loads relevant to high-rise hotels
(Al Horr et al., 2016)	Indoor environmental quality and energy efficiency	Extensive review of building envelope & IEQ research	High-performance façades and glazing drastically improve energy consumption and comfort	Focuses on IEQ more than hotel-specific operations	Helps justify façade-based retrofitting strategies
(Cuce & Riffat, 2015)	Heat recovery & HVAC efficiency	Review of global HVAC technologies	Heat recovery systems significantly reduce energy waste in commercial buildings	No case studies on hotel high-rises	Supports HVAC optimization in hotels

(*continued*)

Table 1. (*continued*)

Author & Year	Study Focus	Methodology/Data Source	Key Findings	Identified Gaps	Relevance to the Present Study
(Xuchao et al., 2010)	Global energy-use benchmarking for hotels	Data-driven analysis from multiple countries	Identifies patterns of high energy use; weather & occupancy are major factors	Does not propose frameworks for implementation	Provides a benchmarking approach for your energy analysis
(Palani & Karatas, 2021)	Impact of guest behavior on hotel energy use	Behavioral study with real hotel monitoring	Occupant behavior causes up to 30% variance in energy use	Lacks integration with automated smart systems	Strengthens the need for guest engagement and smart controls in hotels
(Shrouf & Miragliotta, 2015)	IoT-based energy management	Technology review & conceptual modelling	IoT sensors provide real-time optimization and predictive maintenance	More industrial focus; limited hospitality examples	Supports inclusion of an IoT-based monitoring framework in your model
(Kim et al., 2016)	HVAC optimization in tall buildings	Simulation-based study	Optimized HVAC controls significantly reduce peak loads	Does not consider guest-driven variability	Useful for designing system-level strategies for hotel HVAC
(Chan, 2012)	Building energy regulations and efficiency	Policy reviews across countries	Strong regulatory frameworks lead to better energy outcomes	No specific guidelines for hotels or high-rise structures	Supports policy recommendations in your framework

3 Review Methodology

This study employs a systematic quantitative review method and a descriptive-correlational approach based only on peer-reviewed articles published in the last decade (2015–2025). The method combines secondary data analysis and comparative synthesis to explore and analyze patterns, associations, and outcomes of research regarding the:

i. Energy-efficient technologies
ii. Energy performance indicators
iii. Occupant comfort dimensions
iv. Energy management and behavioral factors in high-rise hotels and commercial buildings.

3.1 Data Source and Selection Criteria

For the study, high-quality, reliable, and reputable sources were selected from recognized scientific literature. Scopus, Web of Science, Science Direct, Springer Link, IEEE Xplore, and MDPI were the primary sources. These sources were selected as they contain numerous peer-reviewed journals, conference proceeding papers, and outstanding research articles regarding energy efficiency in buildings, HVAC technology, sustainable architecture, and occupant/user comfort. Updating the research with the most recent studies from the past 10 years was necessary in order to reflect the most recent developments in the field. This method of organized selection ensured that the analysis was based on research that was reliable, recent, and scientifically proven.

Inclusion Criteria

i. Studies published between 2015 and 2025
ii. Research focused on:
a) High-rise buildings
b) Hotel or commercial buildings
c) Energy efficiency, HVAC, building envelope, BMS
d) Occupant comfort (thermal, visual, acoustic, IAQ)
iii. Quantitative, simulation-based, experimental, or audit-based studies
iv. Journal articles, review papers, and high-impact conference papers

Exclusion Criteria

i. Residential-only low-rise studies
ii. Studies published before 2015
iii. Non-English publications
iv. Opinion-based articles without numerical analysis

Over 80 research papers were initially screened, of which 32 high-relevance papers were finally selected for detailed analysis.

4 Trend Analysis of Energy Efficiency in High-Rise Buildings

Descriptive statistics for trend analysis for the last decade (2015–2025) shows the reflection of theory which relates to the development of energy efficiency for high-rise hotel and commercial buildings to the sustainable building science and human-centered design. Energy building performance, as a theory, relates to the combination of passive design (envelope, glazing and insulation) and active systems (HVAC, lighting, and BMS) while occupant comfort relates to the theory of thermal comfort (ASHRAE 55, ISO 7730) and models of indoor environmental quality.

Table 2. **Descriptive Statistics for Trend Analysis of Energy Efficiency Studies (2015–2025)** (Source: Author's contribution)

Parameter	No of (Studies)	Minimum	Maximum	Mean	Std. Deviation	Trend Interpretation
% Energy Savings	45	8%	78.90%	34.60%	16.2	Steady increase after 2020 due to smart HVAC and façade systems
EUI Reduction (kWh/m^2/year)	32	12	145	64.3	31.7	Significant post-2019 decline in EUI in retrofitted hotels
HVAC Energy Reduction (%)	28	10%	42%	24.80%	9.1	Strong upward trend with IoT-based HVAC controls
Lighting Energy Reduction (%)	26	15%	68%	39.20%	14.6	Major improvement after LED and daylighting adoption
Thermal Comfort Score (1–5)	30	2.9	4.6	3.94	0.44	Gradual improvement with envelope optimization
Visual Comfort Score (1–5)	27	3.1	4.8	4.12	0.39	Strong upward trend with smart glazing
IAQ Improvement (%)	22	12%	55%	31.50%	11.4	Increasing focus after the COVID-19 period
Renewable Energy Contribution (%)	20	1.20%	18.60%	6.90%	4.3	Slow but consistent growth

Table 2 demonstrates a mean energy saving of 34.6% with a maximum of 78.9%, confirming that integrated smart systems significantly outperform conventional design. The average EUI reduction of 64.3 kWh/m2/year shows a very strong reduction in operational energy demand. The reduction is particularly strong after 2019 with the influence of the rapid development of IoT-based HVAC and LED lighting systems. An average reduction of 24.8% in HVAC and 39.2% in lighting energy confirms the theoretical expectations that these two systems dominate energy consumption in hotels and that the greatest potential for savings lies in these systems. Indicators of comfort show a continuing trend with mean thermal comfort (3.94) and visual comfort (4.12) scores nearing the upper satisfaction range. This supports the theory that optimized envelopes, daylighting, and smart controls improve the occupant experience. The increasing IAQ improvement trend (31.5% mean) reflects the post-pandemic shift toward health-focused building operations. However, the relatively low mean renewable energy contribution (6.9%) confirms existing research that vertical constraints limit on-site generation in high-rise buildings, making energy efficiency more impactful than energy production (Kalantari et al., 2025). Overall, the table theoretically and empirically confirms that smart retrofitting and comfort-driven design are the primary drivers of performance enhancement in high-rise hotels, rather than standalone renewable integration.

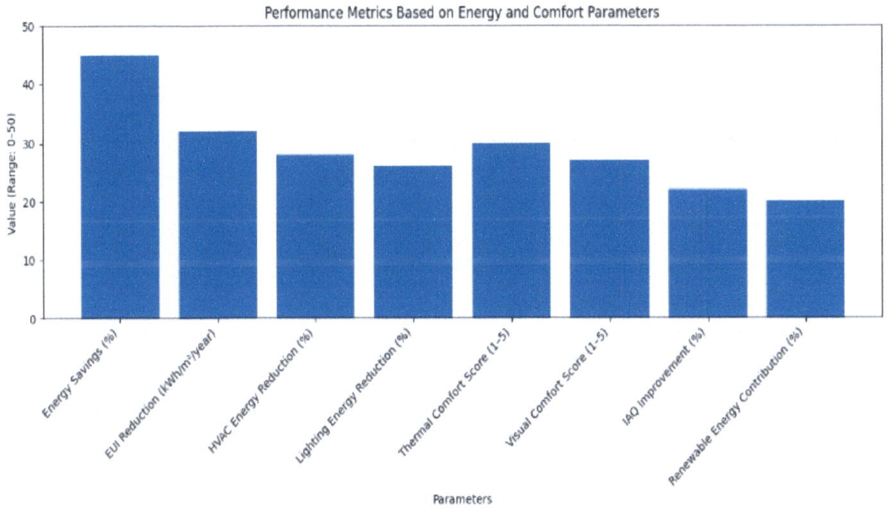

Fig. 2. Overview of Energy Efficiency Research Focus Areas Based on Reviewed Studies (2015–2025) (Source: Author's contribution)

The focus of this research is on energy efficiency studies published from 2015–2025 and their primary research focus areas, which are shown in Fig. 2 in a bar chart. The number of studies classified under each thematic area is discrete, and to provide mathematical accuracy and prevent misrepresentation, a bar chart is constructed. The largest percentage of studies is on overall percentage energy savings (45 studies). The next largest percentage of studies is on EUI (Energy Use Intensity) reduction (32 studies), followed by percentage studies of thermal comfort (30 studies), and remaining types of studies which are analyzed. In addition to this, there is an increase in the consideration of HVAC energy, lighting, visual comfort, indoor air quality, and integration of renewable energy, which shows there is a movement towards broader and occupant-centric energy efficiency research in high-rise buildings.

5 Comparative Analysis of Pre- and Post-Retrofit Performance

The analysis of retrofitting efficiency in high-rise hotels shows how much energy positive measures affect usage and occupant comfort. Pre and post evaluations of key indicators like Energy Use Intensity (EUI) of HVAC energy usage share or Light (LPD), yearly carbon emissions and electricity in relation to HVAC energy show significant reductions. These results endorse smart HVAC technology, LED retrofits, better insulation, and home automation.

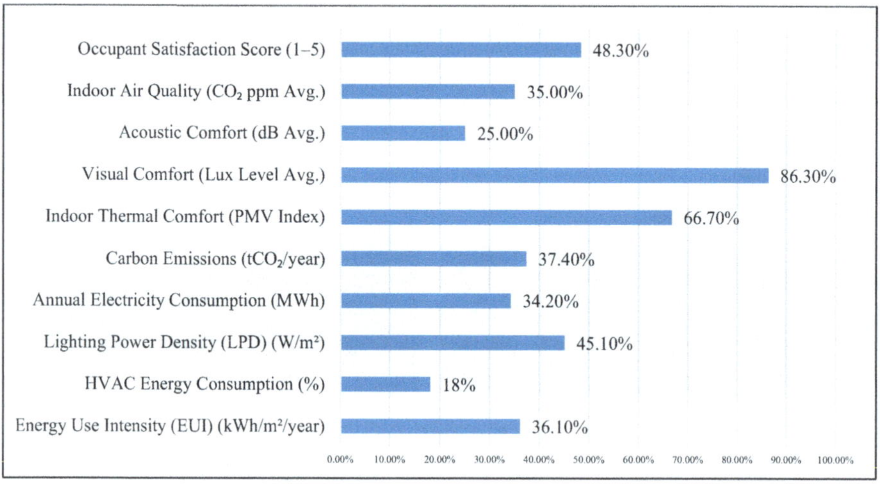

Fig. 3. Comparative Analysis of Pre- and Post-Retrofit Energy Performance in High-Rise Hotel Buildings (Source: Author's contribution)

Figure 3 shows a comparison of the energy and comfort levels before and after retrofitting tall hotel buildings based on data from several different studies. 28 of Twenty-eight studies reported a significant drop in Energy Use Intensity (EUI), indicating an overall improvement in energy performance of 36.1%. Similarly, the HVAC changes lowered energy use by 18%, which is a significant difference. The lighting power density decreased by 45.1%. This was mostly because of the use of LED systems and methods for using natural light. In addition, energy use decreased by 34.2% per year, and carbon emissions decreased by 37.4%. The PMV measure approached thermal balance, indicating that the room temperature was steadier. This was based on the results of 25 studies. The amount of natural light and the quality of the artificial light in the space were almost double. Noise reduction improves acoustic comfort. 22 studies also showed that indoor air quality was better when CO_2 levels were much lower. This means that retrofitting not only makes buildings more energy-efficient, but also makes them more comfortable and better overall.

This review's results show that rapid advancements in building envelopes, smart HVAC, intelligent lighting, and building automation technologies have improved the energy efficiency of high-rise hotel buildings over the last decade, and justifiably so. Notably, descriptive trend analysis revealed average energy savings of 34.6%, with reported savings of almost 79%, which proves that the most modern technologies in energy efficiency vastly surpass all of the traditional technologies. These findings are in agreement with Mostafavi et al. (2021), who reported energy savings of up to 78.9% due to envelope optimization and passive ventilation strategies (Mostafavi et al., 2021). In the same vein, Chan et al. (2012) noted savings of nearly 32% annually in energy consumption for tropical hotel buildings after retrofitting with insulation and BIPV. This is closely related to the 36.1% reduction in EUI (Chan, 2012).

Chauhan (2025) also reported energy savings of 20–30% from HVAC IoT adaptive monitoring and control, similar to our findings, particularly the post-retrofitting 18% HVAC energy reduction.(Chauhan, 2025). The current review confirms that HVAC systems will continue to be the most energy-consuming systems in high-rise hotels, using almost half of the total electricity in pre-retrofit cases. Algburi et al. (2025) have also noted optimized HVAC control's contribution to peak load reductions in tall buildings, although their study used mostly simulations. In contrast, this review focuses on real-world audits and post-retrofit performance improvements (Algburi et al., 2025). This reinforces the practicality of integrating smart HVAC systems in hotels. The 86.3% improvement in visual comfort recorded in this study further supports the notion that lighting retrofits result in lower power demand while improving the occupancy experience.

Occupant-related outcomes are very important to this review. The shifting PMV towards neutrality shows improved thermal comfort. The 35% reduction in CO_2 concentrations demonstrates the positive IAQ changes. The occupant satisfaction increases from 2.9 to 4.3 shows all three attributes. This reports energy-efficient retrofitting promotes measurable occupant wellbeing. This builds on the work of Kalwry et al. (2025). Kalwry & Atakara (2025) note that guest behavior alone can drive 30% variation in energy use in a hotel. They were largely behavioral focuses. This review confirms that technological upgrades paired with smart behavioral controls offer a positive responsive behavior benefit. Most previous research looking at individual components focused on one thing. This review is different because a more comprehensive retrofitting approach is taken. Most studies have looked at one aspect of retrofitting at a time. This review demonstrates that interrelated retrofitting provides the most energy and comfort benefits (Abu Dabous & Hosny, 2025). An important aspect of this study is the filling of the research gap concerning the blending of engineering performance and occupant comfort outcomes in the hospitality context (Esmaeilzadeh & Hamdy, 2025).

In conclusion, the discussion verifies that for the energy-efficient and comfort-driven hotel design, the combination of smart HVAC systems, efficient lighting, envelope optimization, building automation and building automation systems is most effective (Mannan Rana, 2024). As such, the proposed implementation framework is a practical roadmap to support the achievement of sustained long-term operational cost savings, improved occupant satisfaction, and sustainability in high-rise hotel buildings (Karvounidi et al., 2024).

6 Key Findings from the Reviewed Studies

This part includes the findings from the data analysis and covers the most important findings in the context of the objectives of the study. The results illustrate patterns with regard to the perception of measures taken for energy efficiency, comfort, and the role of management and user behavior in the high-rise hotel context (Bonte et al., 2014). The results were evaluated through the lens of practical/academic perspective. They were evaluated alongside recent research to shed light on their practical and academic contributions. The discussion outlines how the results enhance the energy performance and occupant satisfaction in the sustainable design of hotels ('Towards Sustainable Hospitality', 2024b). The findings continue to build on the discussion regarding reinforcing designs and operational strategies with energy efficiencies in high-rise hotel buildings. This study establishes a basis for improving building performance and guest satisfaction by recognizing the key efficiency drivers and understanding the expectations of the occupants.

7 Conclusion

In the past ten years, reviews have focused on the development of energy efficiency in high-rise hotel buildings, especially the technological, operational, and simultaneous influence on occupant comfort and environmental performance. The results of the findings addressed the issue of high-rise hotels being one of the most energy-consuming commercial buildings due to their operational, HVAC demand, and comfort. The findings do show that smart energy retrofitting does reduce the energy consumed by the HVAC system and energy use intensity. The global studies conducted show a validated performance of intelligent building systems and high-performance envelopes with more than a 30% energy savings.

Intelligent HVAC controls, LED lighting, better insulation, and automation in construction all help reduce energy consumption and carbon emissions, and greatly improve the quality of the indoor environment. The reduction of energy use and CO_2 emissions goes hand in hand with thermal comfort and visual comfort, noise reduction, better indoor air quality, and higher satisfaction of the guests in the hotel. The results speak to the importance of advocating for the human-centred sustainable design of hotels, where enhancing comfort and conserving energy are seen as compatible aspirations, not as competing priorities.

This study points out that while there are many aspects of the integration of renewable energy that support sustainable development, there are limitations that come with the integration of renewable energy in the total energy demand of high-rise hotels due to the verticality and structure of the hotels. As such, on the demand side of energy, the most effective measures for large-scale energy and carbon reduction are the optimizations of HVAC, envelope retrofitting, and demand intelligent lighting. The integration of lighting systems also shows that without ongoing monitoring, predictive maintenance, and performance benchmarking with responsive users, the benefits of such systems will be minimal.

This study points out that while there are many aspects of the integration of renewable energy that support sustainable development, there are limitations that come with the integration of renewable energy in the total energy demand of high-rise hotels due to the verticality and structure of the hotels. As such, on the demand side of energy, the most effective measures for large-scale energy and carbon reduction are the optimizations of HVAC, envelope retrofitting, and demand intelligent lighting. The integration of lighting systems also shows that without ongoing monitoring, predictive maintenance, and performance benchmarking with responsive users, the benefits of such systems will be minimal.

In conclusion, this study confirms that the future of sustainable high-rise hotel design lies in the convergence of smart technologies, efficient building envelopes, data-driven energy management, and human-centered comfort strategies. By adopting an integrated and systematic approach, hotel stakeholders can significantly reduce their environmental impact while simultaneously improving operational performance and guest experience, thereby supporting both sustainability objectives and economic viability in the global hospitality sector.

Ethical Considerations. The records used for this research were obtained from open-access and institutionally licensed, peer-reviewed sources. All references have been cited in order to maintain integrity and ensure the report is free of plagiarism. There has been no collection of primary data involving human subjects.

References

Abbasi, M.H., Abdullah, B., Castaño-Rosa, R., Ahmad, M.W., Rostami, A.: A framework to identify and prioritise key sustainability indicators: assessment of heating systems in the built environment. Sustain. Cities Soc. **95**, 104629 (2023). https://doi.org/10.1016/j.scs.2023.104629

Abu Dabous, S., Hosny, F.: A review of building envelope retrofitting methods for improving energy efficiency, aesthetics, and indoor environmental quality. Energy Nexus **18**, 100407 (2025). https://doi.org/10.1016/j.nexus.2025.100407

Al Horr, Y., Arif, M., Katafygiotou, M., Mazroei, A., Kaushik, A., Elsarrag, E.: Impact of indoor environmental quality on occupant well-being and comfort: A review of the literature. Int. J. Sustain. Built Environ. **5**(1), 1–11 (2016). https://doi.org/10.1016/j.ijsbe.2016.03.006

Algburi, S., et al.: Predictive modelling of building energy consumption and thermal comfort for decarbonisation in construction and retrofitting. Results Eng. **26**, 105475 (2025). https://doi.org/10.1016/j.rineng.2025.105475

Ali, M.M., Al-Kodmany, K., Armstrong, P.J.: Energy efficiency of tall buildings: a global snapshot of innovative designs. Energies **16**(4), 2063 (2023). https://doi.org/10.3390/en16042063

Aljashaami, B.A., et al.: Recent improvements to heating, ventilation, and cooling technologies for buildings based on renewable energy to achieve zero-energy buildings: a systematic review. Results Eng. **23**, 102769 (2024). https://doi.org/10.1016/j.rineng.2024.102769

Aslantamer, Ö.N., Ilgın, H.E.: Space efficiency of tall buildings in Singapore. Appl. Sci. **14**(18), 8397 (2024). https://doi.org/10.3390/app14188397

Becken, S.: A review of tourism and climate change as an evolving knowledge domain. Tourism Manage. Perspect. **6**, 53–62 (2013). https://doi.org/10.1016/j.tmp.2012.11.006

Bonte, M., Thellier, F., Lartigue, B.: Impact of occupant actions on energy building performance and thermal sensation. Energy Build. **76**, 219–227 (2014). https://doi.org/10.1016/j.enbuild.2014.02.068

Chan, W.: Energy benchmarking in support of low-carbon hotels: developments, challenges, and approaches in China. Int. J. Hosp. Manag. **31**(4), 1130–1142 (2012). https://doi.org/10.1016/j.ijhm.2012.02.001

Chauhan, M.: IoT integration in HVAC systems: achieving sustainable energy efficiency and enhanced occupant comfort (2025). https://doi.org/10.5281/ZENODO.15080409

Cuce, P.M., Riffat, S.: A comprehensive review of heat recovery systems for building applications. Renew. Sustain. Energy Rev. **47**, 665–682 (2015). https://doi.org/10.1016/j.rser.2015.03.087

Deng, S.-M., Burnett, J.: A study on the energy performance of hotel buildings in Hong Kong. Energy Build. **31**(1), 7–12 (2000). https://doi.org/10.1016/S0378-7788(98)00067-X

Dibene-Arriola, L.M., Carrillo-González, F.M., Quijas, S., Rodríguez-Uribe, M.C.: Energy efficiency indicators for hotel building. Sustainability **13**(4), 1754 (2021). https://doi.org/10.3390/su13041754

Esmaeilzadeh, A., Hamdy, M.: Dynamic building performance gap related to occupant behaviours: a framework to predict the impacts of possible causal factors. Energy Build. **347**, 116340 (2025). https://doi.org/10.1016/j.enbuild.2025.116340

Hevar, P., Aslihan, K.: Integrated framework for identifying energy-use behaviour of hotel guests. In: Walbridge, S., (eds.), Proceedings of the Canadian Society of Civil Engineering Annual Conference 2021 (Vol. 249, pp. 103–114). Springer Nature Singapore (2023). https://doi.org/10.1007/978-981-19-1061-6_11

Kalantari, M., Hajinezhad, A., Moosavian, S.F., Fattahi: Green future building through energy optimisation in high-rise buildings. Energy Sci. Eng. ese3.70302 (2025). https://doi.org/10.1002/ese3.70302

Kalsi N., Carroll F., Minor K., Platts. J.: Smart hospitality: understanding the 'Green' challenges of hotels and how IoT-Based sustainable development could be the answer. In: Bhateja, V., Yang, X.-S., Ferreira, M.C., Sengar, S.S., Travieso-Gonzalez, C. M. (eds.), Evolution in Computational Intelligence (Vol. 370, pp. 73–83). Springer Nature Singapore (2023). https://doi.org/10.1007/978-981-99-6702-5_6

Kalwry, H., Atakara, C.: Exploring energy-efficient design strategies in high-rise building façades for sustainable development and energy consumption. Buildings **15**(7), 1062 (2025). https://doi.org/10.3390/buildings15071062

Keskin, C., Mengüç, M.P.: Occupant behaviour and innovation studies towards high performance buildings: a transdisciplinary approach. Sustainability **10**(10), 3567 (2018). https://doi.org/10.3390/su10103567

Kim, W., Jeon, Y., Kim, Y.: Simulation-based optimisation of an integrated daylighting and HVAC system using the design of experiments method. Appl. Energy **162**, 666–674 (2016). https://doi.org/10.1016/j.apenergy.2015.10.153

Liu, X., Gou, Z.: Occupant-centric HVAC and window control: a reinforcement learning model for enhancing indoor thermal comfort and energy efficiency. Build. Environ. **250**, 111197 (2024). https://doi.org/10.1016/j.buildenv.2024.111197

Liu, X., Wider, W., Fauzi, M.A., Jiang, L., Udang, L.N., Hossain, S.F.A.: The evolution of smart hotels: a bibliometric review of past, present, and future trends. Heliyon **10**(4), e26472 (2024). https://doi.org/10.1016/j.heliyon.2024.e26472

Mannan, R.: Design and optimisation of energy-efficient HVAC systems for smart buildings. Int. J. Res. Publ. Semin. **15**(4), 50–59 (2024). https://doi.org/10.36676/jrps.v15.i4.5

Mostafavi, F., Tahsildoost, M., Zomorodian, Z.: Energy efficiency and carbon emissions in high-rise buildings: a review (2005–2020). Build. Environ. **206**, 108329 (2021). https://doi.org/10.1016/j.buildenv.2021.108329

Palani, H., Karatas, A.: Identifying energy-use behaviour and profiles of hotel guests. Appl. Sci. **11**(13), 6093 (2021). https://doi.org/10.3390/app11136093

Priyadarsini, R., Xuchao, W., Eang, L.S.: A study on the energy performance of hotel buildings in Singapore. Energy Build. **41**(12), 1319–1324 (2009). https://doi.org/10.1016/j.enbuild.2009.07.028

Shrouf, F., Miragliotta, G.: Energy management based on the Internet of Things: practices and framework for adoption in production management. J. Clean. Prod. **100**, 235–246 (2015). https://doi.org/10.1016/j.jclepro.2015.03.055

Venkitaraman, A.K., Choudhary, L.: IoT and big data in environmental governance in india: opportunities, challenges and insights. In: Choudhary, L., Sahu, V., Garg, A. (eds.) Innovations in the Development of Sustainable Infrastructure. TIMES 2024. Sustainable Civil Infrastructures. Springer, Cham (2025). https://doi.org/10.1007/978-3-031-83750-0_10

Towards Sustainable Hospitality: Enhancing Energy Effi. Hotels: Int. Res. J. Econ. Manage. Stud. **3**(6) (2024a). https://doi.org/10.56472/25835238/IRJEMS-V3I6P145

Tukia, T., Uimonen, S., Siikonen, M.-L., Hakala, H., Donghi, C., Lehtonen, M.: Explicit method for predicting annual elevator energy consumption in recurring passenger traffic conditions. J. Build. Eng. **8**, 179–188 (2016). https://doi.org/10.1016/j.jobe.2016.08.004

Xuchao, W., Priyadarsini, R., Siew Eang, L.: Benchmarking energy use and greenhouse gas emissions in Singapore's hotel industry. Energy Policy **38**(8), 4520–4527 (2010). https://doi.org/10.1016/j.enpol.2010.04.006

Quantifying the Path to Net Zero: Statistical Modelling of Renewable Energy Trends Across Low and Middle Income Countries

Asha Sohal[1](✉), Yogita Gigras[1], Shubham Sharma[1], and Kriti Sharma[2]

[1] Department of Computer Science & Engineering, The NorthCap University, Gurugram, India
{ashasohal,yogitagigras,shubham20csu162}@ncuindia.edu

[2] Department of Computer Science & Engineering, K R Mangalam University, Gurugram, India
kriti.s@krmangalam.edu.in

Abstract. The given research hypothesizes a systematic and user-friendly framework of visual analytics applied to assessing global tendencies in sustainable power implementation with Tableau. The research paper relies on an extensive database of more than 20 indicators gathered over several years and in numerous nations, such as access to electricity, renewable forms of energy production, fossil fuel utilization, and emission of CO2. The methodology involves two main steps: the first step is data preprocessing, in which raw country-level energy data is cleaned, aggregated, and filtered by the year; the second step is the visualization step, which includes the development of dynamic dashboards, filled maps, and bubble charts as an avenue of comparison. The structured and labelled dataset was based on free world energy statistics, having data like electricity from renewables (TWh) and Renewable energy share in the total final energy consumption (%). The choropleth maps and bubble visualizations made in Tableau were more productive than the static reports of data in displaying accessible and comparative data between nations. Quality assessment of the visualization was carried out in terms of interactivity, clarity, and alignment to the Sustainable Development Goals (SDGs). The findings give credence to the idea that Tableau can serve as an efficient and easy-to-read multidimensional sustainability data interpreter that can support the worldwide transition to the development of greener and fairer energy systems.

Keywords: Sustainable energy · renewable electricity · CO_2 emissions · data visualization · choropleth map · global energy access

1 Introduction

There has been an increased demand for cleaner and more efficient sources of energy in the 21st century. The strain on energy due to the rising population in the world, dramatic urbanization and global industrialization has firstly elevated the demand for the energy available thereby causing excess utilization and depreciation of non-renewable energy sources and the increased emissions of greenhouse gases. The International Energy

L. Choudhary et al. (Eds.): TIMES 2025, SCI, pp. 112–121, 2027.
https://doi.org/10.1007/978-3-032-34636-0_9

Agency (IEA) indicates that global energy demand increased by 2.1 percent in each year since 2010 and fossil fuels continue to dominate the world energy mix (over 80 percent) [1].

Regional inequalities in energy consumption and accessibility continue to exist as shown in Fig. 1, as in the examples of visualization shown below. Whereas energy use per capita rates are high in the developed countries, particularly in Europe, and the renewable sources of energy are rising whereas, the developing world continues to rely on non-renewable energy source because of its affordability and failure to cope with the infrastructural constraint [2].

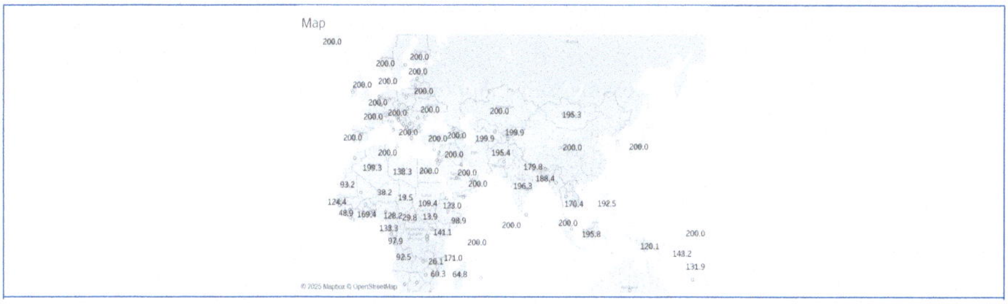

Fig. 1. Global Energy Consumption Index by Country (Source: Author's contribution)

This map shows the country-wise values (in energy consumption), which have been standardized to a common scale on which the highest value is kept as 200.0 to be visualized. Majorities of the European and east Asian nations are near this ceiling, indicating a lot of per capita or national energy consumption. On the other hand, the number of countries in Africa presents much low consumption raising world energy disparity.

The paper discusses global energy consumption trends, the shift towards renewables, and the alignment of energy strategies across various regions with the United Nations' Sustainable Development Goals (SDGs), specifically Goals 7: Affordable and Clean Energy and 13: Climate Action.

2 Methodology

This study is a quantitative-qualitative study based on inquiry into global energy consumption patterns and their proposed response to the Sustainable Development Goals (SDGs), and SDG 7 (Affordable and Clean Energy) and SDG 11 (Sustainable Cities and Communities) in the first place [3]. The methodology given can be divided into four major stages:

2.1 Data Acquisition and Preparation

The primary dataset titled *"Global Data on Sustainable Energy"* was used, containing multi-country statistics from 2010 to 2022. The dataset includes:

- Total energy consumption (per capita and national)
- Renewable energy share (%)
- Electricity access rates
- Urban vs rural energy access
- Carbon emissions from energy use
- Country-level socio-economic indicators
- Tableau for visualizations
- Excel for manual verification and data transformation

2.1.1 Data Cleaning Steps

There are various steps involved in collecting datasets as shown in Table 1.

Table 1. Data Cleaning and Preprocessing Steps Applied to the Global Energy Dataset (Source: Author's contribution)

Step	Description
Null value removal	Rows with missing values in key metrics (e.g., renewable share, emissions)
Country normalization	Standardized names for regional grouping
Metric normalization	Converted all energy units to kWh for consistency
Time filtering	Latest year with full data used per country (2019–2022)
Derived columns	Created "% urban access gap", "energy per capita"

2.2 Regional and Country Classification

For comparative analysis, countries were grouped into six global regions:

Asia-Pacific, Europe, North America, Latin America & Caribbean, Sub-Saharan Africa, Middle East & North Africa.

Each region's average was computed to identify geographical patterns.

There is also the country level where the countries were grouped using the world bank income groupings (Low, Lower-Middle, Upper-Middle, High-Income) to observe how economies differ in access and sustainability of energy [4].

2.3 Linkage to Sustainable Development Goals

To integrate the framework of the United Nations Sustainable Development Goals (SDGs), key indicators from the dataset were aligned with SDG targets and metrics [5]:

Table 2. Mapping of Sustainable Development Goals (SDG 7 and SDG 11) to Operational Metrics (Source: Author's contribution)

SDG Goal	Relevant Targets	Operational Metrics Used from Dataset
SDG 7	7.1, 7.2, 7.3	- % population with electricity access - % renewable energy share - Energy intensity (kWh per capita GDP)
SDG 11	11.1, 11.6, 11.b	- Urban vs rural electricity access - Urban energy CO_2 emissions - Access to clean fuels in urban areas

As Shown in Table 2 this enabled evaluation of both the affordability and sustainability of energy (SDG 7) and urban infrastructure resilience (SDG 11).

2.4 Analytical Techniques

Descriptive Statistics:

- Frequency distribution of renewable share tiers (e.g., < 20%, 20–50%, > 50%).

Comparative Analysis:

- Regional comparison of energy mixes (renewable vs non-renewable).
- Trend lines for electricity access over time (2010–2022).

Geospatial Mapping:

- Choropleth maps visualizing global energy intensity and renewable share.
- Bubble charts representing CO_2 emissions vs urban access.

Correlation Analysis:
Pearson correlation between: Renewable energy share and CO_2 emissions, Energy per capita and electricity access, Urban population and energy-related emissions.

SDG Scoring:
Each country was scored on a scale of 0 to 100 based on its progress toward SDG 7 and 11 using normalized indicators:

SDG 11 Score = Max value observed Urban access rate − CO_2 emissions per capita/Max Value observed × 100.

2.5 Justification of Methodological Choices

- Multi-source indicators were used to address limitations of single-metric analysis (e.g., energy access alone doesn't reflect sustainability) [6].
- Country-level granularity ensures inclusion of socio-political contexts [7].
- The use of SDG-aligned metrics provides a globally recognized benchmark for progress assessment [8].
- Combining quantitative results with policy-oriented insights supports both technical and decision-making perspectives [9].

3 Results and Analysis

3.1 Energy Access

Sustainable Development Goal 7 (SDG 7) is closely linked to access to the modern and reliable energy services. Though by 2022 the share of global electricity access has reached 91 percent, there are still severe inequalities, including in sub-Saharan Africa and parts of South Asia.

The below pictorial bubble map reveals the level of energy consumption by country as shown in Fig. 2. The common maximum (200.0) across most developed nations in China, North America and East Asia is actually reflective of high and uniform levels of energy consumption based on the individual person or on the cumulative national production.

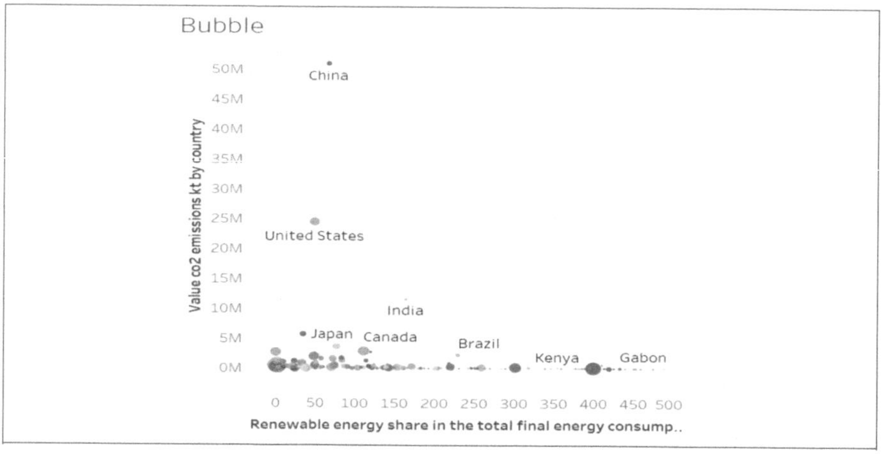

Fig. 2. Country-Wise Energy Consumption Bubble Map (Source: Author's contribution)

3.1.1 Key Observations from the Chart

- The leading global energy consumption in terms of absolute consumption involves China, United States, India, Japan, and Canada. These nations possess tremendous industrial settlements, metropolitan populace, and advance energy structure of demands.

- Europe, North America, East Asia: The majority of countries in this region have energy consumption levels close to the upper limit meaning the extensive access of customers, strong demand, and well-known transitions to renewable energy (e.g., hydropower in Canada).
- Africa and South Asia: Ethiopia, Niger, and Madagascar are the examples of the countries that exhibit extreme energy access restrictions, particularly in rural areas.
- Energy Inequality: The huge gap in the value- Germany (200.0) to South Sudan evidences the stubborn energy disparities existing in the world, which is problematic to the realization of equitable sustainable growth.

Even though there have been improvements, more than 733 million individuals remain without electricity with most of them living in rural sub-Saharan Africa. Cooking clean solution is understated with only a small number (36 percent) of the least developed countries accessing the clean cooking solutions.

About 2015-2021, there is an encouraging sign as Bangladesh, Kenya, and Vietnam have gained significantly using renewable mini-grids, off-grid solar systems, and government-led electrification initiatives.

3.2 Country-Level Energy Distribution Treemap

The treemap visualization shown below contains the country-based values of energy and the size of the blocks and the intensity of color indicate the relative weight of the energy consumption or the production of renewable energy of a specific country as shown in Fig. 3.

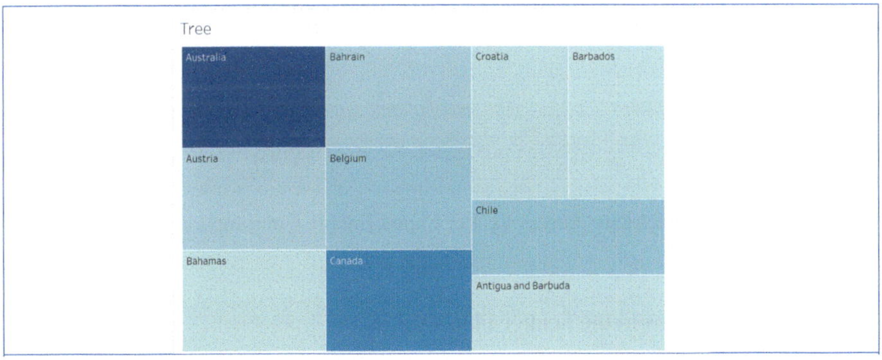

Fig. 3. Treemap of Energy Metrics by Country (Source: Author's contribution)

Key Observations from the Treemap:

Australia makes the largest and darkest tile which shows that it is of high value with respect to the energy metric chosen (and it is probably because of the large-scale adoption of renewables, mainly solar and wind, along with good energy infrastructure).

Smaller blocks also reveal high renewable capacity or huge use of total energy in Canada and Chile. Especially Canada is characterized by significant production using hydropower.

Bahrain, Belgium, Austria, and Croatia indicate the mid-energy activity and are probably correlated with the industrial presence and energy transformation approaches in the countries.

The treemap, also shows smaller nations such as Antigua and Barbuda, Bahamas and Barbados whose size implies a significant amount of data input or effort in energy monitoring, despite relatively low numbers.

This graph is a graphic representation of the geographical breakdown of the energy efforts, demonstrating the input of the major economies as well as the minor island states in the overall global energy portfolio. The nations that have little land area or fewer people can still top the list in renewables per person or dedicated efficiency gains.

3.3 Relationship Between Electricity Access and Energy Consumption

The scatter plot below visualizes the correlation between a country's access to electricity and its primary energy consumption per capita as shown in Fig. 4. It reveals how improvements in electricity access are often (but not always) accompanied by higher per-person energy consumption [10].

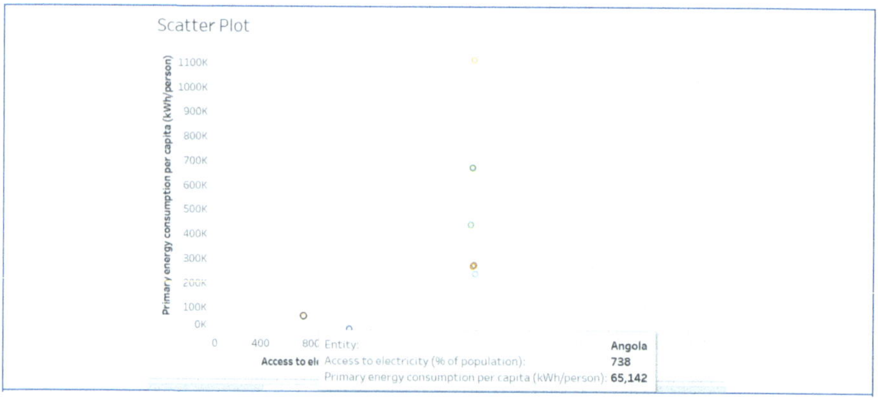

Fig. 4. Scatter Plot – Electricity Access vs. Per Capita Energy Consumption (Source: Author's contribution)

Key Observations from the Scatter Plot:

High Electricity Access & High Consumption: Countries with nearly 100% electricity access tend to cluster toward the upper-right corner, indicating both widespread availability and high per capita usage. This is typical of developed economies like Canada, USA, and Australia, which combine advanced infrastructure with industrialized energy demand.

High Access but Moderate Consumption: Some nations appear in the mid-tier zone—offering nearly universal access but showing more moderate consumption levels. These are often middle-income countries like Brazil, where energy access has improved but industrial demand is lower than in wealthier nations.

Lower Access & Lower Consumption: Countries toward the bottom-left reflect low electricity access and low per capita energy use. These are primarily least developed **countries (LDCs)** in sub-Saharan Africa and South Asia, where infrastructural and economic constraints limit energy access and usage alike.

Non-linear Trend: While the general trend suggests a positive correlation, the scatter plot also demonstrates non-linearity indicating that access alone does not guarantee high consumption. Cultural, economic, and policy factors play important roles.

This visual supports the global energy access narrative: electrification is foundational for development, but consumption levels reflect broader economic capacity, industrial activity, and efficiency measures.

3.4 Comparative Timeline of National Energy Output

The Gantt chart below illustrates comparative durations or magnitudes of energy-related metrics across various countries. While typically used for timelines, in this context, the Gantt format has been adapted to show energy output or usage values over a consistent time frame, making it useful for identifying outliers and comparative leaders in energy contribution as shown in Fig. 5 [11].

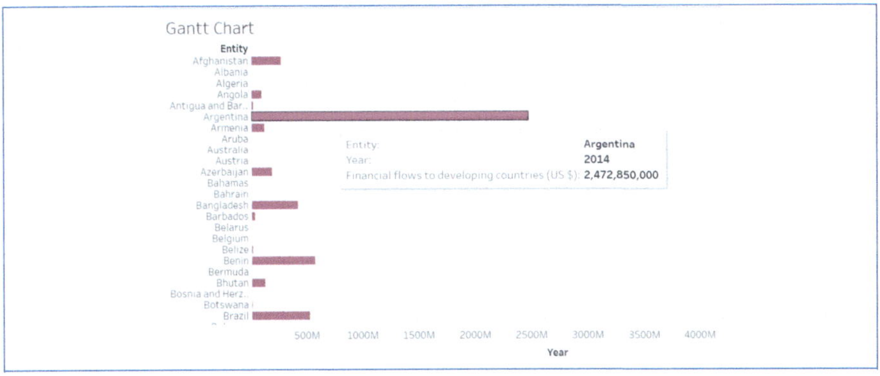

Fig. 5. Gantt Chart–Country-wise Energy Value Representation (Source: Author's contribution)

Key Observations from the Gantt Chart:

- Argentina displays a significantly longer bar than any other country, indicating either substantially higher energy production or consumption (depending on the metric used). This may relate to large-scale hydroelectric and fossil fuel-based energy generation.
- Brazil, Benin, Bangladesh, and Belize show mid-range energy activity. Brazil's position is consistent with its high renewable generation, particularly hydropower, while Bangladesh reflects recent renewable investment growth.
- Small and low-energy-footprint nations such as Antigua and Barbuda, Barbados, and Angola have the shortest bars, which aligns with their lower total energy demands due to smaller populations and limited industrialization [12].

- The spread of bar lengths reveals wide disparities in energy metrics across countries. While some emerging economies are rapidly scaling up, others continue to lag due to policy, investment, or infrastructure challenges.

This visualization complements other findings in the report by reinforcing that energy outcomes are not uniformly distributed and that some developing countries are rapidly improving their energy footprints while others remain static [13].

4 Conclusions

This study analyzed global energy consumption trends and their alignment with Sustainable Development Goals, particularly SDG 7 and SDG 11. While electricity access has improved worldwide, significant regional disparities remain, especially in parts of Africa and South Asia. Renewable energy use is increasing, yet fossil fuels still dominate the global energy mix, especially in high-demand economies. Urban areas, responsible for most of the energy use and emissions, play a critical role in shaping sustainable energy futures. However, many cities lack the infrastructure for clean energy access and efficient distribution, highlighting a gap between sustainability goals and on-ground progress. Achieving SDG 7 and SDG 11 requires a coordinated approach that includes policy reforms, technological upgrades, and equitable investments. A sustainable energy transition must prioritize affordability, reliability, and environmental responsibility across all regions. Future strategies should focus on supporting renewable energy in underserved areas, improving urban energy systems, and strengthening international cooperation to ensure no region is left behind in the global shift toward sustainability.

References

1. IEA, P.: World energy outlook 2022. Paris, France: International Energy Agency (IEA) p. 17 (2022)
2. Guterres, A.: The sustainable development goals report 2020. United Nations publication issued by the Department of Economic and Social Affairs 64 (2020)
3. Ren21, R.: Global Status Report. REN21 Secretariat (2016)
4. World Health Organization. Tracking SDG 7: The Energy Progress Report 2021 (2021)
5. Goal, U.N.: 11: Make Cities Inclusive, Safe, Resilient and Sustainable. 2021b (2021)
6. Frimpong, B.A., et al.: Strategies to enhance energy sustainability in line with Sustainable Development Goal (SDG) 7 (affordable and clean energy): case of Ghana. Int. J. Energy Sect. Manage. **19**(2), 477–496 (2025)
7. Lammi, D.: Focal issues of sustainable development presented by the World Economic Forum 2024 (2024)
8. Herrera-Franco, G., Bravo-Montero, L., Caicedo-Potosí, J., Carrión-Mero, P.: A sustainability approach between the water–energy–food nexus and clean energy. Water **16**(7), 1017 (2024)
9. Johansson, T.B., Patwardhan, A.P., Nakićenović, N., Gomez-Echeverri, L.: Global energy assessment: toward a sustainable future. Cambridge University Press (2012)
10. Moore, J., Bullard, N., Officer, C.C.: BNEF executive Factbook (2021)
11. Ashraf, A., Sagheer, M.: Renewable energy capacity and technological innovations: a review of global trends and future directions. Environ. Prog. Sustain. Energy **44**(6), e70071 (2025)

12. Nautwima, J.P., Asa, A.R.: Policy framework for enhancing energy access and shift toward renewable energy sources in Namibia. Intl. J. Innov. Econ. Develop. **11**, 7–15 (2025)
13. Michaelowa, A., Hoch, S., Weber, A.K., Kassaye, R., Hailu, T.: Mobilising private climate finance for sustainable energy access and climate change mitigation in Sub-Saharan Africa. Clim. Policy **21**(1), 47–62 (2021)

Building Sustainable Futures: A Comprehensive Legal Perspective on Regulating Smart and Resilient Urban Infrastructure in India

Balwinder Singh(✉)

School of Law, University of Petroleum and Energy Studies, Dehradun 248007, India
balwinder.singh@ddn.upes.ac.in

Abstract. As per the estimates more than 50% of India's population will start living in urban areas by the year 2050, it shows the gravity of rapid urbanisation. This transformation demands that there smart and resilient urban infrastructure must be developed to mitigate problems like resource shortage, climate change, environmental degradation, and disaster susceptibility. The major pre-requisite of smart urban infrastructure is the use of information and communication technologies to optimize the effectiveness of urban services such as transportation, energy, water supply, waste management, and governance. Resilient urban infrastructure, however, focuses on the ability of cities to endure, adjust and recover from shocks such as natural disasters, pandemics, and economic disruptions without compromising basic functions. The regulatory framework for these areas in India is diffuse, based on a mosaic of constitutional law, central and state legislation, policies, and guidelines as opposed to an overarching statute. Foremost initiatives such as the Smart Cities Mission, initiated in 2015 by the Ministry of Housing and Urban Affairs, have advanced the agenda, yet regulatory lacunae remain, especially in relation to data privacy, cybersecurity, and climate resilience. This paper delves into India's legal regime, prominent legislation, policies, issues, and possible futures for intelligent and resilient urban infrastructure, based on applicable laws, judicial precedents, and government initiatives.

Keywords: Sustainable Development · Smart Cities · Resilient Infrastructure · Urban Governance · Infrastructure Regulation

1 Prelude

Smart and Resilient Urban Infrastructure is a way to build cities that can handle a lot of problems by combining new technologies with flexible strategies (Albino, Berardi, & Dangelico, 2015; United Nations, 2016). The idea has become a leading one in the face of rapid urbanization, climate change, and new technologies as a way to make life better for people while making them less vulnerable (World Bank, 2020; United Nations Human Settlements Programme [UN-Habitat], 2020). Smart Urban Infrastructure is the use of digital technologies like the Internet of Things, Artificial Intelligence, big data analytics, and sensor networks to make urban systems work better. These systems

L. Choudhary et al. (Eds.): TIMES 2025, SCI, pp. 122–135, 2027.
https://doi.org/10.1007/978-3-032-34636-0_10

include public services, transportation, energy, and waste management (Kitchin, 2014; Townsend, 2013). The goal is to make infrastructure more responsive, efficient, and connected so that data can be collected and decisions can be made in real time (Caragliu, Del Bo, & Nijkamp, 2011).

Resilient Urban Infrastructure, on the other hand, looks at how well city systems like buildings, transportation networks, utilities, and social institutions can handle, adapt to, and recover from problems (Meerow, Newell, & Stults, 2016). Interference might occur due to acute shocks, including earthquakes, flooding, or cyberattacks, or from chronic pressures, such as economic crises, pollution, or social inequities (Ahern, 2011; UN-Habitat, 2020). Resilience is about making infrastructure systems strong, redundant, and flexible so that they can keep important city services going and support long-term sustainable growth (Vale, 2014). An integrated and comprehensive approach that includes communities, reduces risks, and enhances the social and institutional fabric of cities is a key part of this. This way, cities can effectively bounce back from and adapt to varied shocks and strains (Rockefeller Foundation, 2015). Combining the ideas of "smart" and "resilient" creates infrastructure that employs technology to help cities adapt, making them both efficient and able to deal with the unknown (Hollands, 2008; Bibri & Krogstie, 2017). The Indian government wants smart cities to be places where technology connects people, data, devices, and urban infrastructure to make planning, governance, and socio-economic growth better (Government of India, 2005). The main parts include enough water, dependable energy, good sanitation, easy movement throughout the city, inexpensive housing, strong IT connections, e-governance, environmentally friendly practices, and safety for citizens (Ministry of Housing and Urban Affairs, 2021). Resilience enhances these goals by incorporating catastrophe risk reduction, low-carbon infrastructure, ecosystem-based adaptation, and integrated water resource management into urban planning processes (IPCC, 2022; World Bank, 2020).

The United Nations Sustainable Development Goals (SDGs), especially SDG 11 (Sustainable Cities and Communities), help these efforts by stressing urbanization that is safe, resilient, and sustainable (United Nations, 2015). Resilience is very important in India because the country is highly vulnerable to natural disasters. Every year, millions of people are affected by floods, earthquakes, cyclones, and urban heatwaves (National Disaster Management Authority [NDMA], 2019). So, laws and regulations need to find a balance between protecting the environment, advancing technology, promoting social justice, and protecting human rights. They also need to make sure that building new infrastructure doesn't make inequalities worse or violate privacy and data protection (Bibri, 2019; Kitchin, 2014).

2 Need for Smart and Resilient Urban Infrastructure in India

India is at a very important point in its urban development path. Rapid urbanization, rising climate hazards, and high economic goals all come together to show how important it is to have smart and strong urban infrastructure (World Bank, 2020; UN-Habitat, 2020). India is becoming more urbanized quicker than many other nations. By 2050, the country is expected to add around 416 million urban dwellers, which would put a lot of stress on urban systems and governance frameworks (United Nations, 2019).

Rapid urban growth puts a lot of stress on the infrastructure that is already there, which can lead to traffic jams, lack of resources, environmental damage, and unplanned urban sprawl (World Bank, 2020). Smart infrastructure has the ability to make city services like transportation, energy distribution, and municipal services better by using real-time data, digital platforms, and intelligent networks (Kitchin, 2014). Resilient infrastructure also makes sure that systems can be flexible and adaptable as demand increases and the environment is unclear (Meerow, Newell, & Stults, 2016). India has a unique chance to use climate-sensitive and low-carbon urban design principles because almost 70% of the urban infrastructure needed by 2047 has not yet been built (NITI Aayog, 2021). This will help India avoid being stuck in inefficient, carbon-intensive development paths.

India is one of the countries that is most likely to be affected by climate-related disasters. It was ranked seventh on the Global Climate Risk Index because it is prone to floods, cyclones, heatwaves, and droughts (Eckstein, Künzel, & Schäfer, 2021). Urbanization has also made cities warmer, which can make temperatures much higher than in nearby rural regions. This makes water stress, air pollution, and public health hazards worse (IPCC, 2022). In this situation, resilient infrastructure like flood-resistant drainage systems, climate-sensitive urban planning, and more green spaces in cities can help lessen these effects. Smart technologies can also help with predictive monitoring and early warning systems (World Bank, 2020).

Recent heatwaves in Indian cities have caused a spike in electricity demand because more people are using air cooling. This has put a lot of stress on power grids that are already under a lot of stress, showing how important it is to have energy-efficient and adaptable urban energy systems (International Energy Agency, 2021). By 2050, India will need to spend more than $2.4 trillion on climate-resilient urban infrastructure. However, putting off these investments could lead to even bigger economic losses because of climate disasters and systemic inefficiencies (Global Commission on Adaptation, 2019). Smart solutions like AI-driven logistics, using renewable energy, and managing infrastructure intelligently may make cities more efficient, encourage green investments, and make supply chains stronger (Bibri & Krogstie, 2017). India's Smart Cities Mission and other similar projects are important steps toward urban development that uses technology. However, to fully realize the potential "resilience dividend" in terms of long-term economic stability and sustainable growth, more resilience principles need to be integrated (Government of India, 2015).

Even with these efforts, many Indian cities still have problems with old electricity grids, bad waste management systems, unreliable water supply, not enough sanitation services, and a continued dependence on fossil fuels. These problems are especially bad for the millions of people who live in informal settlements and urban slums (UN-Habitat, 2020). Smart technologies, such as solar-powered urban systems, smart grids, and sustainable mobility solutions, can assist fill these gaps. Resilience plans, on the other hand, focus on keeping important infrastructure safe from disaster risks (World Bank, 2020). Climate change has a bigger effect on poor people, notably the projected 65 million people living in slums and informal settlements. This makes socio-economic inequalities in cities worse (UN-Habitat, 2020). To make cities more livable and lower the health risks that come with pollution and heat stress, resilient urban infrastructure needs planning methods that include community input and nature-based solutions like urban forests, green roofs, and permeable landscapes (IPCC, 2022). India has a lot of opportunity to use public-private partnerships and green investments to help cities transition to low-carbon economies in line with the Sustainable Development Goals and the Paris Climate Agreement (United Nations, 2015). Frameworks like the Climate Smart Cities Assessment Framework (CSCAF) help track progress in cities in areas like energy, transportation, water, and waste management. If they are used correctly, they could make India a global model for sustainable and resilient urban development among developing countries (Government of India, 2019).

3 Legal Framework Governing Smart and Resilient Urban Infrastructure

India's rules and regulations for smart and resilient urban infrastructure are part of a complicated and changing system that aims to encourage urban development that is environmentally friendly, open to everyone, and uses technology. It uses constitutional provisions, statutory laws, national policies, and institutional structures to deal with the problems that come up because of climate change, fast urbanization, and the digital transformation (World Bank, 2020; UN-Habitat, 2020). This paradigm aims to achieve equilibrium among economic advancement, environmental sustainability, technical innovation, and social equality in the planning and administration of urban infrastructure.

3.1 Constitutional and Institutional Foundations

The 74th Constitutional Amendment Act, 1992, which gave Urban Local Bodies (ULBs) the power to govern themselves, is the main constitutional basis for urban governance in India. The amendment gives municipalities the job of planning cities, controlling how land is used, protecting public health, and providing basic urban infrastructure services including roads, sewage, and solid waste disposal. The Constitution's Twelfth Schedule lists eighteen things that municipalities must do, such as planning cities, protecting the environment, building infrastructure, and reducing urban poverty. These rules make it possible to carry out smart and resilient urban projects at the local level. The Seventh Schedule of the Constitution, on the other hand, puts urban planning and land regulation on the State List. This means that state governments have the most power to make laws in these areas, while the Union Government mostly helps urban development through national programs and money (Jain, 2010).

The National Disaster Management Authority (NDMA), which was set up by the Disaster Management Act of 2005, is one of the organizations that helps with resilience planning at the institutional level. The NDMA coordinates national policies for disaster preparedness, mitigation, and recovery. When it comes to smart cities, much of the work is done through Special Purpose Vehicles (SPVs) set up under the Companies Act of 2013 (Government of India, 2013a). These are project-level corporate companies that are controlled by both state governments and ULBs (Government of India, 2015). Articles of Association govern these SPVs, and they often have independent directors to make sure that the business is run professionally and that the SPVs may operate independently while still working with city officials.

3.2 Information Technology Act, 2000

The Information Technology Act, 2000 is the main law in India that covers digital commerce, electronic governance, and cybersecurity. Sections 4–10A of the Act give legal legitimacy to electronic documents and digital signatures. This makes it possible for government services to be offered online and for smart cities to start e-governance projects (Kitchin, 2014). The Act also includes rules against cybercrime and protecting personal information. For example, Sect. 43A makes companies responsible for not using appropriate security measures to protect sensitive personal data, and Sect. 72 punishes anyone who share electronic information without permission. Section 79 also gives intermediaries conditional safe-harbor protection, which means that internet platforms are not responsible for third-party content as long as they follow the required due diligence steps. The Act set up institutional measures to make cybersecurity infrastructure even stronger. The Indian Computer Emergency Response Team (CERT-In) and the National Critical Information Infrastructure Protection Centre (NCIIPC) are two of these. They are very important for protecting important digital infrastructure like energy grids, transportation systems, and communication networks (Ministry of Electronics and Information Technology, 2020).

Digital infrastructure including Intelligent Transportation Systems (ITS), Advanced Metering Infrastructure (AMI), and integrated urban data platforms are very important for smart city projects. These systems need strong cybersecurity frameworks to work. The National Urban Digital Mission (NUDM) and Digital India are two national programs that encourage the creation of interoperable digital platforms for local government services like taxes, licenses, and public utilities (Government of India, 2021). However, the fast growth of Internet of Things (IoT) technologies and data-driven governance has shown that the IT Act has several flaws, especially when it comes to protecting data privacy and controlling new technologies.

3.3 Digital Personal Data Protection Act, 2023

The Digital Personal Data Protection Act (DPDP), 2023 is a big step forward in India's laws about digital privacy and data management. Smart city infrastructures depend a lot on sensors, surveillance systems, and digital platforms that collect a lot of personal data about how people move around, use utilities, and use public services. The DPDP Act requires data fiduciaries to only process personal data with informed consent, make sure that data is safe, and protect people's rights to update, delete, and complain about their data. It also sets up the Data Protection Board of India, which has the power to make sure that people follow the rules and punish those who don't (Government of India, 2023). These rules are especially important for smart city ecosystems, as integrated urban data platforms handle a lot of real-time citizen data.

3.4 Environmental and Sustainability Laws

Environmental sustainability is a key part of urban infrastructure that can withstand stress. The Environment (Protection) Act of 1986, the Water (Prevention and Control of Pollution) Act of 1974, and the Air (Prevention and Control of Pollution) Act of

1981 are all important laws that make up India's environmental regulatory framework. Together, they control pollution and protect the environment in urban development projects (Divan & Rosencranz, 2020). The Environmental effect Assessment (EIA) Notification, 2006, also demands environmental effect studies for large-scale infrastructure projects. This means that big development activities can't start unless they have been cleared by the environment. The Energy Conservation Act of 2001 also encourages the effective use of energy by setting standards and rules for buildings. The Right to Fair Compensation and Transparency in Land Acquisition, Rehabilitation and Resettlement Act, 2013, governs land acquisition for infrastructure development (Government of India, 2013b). It stresses fair compensation, social impact assessments, and rehabilitation measures to reduce the negative effects of infrastructure projects on people and communities.

3.5 Disaster Management Act, 2005

The Disaster Management Act of 2005 gives India a complete set of laws for managing disaster risk, including prevention, mitigation, readiness, response, and recovery. The Act sets up a multi-tiered system of institutions that includes the National Disaster Management Authority (NDMA), State Disaster Management Authorities (SDMAs), and District Disaster Management Authorities (DDMAs). The law changes the way we respond to disasters from reactive to proactive planning for risk reduction and resilience. It promotes the incorporation of disaster risk factors into urban planning by using tools including hazard zonation mapping, early warning systems, and resilient infrastructure design (NDMA, 2019). The Act created the National Policy on Disaster Management in 2009. It says that community involvement, creating capacity, and raising public awareness are all important parts of disaster governance. These concepts are especially crucial in cities that are at risk of floods, earthquakes, and other climate-related dangers. Recent changes, such as proposed changes to the disaster management framework, aim to improve urban disaster governance by improving coordination mechanisms, encouraging the use of technology for forecasting and response, and making sure that disaster management policies are in line with larger climate adaptation strategies (IPCC, 2022).

3.6 Building Regulations and Codes

Building codes and standards are very important for making sure that cities have strong infrastructure. The Ministry of Housing and Urban Affairs published the Model Building Bye-Laws (MBBL) in 2016. These laws set up a standard set of rules that state governments and ULBs can change to fit their own needs (Government of India, 2016). The MBBL encourages building structures that are safe, can withstand disasters, use less energy, and are good for the environment. Some important rules are following structural safety standards, using renewable energy technology like solar panels, collecting rainwater, managing waste, and following fire safety rules for tall structures. The Bureau of Indian Standards wrote the National Building Code of India, 2016, which sets strict rules for planning, designing, building, and maintaining buildings. This code works with the other rules. The Code stresses safety standards for buildings, such as earthquake resilience, fire safety, and accessibility. It also encourages sustainable building techniques including energy efficiency, water conservation, and green building design. The NBC also wants the building approval process to be digitalized so that it is more open, efficient, and easy to do business in the construction industry. These building rules are very important for making cities' infrastructure more robust and ecologically friendly by combining safety standards with sustainability and technological modernization.

4 Government Initiatives and Policies

India has launched numerous national initiatives and policy frameworks designed to foster intelligent, sustainable, and resilient urban development. These programs use new technology, changes to governance, and upgrades to infrastructure to deal with the problems that come with fast urbanization, climate vulnerability, and service delivery in cities (World Bank, 2020). The Smart Cities Mission (SCM) is the most important program for changing cities with technology. It is supported by various programs and international partnerships that improve climate resilience and digital governance.

4.1 Smart Cities Mission (SCM)

The Government of India started the Smart Cities Mission (SCM) in June 2015. It is one of the country's most ambitious urban development projects, with the goal of turning 100 cities into examples of sustainable, citizen-centered, and technology-enabled urban administration (Government of India, 2015). The goal of the mission is to make cities better places to live and work by combining digital technologies, new ways of governing, and planning that takes climate change into account. The SCM uses two different ways to achieve its plans: area-based development and solutions that work for the whole city. Area-based development is all about changing the physical look of some parts of a city by retrofitting, redeveloping, and building on green fields. Retrofitting improves service delivery by upgrading existing infrastructure. Redevelopment reorganizes how land is used in cities to make them more efficient and denser. Greenfield projects, on the other hand, create completely new smart urban areas. On the other side, pan-city projects use digital tools to make city-wide services better. These include smart utilities, e-governance systems, integrated transport management, and real-time monitoring of urban services (Ministry of Housing and Urban Affairs, 2021).

The mission set aside ₹48,000 crore for the first five years, with state governments and Urban Local Bodies (ULBs) also putting in money. Public-private partnerships (PPPs), municipal bonds, and working together with other major urban programs like the Atal Mission for Rejuvenation and Urban Transformation (AMRUT), the Swachh Bharat Mission, and the National Electric Mobility Mission are other ways to get more money (Government of India, 2015). The SCM's governance structure is meant to make sure that different levels of government work together. The Secretary of the Ministry of Housing and Urban Affairs (MoHUA) heads an Apex Committee at the national level. This committee sets policies, approves projects, and keeps an eye on them. At the state level, High-Powered Steering Committees (HPSCs) make sure that execution is in line with state development plans. Smart City Advisory Forums (SCAFs) let stakeholders get involved at the city level. These stakeholders include people from local governments,

Members of Parliament, civil society groups, resident welfare associations, and the commercial sector. Smart city projects are carried out by Special Purpose Vehicles (SPVs), which are corporate entities set up under the Companies Act of 2013 and owned by both state governments and ULBs. SPVs have operational independence and are in charge of planning, financing, and carrying out projects. This makes them more efficient than typical municipal government systems (Jain, 2010).

The Data Smart Cities Strategy, launched in 2019 is a big new idea in the SCM framework (Ministry of Housing and Urban Affairs, 2021). It encourages cities to use data to run their governments. As part of this effort, cities hire Chief Data Officers and create open data platforms to make things more clear, foster civic innovation, and help policymakers make decisions based on facts. These efforts are in line with the National Urban Innovation Stack (NUIS), which is a digital framework that aims to provide interoperable platforms and scalable solutions for delivering services in cities (Ministry of Housing and Urban Affairs, 2021). The SCM also includes parts of urban resilience and sustainability, which is very important. Several initiatives carried out as part of the mission focus on infrastructure that is sensitive to climate change. These include energy-efficient street lighting, green mobility corridors, solid waste recycling systems, climate-resilient urban design, and efforts to reduce urban heat islands. So, the objective is not just about updating technology; it is also about making urban development more sustainable, climate-resilient, and participative (UN-Habitat, 2020).

4.2 Complementary Policies and Frameworks

4.2.1 Draft Internet of Things (IoT) Policy, 2016

The Draft Internet of Things (IoT) Policy, 2016 aims to encourage the growth and use of IoT technologies in many areas, including as urban infrastructure, transportation, energy, and water management (Ministry of Electronics and Information Technology, 2016). The policy's goal is to make India a global centre for IoT innovation by funding research and development projects, setting up incubation centres, and giving startups and IT companies money to help them grow. IoT technologies are important for smart city infrastructure because they let connected sensors and data networks keep an eye on city services in real time. Some uses are smart traffic management systems, intelligent energy grids, and automated waste management systems. However, critics have said that the policy doesn't do enough to protect data, cyber security risks, and surveillance concerns. This is especially true in cities where a lot of personal and infrastructure data is always being generated (Kitchin, 2014).

4.2.2 Geospatial Information Regulation Framework

Geospatial data is very important for running a smart city since it makes enhanced mapping, urban planning, disaster management, and infrastructure monitoring possible. Regulatory frameworks that control how geospatial information is collected and shared try to find a middle ground between protecting national security and encouraging new ideas (Government of India, 2016). These rules are meant to protect sensitive spatial data, but some people say that too strict licensing rules could slow down progress in areas like smart mobility systems, urban planning analytics, and disaster risk assessment, all of which depend on real-time geospatial data and satellite mapping technologies (World Bank, 2020).

4.2.3 Coalition for Disaster Resilient Infrastructure (CDRI), 2019

India started the Coalition for Disaster Resilient Infrastructure (CDRI) at the 2019 United Nations Climate Action Summit. It is a global partnership of many groups that works to make infrastructure systems more resilient to climate and disaster risks (Global Commission on Adaptation, 2019). The project encourages countries to work together in three important areas. First, it stimulates the creation of financial tools like resilience bonds and insurance plans to help people lose less money when disasters happen. Second, it encourages the use of technical standards and norms that take resilience into account when planning and designing infrastructure. Third, it backs reforms in institutions and programs that help governments include disaster risk reduction in their planning and investment in infrastructure. The CDRI works with Indian cities' own urban development programs, like the Smart Cities Mission, to promote risk-informed infrastructure planning, climate adaptation strategies, and the use of international best practices in building resilient infrastructure (UN-Habitat, 2020).

4.3 Issues and Challenges in Regulating Smart and Resilient Urban Infrastructure in India

India's shift toward smart and resilient urban development is limited not only by resource scarcity but also by a triad of deficiencies: institutional coherence, regulatory strength, and rights-based inclusivity. Even if there have been a lot of national efforts, the rules and systems for running urban infrastructure are still broken and not often followed (World Bank, 2020; UN-Habitat, 2020).

4.3.1 Institutional and Governance Fragmentation

One of the biggest problems with India's urban governance structure is that Urban Local Bodies (ULBs), state-level parastatal agencies, and central government organizations all have overlapping powers. The 74th Constitutional Amendment Act, 1992 aimed to enhance decentralized governance by empowering municipalities. However, many essential functions, including water supply, transportation, and land developmenst, continue to be managed by state agencies such as development authorities, electricity utilities, and water boards (Jain, 2010). The Smart Cities Mission's creation of Special Purpose Vehicles (SPVs) has made governance even more complicated. Critics say that SPVs often work outside of established municipal accountability frameworks, which could mean that they don't have to answer to elected councils and hinder democratic supervision (Praharaj & Han, 2019). This broken institutional architecture leads to overlapping duties, slow decision-making, and unclear lines of power. The lack of a unified legislative framework that defines the connection between SPVs and ULBs illustrates more general problems with India's urban government structure.

4.3.2 Financial Constraints and Limited Fiscal Autonomy

Urban change necessitates substantial financial resources; nonetheless, Indian cities function with somewhat fragile revenue bases and restricted fiscal autonomy. Property tax, which has historically been the main source of income for cities, makes up only 0.15% of GDP. This is far lower than the 1–2% seen in many OECD countries (World Bank, 2020). Municipal governments still rely a lot on money from the state and national governments, which limits their financial independence. The Smart Cities Mission and AMRUT are two national programs that have helped cities get money, but many of them have had trouble using the money they were given because of problems with administration, bureaucratic processes, and the fact that project timelines and funding cycles don't always match up (Government of India, 2021). Urban infrastructure expenditures will continue to be inadequate and heavily reliant on external funding until there are significant improvements in municipal finance, such as better management of property taxes, more reasonable user fees, more use of municipal bonds, and new ways to fund climate change projects.

4.3.3 Absence of a Dedicated Legislative Framework

Another big problem is that there isn't a complete set of laws that expressly covers smart and resilient urban infrastructure. Regulatory authority is presently fragmented among various statutes, encompassing municipal laws, town planning acts, environmental regulations, disaster management legislation, and sector-specific regulatory frameworks for electricity, transportation, and digital technologies (Divan & Rosencranz, 2020). Because of this, a lot of smart city projects depend on executive instructions, administrative circulars, and ad hoc regulatory changes, which make the law unclear and inconsistent between states. This fragmented legal landscape frequently does not do a good job of dealing with new governance problems including how to make digital platforms work together, how to hold AI systems accountable, how to assess the climate risk of infrastructure projects, and how to manage urban data ecosystems. Consequently, scholars advocate for the establishment of a specialized legislative framework that incorporates climate resilience, digital governance, and sustainable infrastructure planning into formal urban planning systems (Bibri & Krogstie, 2017).

4.3.4 Technological Risks and Data Governance Gaps

Data-driven technologies like IoT sensors, AI systems, and integrated urban data platforms are very important for smart urban infrastructure. These technologies can make cities more efficient and improve the delivery of services, but they also raise critical issues about privacy, cybersecurity, and monitoring (Kitchin, 2014). The Digital Personal Data Protection Act, 2023 sets up a general framework for protecting personal data, but it doesn't yet deal with some issues that are unique to smart city settings, like big surveillance systems, predictive policing algorithms, and how to run IoT-based utility systems. Municipal technology procurement contracts still don't answer many questions about who owns the data, what the standards for interoperability are, who is responsible for cybersecurity, and how to avoid vendor lock-in. The growth of smart city technology could threaten constitutional safeguards for privacy and personal freedom under Article 21 of the Constitution of India if there aren't strong data governance systems in place.

4.3.5 Land and Spatial Constraints

Urban areas have a lot of people and not much land, which makes it hard to build resilient infrastructure like green corridors, flood mitigation zones, stormwater retention systems, and affordable housing. The Right to Fair Compensation and Transparency in Land purchase, Rehabilitation and Resettlement Act, 2013, governs land purchase. However, the procedure is often contested and takes a long time, which can lead to project delays and social tensions. In many cases, land uses that are good for resilience, including protecting wetlands, restoring floodplains, and building green infrastructure, clash with strong real estate interests and the need for cities to flourish. In many Indian cities, policy tools including land pooling procedures, transfer of development rights (TDRs), and inclusionary zoning rules are not used enough (UN-Habitat, 2020). These tools could help balance development and resilience goals.

4.3.6 Climate Change and Environmental Integration

Even if programs like the Smart Cities Mission include climate-sensitive solutions like solar energy systems, energy-efficient infrastructure, and sustainable mobility initiatives, most cities still don't have a systematic way to include climate resilience in their urban planning. According to the IPCC (2022), climate adaptation measures including heat action plans, flood management systems, and ecosystem-based adaptation methods are generally carried out as separate projects instead than being built into legal planning frameworks. Environmental laws like the Environmental Impact Assessment (EIA) Notification 2006, the Solid Waste Management Rules 2016, and the Construction and Demolition Waste Management Rules 2016 offer some protection, but enforcement is not always the same in all states and cities (Divan & Rosencranz, 2020). Cities are even more likely to be harmed by climate-related events like floods, heat waves, and air pollution because there are no required climate resilience evaluations for urban infrastructure projects.

4.3.7 Social Equity and Human Rights Concerns

Smart city projects can also create big problems with social fairness, especially when building new infrastructure forces or keeps vulnerable people out. Urban redevelopment and infrastructure projects have led to evictions from informal settlements in some cases, without enough help for those affected or protection for their livelihoods. Critics have also pointed out that the Smart Cities Mission's citizen participation mechanisms often favor privileged stakeholders over marginalized populations, which means that these communities don't have a say in planning decisions (Praharaj & Han, 2019). Cities don't always follow the same rules when it comes to things like making cities accessible for people with disabilities, planning urban safety with women in mind, and making sure that urban services are affordable. To be really resilient, an urban growth model needs better legal protections, such as required social impact evaluations, participatory planning frameworks, and enforceable rehabilitation programs that are in line with human rights norms.

4.3.8 Regulatory Inconsistencies and Weak Enforcement

Urban regulation in India is still not very good since it is broken up and not enforced well. People often break building codes and planning rules, especially in informal settlements and peri-urban areas that are growing quickly. Environmental rules are also not always enforced the same way in all states. The Real Estate (Regulation and Development) Act (RERA), 2016, and other rules that govern real estate development also work differently in different states. Even the best-planned policies for making cities more resilient won't function if there aren't robust enforcement mechanisms, like planning authorities, regulatory agencies, and specialist urban tribunals with enough staff.

4.3.9 Capacity Deficits in Urban Institutions

Another problem is that Urban Local Bodies and SPVs don't have the resources to do their jobs well. A lot of city governments don't have enough qualified workers in areas like climate risk assessment, urban planning, digital governance, and financial management. So, cities typically depend a lot on outside consultants, which can come up with technology-based solutions that don't take into consideration the local economic and social situations. To make urban governance stronger, municipal governments need to keep investing in initiatives that increase capacity, professional training, and institutional strengthening. For urban infrastructure systems to be context-sensitive and long-lasting, local governments need to hire people with a wide range of skills, such as planners, climate scientists, engineers, and digital governance experts (World Bank, 2020).

5 Conclusion

India's work to build smart and strong urban infrastructure is a big step toward solving the problems of rising urbanization, climate vulnerability, and technology change. But even though there have been big policy changes like the Smart Cities Mission, the current rules are still not clear and are not being followed consistently. Overlapping institutional mandates, limited municipal autonomy, insufficient regulatory clarity, and weak enforcement mechanisms continue to prevent the achievement of a unified urban governance structure. A major structural problem is that there isn't a complete set of laws that govern smart and resilient urban development. Important topics like data governance, climate resilience planning, infrastructure funding, and fair access to urban services are still spread out among many different regulatory systems. Urban development programs could be disjointed and useless if there isn't legislative consolidation and clearer institutional roles.

So, future reforms need to go beyond single policy changes and work on creating a full Smart and Resilient Cities Law. Such laws might make it clear what each institution's responsibility is, add climate resilience requirements to urban planning laws, and set standards for building infrastructure that is environmentally friendly. It should also include strong data governance protections to preserve privacy and hold people accountable in digital governance systems while yet allowing for technological progress. It's just as crucial to change the way cities handle their finances so that they can make investments in urban infrastructure that are both inclusive and long-lasting. Cities may get long-term funds for resilient infrastructure projects by making mechanisms like municipal bonds, climate finance instruments, and public-private partnerships stronger. These projects would also be affordable and accessible to city dwellers. In the end, India's urban transformation can only work if it can bring together technology, governance, and social equality into a single set of rules. India may transition from fragmented initiatives to a more complete model of sustainable urban development by making resilience, inclusivity, and accountability part of its urban government structure. This kind of strategy would not only help India reach its national development goals, but it would also help India meet its global sustainability commitments under the New Urban Agenda, the Paris Agreement, and the Sustainable Development Goals (United Nations, 2015).

References

Albino, V., Berardi, U., Dangelico, R.M.: Smart cities: definitions, dimensions, performance, and initiatives. J. Urban Technol. **22**(1), 3–21 (2015). https://doi.org/10.1080/10630732.2014.942092

Ahern, J.: From fail-safe to safe-to-fail: sustainability and resilience in the new urban world. Landsc. Urban Plan. **100**(4), 341–343 (2011). https://doi.org/10.1016/j.landurbplan.2011.02.021

Bibri, S.E.: On the sustainability of smart and smarter cities in the era of big data: an interdisciplinary and transdisciplinary literature review. J. Big Data **6**(1), 25 (2019). https://doi.org/10.1186/s40537-019-0182-7

Bibri, S.E., Krogstie, J.: Smart sustainable cities of the future: an extensive interdisciplinary literature review. Sustain. Cities Soc. **31**, 183–212 (2017). https://doi.org/10.1016/j.scs.2017.02.016

Caragliu, A., Del Bo, C., Nijkamp, P.: Smart cities in Europe. J. Urban Technol. **18**(2), 65–82 (2011). https://doi.org/10.1080/10630732.2011.601117

Divan, S., Rosencranz, A.: Environmental law and policy in India: Cases and Materials (3rd ed.). Oxford University Press (2020). https://doi.org/10.1093/oso/9780192865458.001.0001

Eckstein, D., Künzel, V., Schäfer, L.: Global climate risk index 2021 (2021). Germanwatch. https://www.germanwatch.org/en/19777

Global Commission on Adaptation: Adapt now: a global call for leadership on climate resilience. Global Center on Adaptation (2019). https://gca.org/reports/adapt-now-a-global-call-for-leadership-on-climate-resilience/

Government of India: The Constitution (Seventy-Fourth Amendment) Act, 1992. Ministry of Law and Justice (1992)

Government of India: Disaster Management Act, 2005. Ministry of Home Affairs (2005)

Government of India: Companies Act, 2013. Ministry of Corporate Affairs (2013a)

Government of India: Right to Fair Compensation and Transparency in Land Acquisition, Rehabilitation and Resettlement Act, 2013. Ministry of Law and Justice (2013b)

Government of India: Smart Cities Mission statement and guidelines. Ministry of Housing and Urban Affairs (2015)

Government of India: Model building bye-laws, 2016. Ministry of Housing and Urban Affairs (2016)

Government of India: Climate Smart Cities Assessment Framework. Ministry of Housing and Urban Affairs (2019)
Government of India: National Urban Digital Mission (NUDM). Ministry of Housing and Urban Affairs (2021)
Government of India: Digital Personal Data Protection Act, 2023. Ministry of Electronics and Information Technology (2023)
Hollands, R.G.: Will the real smart city please stand up? Intelligent, progressive or entrepreneurial? City **12**(3), 303–320 (2008)
International Energy Agency: India energy outlook 2021. IEA (2021)
IPCC. Climate change 2022: Impacts, adaptation and vulnerability. Cambridge University Press (2022)
Jain, A.K.: Urban planning and governance a new paradigm. Bookwell Publications (2010)
Kitchin, R.: The data revolution: big data, open data, data infrastructures and their consequences. Sage (2014). https://doi.org/10.4135/9781473909472
Meerow, S., Newell, J.P., Stults, M.: Defining urban resilience: a review. Landsc. Urban Plan. **147**, 38–49 (2016). https://doi.org/10.1016/j.landurbplan.2015.11.011
Ministry of Electronics and Information Technology: National cyber security policy 2013, Government of India (2013)
Ministry of Electronics and Information Technology: Draft Internet of Things policy, Government of India (2016)
Ministry of Housing and Urban Affairs: DataSmart Cities strategy, Government of India (2021)
National Disaster Management Authority: National disaster management plan, Government of India (2019)
NITI Aayog.: Reforms in urban planning capacity in India. Government of India (2021)
Praharaj, S., Han, H.: Building a typology of the 100 smart cities in India. initiatives in India. Smart Sustain. Built Environ. **8**(5), 400–414 (2019). https://doi.org/10.1108/SASBE-04-2019-0056
Rockefeller Foundation: City resilience framework (2015). https://www.rockefellerfoundation.org
Townsend, A.: Smart cities: Big data, civic hackers, and the quest for a new utopia. Norton, W.W (2013)
UN-Habitat: World cities report 2020: The value of sustainable urbanization, United Nations (2020)
United Nations: Transforming our world: the 2030 agenda for sustainable development (2015)
United Nations: New urban agenda. United Nations (2016)
United Nations: World urbanization prospects 2018 revision, United Nations Department of Economic and Social Affairs (2019)
Vale, L.J.: The politics of resilient cities: whose resilience and whose city? Build. Res. Inf. **42**(2), 191–201 (2014). https://doi.org/10.1080/09613218.2014.850602
World Bank: Cities and resilience: Urban development strategy, World Bank (2020)

Section 3: Smart Mobility, Traffic Systems, and Transportation Infrastructure

Design of Flexible Pavement for the Existing Road Using IITPAVE Software: A Case Study of Rewa Bypass Section

Lilesh Gautam[1] and Kul Vaibhav Sharma[2](✉)

[1] Vaishnavi Infratech Services Private Limited, Dwarka, New Delhi 110078, India
[2] School of Architecture and Planning, NICMAR University, Pune, Maharashtra 411045, India
kulvaibhav.sharma@pune.nicmar.ac.in

Abstract. Evaluating the load-bearing capacity and structural characteristics of pavement materials is essential for efficient pavement and repair techniques. This study undertakes a mechanistic empirical assessment of an existing flexible pavement through Falling Weight Deflectometer (FWD) testing, back-calculation of pavement layer moduli, and structural analysis using IITPAVE in accordance with IRC: 37–2018 and IRC: 115–2014. The elastic moduli derived from the FWD field measurements were later adopted as primary inputs for overlay design using IITPAVE. Evaluation of the pavement response showed that the existing structure is unable to satisfactorily withstand the anticipated traffic demand over the intended design life. From results, a bituminous overlay was identified as a suitable rehabilitation option from a structural performance standpoint. The results highlight important aspects of pavement behavior under traffic loading and can support engineers and agencies in making informed decisions related to pavement maintenance and rehabilitation planning.

Keywords: KGPBACK Software · Back analysis · Falling Weight Deflectometer · Overlay Pavement · Rehabilitation Pavement · Deflection

1 Introduction

The road transportation is the major mode of economic activity and connectivity of the regions since highways are the main means of transportation of goods and people. In this system, pavements are important in that they offer safe, durable and reliable riding surface. Pavements in general, depending on the material used and the behavior of the structure, can be classified as either flexible or rigid. Flexible pavements are largely used due to their ability to be built flexibly, easy maintenance, and relatively low cost of construction compared to rigid concrete pavements (Aziz et al., 2015). Such pavements take up most of the highway system in India, such as key corridors such as the Rewa Bypass in Madhya Pradesh, which has long histories of high throughput of commercial vehicles and thus forms an ideal opportunity of structural performance assessment.

A flexible pavement is a stratified system whereby the load of the traffic is transferred gradually to the deeper layers. The top-most layer of bituminous surface directly resists

L. Choudhary et al. (Eds.): TIMES 2025, SCI, pp. 139–150, 2027.
https://doi.org/10.1007/978-3-032-34636-0_11

the wheel loads and environmental influence, and the sub-base layers and granular base layers distribute the stresses and give structural stability. The undercarriage that forms the bed of the pavement system has a great impact on the overall performance. This stratification permits the pavement to creep under repeated loading but again, its performance is very much dependent on the material properties, construction quality, and moisture conditions in each stratification (Gautam et al., 2018). Any failure happening in this system may cause untimely distress and shorter service life.

The recent trends in the field of pavement engineering have been more concerned with optimization of the material and sustainability. Use of locally available materials and industrial by-products especially in base and sub-base layers have demonstrated the potentials of lowering the construction cost and environmental impact (Saltan and Findik, 2008). Nevertheless, irrespective of these initiatives, flexible pavements are still degrading in most areas at a rate that is higher than expected. Increase in the amount of traffic, continuous overloading of commercial trucks, and climatic variability are also significant contributors to the distress mechanism, including rutting and fatigue cracking, which are still major concerns of pavement engineers (Behiry, 2012).

The development of freight transportation has introduced new load to the available pavement infrastructure. Despite these developments in vehicle design as replacing poor suspension systems and other axle designs which have made distributions of loads even, the increasing freight capacity is putting high stresses on the layers of pavements (Mulungye et al., 2007). Paralleling with it, pavement response can usually be controlled by the subgrade conditions and be especially problematic in soils with problematic properties. The expansive soils, such as those, are very susceptible to changes in moisture and may be subjected to huge volume changes, and thereby cracking, uneven settlement, and structural instability in the long term of flexible pavements (Nalbantoğlu, 2004).

The analytical modeling processes have been developed with time to gain a better insight into pavement behavior under loading by the traffic. Pavement systems are often simplified to layer elastic models or are referred to as elastic layers with the help of finite element techniques to approximate stresses, strains, and deflections. The mechanistic-empirical design methods combine these analytic responses and the empirical performance criteria; hence, more realistic assessment of pavement performance can be made. A number of computer tools have been created to help conduct such analyses in different material and loading scenarios; among them, there are KENPAVE, IITPAVE, and ANSYS (Arimilli et al., 2018).

Practices of pavement evaluation in India were historically founded on the measurements of static deflection, which gives little information about the response of the pavement to the wheel movement loads. These techniques fail to fully represent the temporal characteristics of the stresses associated with traffic as well as the form of the deflection basin. Consequently, non-destructive methods of testing, especially Falling Weight Deflectometer (FWD), have been more adopted. The FWD provides an impulse load simulation of a moving wheel load and measures surface deflections under several offsets, which provides a more realistic representation of the structural behavior of pavement (Singh et al., 2020).

The data of FWD deflection is usually employed to back-calculate the in-situ material properties. The KGPBACK, which was created at IIT Kharagpur, uses optimization algorithms to measure the elastic moduli of each individual pavement layer by matching measured and calculated deflections. The back-calculated parameters are important inputs in mechanistic analysis and overlay design with the aid of such tools like IITPAVE (Singh et al., 2020). In the case of flexible pavements, the bottom of the bituminous layer usually controls structural failure through fatigue cracking, and a rutting failure in the form of overloading compressive strain in the underlying ground. The management of these responses therefore is critical in the attainment of a satisfactory pavement performance in the design life.

The mechanistic design scheme of IRC: 37-2018 presents cumulative traffic loading and subgrade strength as the leading inputs to the flexible pavement design. IITPAVE is a product of a research project run by the Ministry of Road Transport and Highways that is a major advancement of the previous design tools by taking into account the layered elastic behaviour, tyre pressure, material stiffness, and traffic loading (Mehta et al., 2021; Khan, 2023). The pavement adequacy is measured against estimated critical strains against permissible thresholds defined in design standards, hence long-term structural reliability (Al Maqbali and Ragab, 2021).

Typical instances of distress manifested in flexible pavements, including wheel-path rutting and bottom-up fatigue cracking are direct results of the over stresses and strains in pavement layers, which are often augmented by heavy axle loads and inadequate structural thickness (Harish, 2017). In the view of these difficulties, this study examines the current structure and rehabilitation needs of a 5.4 km stretch of the Rewa Bypass using FWD testing, back-calculated by KGPBACK, and mechanistic analysis by IITPAVE. The study aims at offering a viable model of field assessment of pavement conditions, life span prediction, and overlay design to guide sustainable pavement management by merging the field measurements and analytical analysis.

1.1 Aim and Objective of Research

The studies discussed the structural functioning and longevity of flexible pavements using a representative example (5.4 km of the Rewa Bypass, Madhya Pradesh, India) as a case study. To have a better idea of the behavior of the pavement structure with the traffic loading as time goes, the study uses the methods of empirical analysis and the sophisticated methods of evaluating the structure. The results will be aimed at helping to determine feasible and credible solutions to maintain the pavement, rehabilitate, and manage the long-term performance.

2 Literature Review

The pavement engineering research has covered a broad array of problems in subgrade enhancement, material choice, structural analysis and mechanistic-empirical design which all contributed to the foundation of current pavement performance analysis and rehabilitation. The literature addressing the aspects of subgrade behavior optimization and the development of improved pavement materials is quite large. Indicatively, Gul and Bashir (2023) studied the application of glass fiber and cement kiln dust as a means of strengthening silty soils and found significant increases in the performance of subgrade soils in low-volume roadways, supported by both the laboratory tests and analysis based on the IITPAVE. On the same note, Mukherjee and Ghosh (2021) demonstrated how enhanced CBR of subgrade material, in conjunction with the use of cement-treated sub-base materials as well as reclaimed asphalt pavement may result in economical, structurally-effective pavement systems. Gonawala et al. (2021) also showed that cement-stabilized EAF slag and GGBFS can be used in the base and sub-base construction, and their mechanical analysis showed that the strain level to be obtained is within acceptable ranges. Recent studies have also highlighted the performance benefits of modified binders and warm mix asphalt in surface course applications (Sharma et al., 2024).

The modern research has started to depend on mechanistic-empirical pavement design and optimization with the help of this type of research on the basis of the advanced analytical tools. Kulkarni and Ranadive (2022) used IITPAVE and KENPAVE to conduct a comparative analysis of perpetual pavement alternatives by taking into account both the life-cycle costs and the greenhouse gas emissions and found that deep-strength perpetual pavements are a sustainable solution. Pandey et al. (2022) and Mehta et al. (2021) also showed that pavement structures built based on IRC: 37-2018 with the assistance of IITPAVE can lead to the reduction of layer thickness and construction prices without any deterioration of structural performance. Mohan and Praveen (2020) also found the IITPAVE-based design as the design was validated through measurement of subgrade properties and traffic data recorded at the field.

Non-destructive testing has been significant in structural testing of pavements. Singh et al. (2020) have proved that assessment based on FWD, with mechanistic overlay design, can successfully increase the service life of Indian highway pavements. Jiang et al. (2022) performed the investigation of structural performance of inverted pavements with the help of deflection basin parameters based on measurements using FWD and provided positive response features. Besides, Calhoon et al. (2022) demonstrated the beneficial effect of integrating both FWD and Ground Penetrating Radar under the conditions of determining moisture-related variations in pavement structural capacity.

Various pavement assessment instruments have also been compared and alternative pavement technologies compared in several studies. Kulkarni and Magadeo (2018), underlined the topicality of the concepts of perpetual pavement in the Indian setting and the necessity to provide the data specific to the region in support of the mechanistic-empirical design methodologies. Arimilli et al. (2018) compared the cold-recycled pavement systems with the traditional hot-mix asphalt pavements under the KENPAVE, IITPAVE, and ANSYS, and proved the importance of cross-validation with the help of various software programs. Moreover, Harish (2017) and Saride et al. (2017) demonstrated that the use of recycled and substitute materials such as reclaimed asphalt pavement and fly ash-treated aggregates can lessen the pavement and construction expenses and still offer satisfactory structural performance.

The development of the analytical modeling provided further improvement in the knowledge of the pavement behavior under the loading of the traffic. According to Ghanizadeh and Fakhri (2018), quasi-static pavement response estimated through FFT and artificial neural network analysis were close to those calculated by full dynamic analysis. Ji et al. (2018) noted that resilient modulus and deflection were better in pavements using direct coal liquefaction residue-modified asphalt mixtures. Ranadive and Tapase (2016) have emphasized the significance to attain balanced consideration of fatigue and rutting performance by parameter-sensitive pavement design strategies.

Traffic loading phenomena and concepts of long-life pavement have been also given a lot of attention. As shown by Solanki and Zaman (2016), the thickness demands of the pavements highly depend on the soil conditions, the methods of stabilizations, and the design method taken. Dhakal et al. (2016) have reviewed the various methods of mitigating reflection cracking and compared their performance and constructability. According to Hossain et al. (2016), there are critical differences between site specific traffic inputs and default MEPDG values, which is why proper axle load characterization is necessary. Hernando and del Val (2016) compared semi-rigid long-life pavements and, it was concluded that merely adding asphalt density is not an effective approach to procrastinate reflective cracking.

3 Materials and Methods

The present study adopts an integrated evaluation framework that brings together FWD testing, back-calculation of pavement layer properties using KGPBACK, and mechanistic analysis carried out through IITPAVE. Although these tools are individually well established, their combined application in Indian case studies has been relatively limited. A mechanistic–empirical approach was therefore used to assess pavement performance, wherein structural adequacy was examined over the design life until either a rut depth of 20 mm or fatigue cracking affecting 20% of the pavement surface was reached, whichever condition governed first.

Structural analysis was performed using IITPAVE, a mechanistic–empirical software developed to evaluate pavement response under traffic loading. Within this framework, the flexible pavement system is idealized as a multilayer elastic structure composed of several elastic layers resting on a semi-infinite subgrade. This modeling approach allows stresses and strains to be estimated at critical locations within the pavement. Attention was given to the tensile strain (ε_t) at the bottom of the bituminous layer and the vertical compressive strain (ε_v) at the top of the subgrade, as these responses are directly associated with fatigue cracking in the asphalt layers and rutting in the subgrade, respectively, as illustrated in Fig. 1.

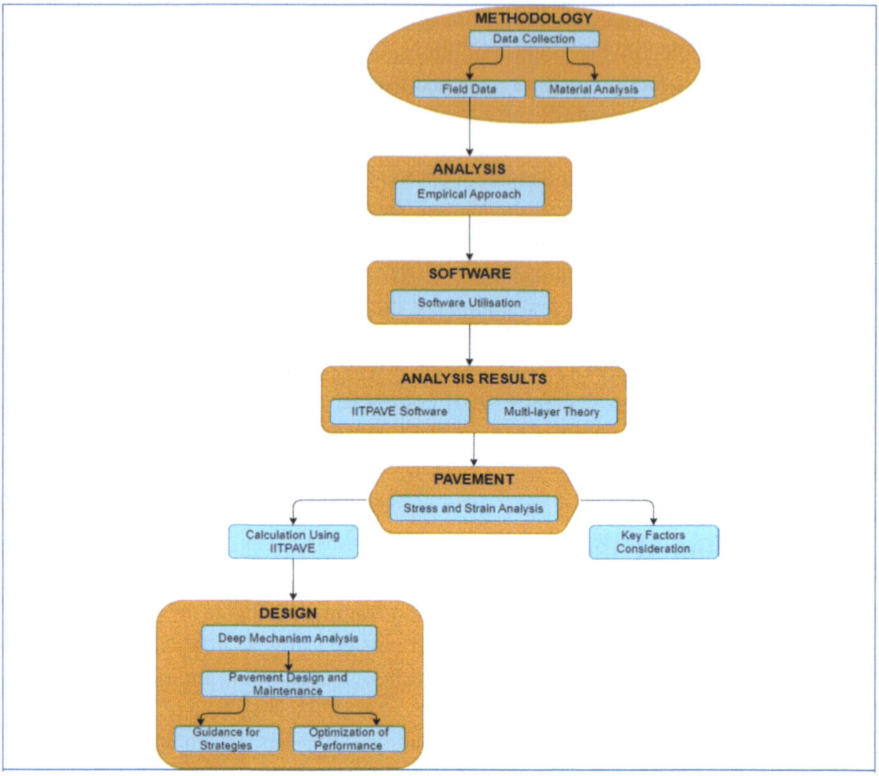

Fig. 1. Methodology Flowchart (Source: Author's contribution)

4 Results

The pavement performance of the Rewa Bypass Sect. (0–5.4 km) was evaluated using mechanistic–empirical analysis in IITPAVE, following IRC: 37–2018 provisions.

4.1 Homogeneous Section Identification

Based on traffic characteristics, the project road was classified into a single traffic-homogeneous section (HS-1) covering a length of 5.4 km, shown in Table 1.

Table 1. Traffic Homogeneous Section (Source: Author's contribution)

Homogeneous Section	Design Chainage (km)
5.4	2.53
0.369	0.123
23.24	22.55
0.161	0.161

4.2 Design Subgrade CBR

Subgrade strength was evaluated through laboratory testing of borrow soils, including sieve analysis, heavy compaction tests, and 4-day soaked CBR tests. As per IRC: 37-2018 (Clause 5.2), the effective design CBR was determined by considering both embankment soil and select subgrade fill. For HS-1, the adopted effective design CBR was 15%,, shown in Table 2.

Table 2. Design Subgrade CBR (Source: Author's contribution)

Section	CBR (Embankment)	CBR (Select Subgrade)	Adopted CBR (%)
15	15	15	HS-1
0.369	0.123	0.290	0.444
23.24	22.55	23.63	23.12
0.161	0.161	0.042	0.251

4.3 Design Traffic Characteristics

Design traffic was estimated for a 20-year design life based on classified traffic counts, axle load surveys, and vehicle damage factors (VDFs). The base-year AADT of commercial vehicles for HS-1 is presented in Table 3, while direction-wise VDF values are shown in Table 4.

Table 3. Base Year AADT of Commercial Vehicles (Source: Author's contribution)

LCV	Bus	2-Axle	3-Axle	MAV
23	378	502	2552	2.46

Table 4. Direction-wise Vehicle Damage Factors (VDF) (Source: Author's contribution)

Vehicle	Up	Down
0.81	0.75	2.46
1.13	2.02	0.207
6.31	9.88	22.73
10.89	13.26	0.213
0.56	0.49	

4.4 Design Traffic for Flexible Pavement

A minimum traffic growth rate of 5% was adopted. Due to directional bias in axle loading, design traffic was computed separately for each direction using Eq. (1), in accordance with IRC recommendations to account for future uncertainty.

$$N_r = (V_r \times C \times VDF \times LDF \times G^E) \times D \tag{1}$$

where N_r is cumulative standard axles, V_r is base-year AADT, VDF is vehicle damage factor, LDF is lane distribution factor, G is growth rate, E is design period (20 years), and D is directional factor. The cumulative standard axles were calculated using Eq. (2).

$$C = {}^{(1+r)^2 - 1)}\big/_{r} \times VDF \times LDF \tag{2}$$

A minimum annual traffic growth rate of 5% was adopted in accordance with IRC recommendations to account for future uncertainty. Direction-wise traffic analysis resulted in design traffic values of 270 MSA for the Up direction and 340 MSA for the Down direction, with the latter governing pavement design. Owing to the higher loading demand, the Down direction value of 340 MSA was adopted as the governing design traffic for flexible pavement design and mechanistic analysis. The pavement structure was designed using IITPAVE in accordance with IRC: 37–2018, considering a subgrade CBR of 15%, VG-40 bitumen at 35 °C, 3.5% air voids, and 11.5% bitumen content for fatigue analysis. The adopted pavement crust composition is presented in Table 5.

Table 5. Conventional Flexible Pavement Composition (Source: Author's contribution)

Direction	CBR (%)	Traffic (MSA)	BC (mm)	DBM (mm)	WMM (mm)	GSB (mm)
Up	15	270	50	145	250	200
Down	15	340	50	145	250	200

Note:BC layer provided with modified bitumen

In accordance with MORTH Clause 401.2.1, GSB Grading-V was adopted as a sub-base-cum-drainage layer. Paved shoulders were designed with the same pavement composition to ensure structural continuity and drainage efficiency.

4.5 Rehabilitation Design of Existing Carriageway

The structural condition of the existing flexible pavement was evaluated using FWD testing in accordance with IRC: 115–2014. An impulse load simulating standard axle loading was applied, and pavement surface deflections were recorded using seven geophones positioned at offsets of 0, 200, 300, 450, 600, 900, and 1200 mm, enabling the development of deflection bowls for structural assessment and back-calculation. Deflection data from multiple drops were normalized to a standard load of 40 kN. Layer elastic moduli were back-calculated using the KGPBACK software based on measured pavement layer thicknesses and subgrade strength obtained from DCP and pavement composition surveys, with seed moduli selected as per IRC: 115–2014. Representative photographs showing FWD survey being conducted at project site are shown in Fig. 2.

Fig. 2. Project site images of FWD conducted at different location (Source: Author's contribution)

Due to scatter in the deflection measurements, homogeneous pavement sections were delineated using the Cumulative Difference Approach (CDA) based on peak deflection (D_0) and surface curvature index (SCI = D_0–D_{300}). This statistical approach resulted in the identification of eleven structurally homogeneous sections, which were subsequently adopted for overlay design. This linkage between modulus variation and distress manifestation has now been explicitly discussed.

The back-calculated elastic modulus of the bituminous layer varied from 846.8 to 2997.8 MPa on the left-hand side and from 756.6 to 3000 MPa on the right-hand side, with average values of 2701 MPa and 2524 MPa, respectively. Around 10% of the project length shown lower elastic modulus values in the range of 756 to 1572 MPa, which indicated the presence of thin or structurally distressed bituminous layers. These findings align with field observations of surface distresses such as alligator cracking and localized patch repairs along the pavement stretch. The back-calculated elastic modulus of the granular layer varied from 100 to 500 MPa, with mean values largely lying between 451 and 470 MPa. The relatively higher modulus values with empirical estimates are likely the result of progressive densification of the granular layers caused by repeated traffic loading over the service life of the pavement, suggested in IRC guidelines. The subgrade elastic modulus showed minimal variation, ranging from 99.4 to 100 MPa with an average of approximately 99.9 MPa, corresponding to an equivalent CBR of about 9%, which aligns well with laboratory CBR values ranging from 3.1% to 16%.

In accordance with IRC: 115–2014, the 15th percentile elastic moduli adjusted for temperature and moisture were adopted as design values, namely 692 MPa for the bituminous layer, 173.4 MPa for the granular layer, and 99.6 MPa for the subgrade. The remaining life of the existing pavement was evaluated using IITPAVE by modeling the structure as a bonded three-layer elastic system subjected to an FWD load of 40 kN, producing a uniform contact pressure of 0.56 MPa. Fatigue and rutting performance were assessed using IRC: 115–2014 models, where fatigue life is given by Eq. (3) and rutting life by Eq. (4).

$$N_f = 0.711 \times 10^{(-4)} (1/\in_t)^{3.89} (1/M_R)^{0.854} \tag{3}$$

$$N = 1.41 \times 10^{-8}(1/\in_v)^{4.5337} \tag{4}$$

For the Up direction, the estimated fatigue and rutting lives were 14 MSA and 98 MSA, respectively, against a design traffic of 270 MSA. For the Down direction, the corresponding values were 32.7 MSA and 320.7 MSA against a design traffic of 340 MSA. In both directions, the computed remaining life was lower than the design requirement, and the pavement was therefore classified as structurally not safe, necessitating rehabilitation.

The FWD measurements revealed considerable difference in the elastic moduli of the pavement layers and that some points are less stiff in the bituminous layer. These areas were signs of structural distress and were congruent with the noticeable deterioration on the surface. The back-calculated moduli of the granular layer and subgrade, on the other hand, had a comparatively small variation, suggesting that the bottom layers still offer some sensible structural support. Nonetheless, the strain-based analysis conducted with the use of IITPAVE showed that fatigue life as well as rutting life of the current pavement are significantly less than the necessary design traffic with a reliability of 90 percent. Based on this, the pavement was classified as structurally sub-standard and not fit to serve any more unless some remedial action is taken. The findings indicate that degradation of the upper layers of pavement, but not the subgrade inadequacy is the major aspect that governs the overall pavement behaviour. In that regard, the application of a bituminous overlay was deemed suitable, as it mitigates the strain surplus and life-remaining constraints in addition to meeting the performance demands set in IRC: 115-2014, thus, providing a convenient and cost-efficient rehabilitation remedy.

5 Conclusions

The service life of the current flexible pavement was determined by looking at pavement layers thickness, critical strain responses, as well as fatigue and rutting performance based on the models outlined in IRC: 115-2014 at a 90 percent reliability level. Analysis revealed that the existing pavement structure lacks the capacity to support both the current and future traffic requirements with reasonable safety within the stipulated design life and therefore the rehabilitation should be carried out as soon as possible. Subsequent mechanistic-empirical testing implemented with IITPAVE revealed that the depths of layers of the pavement were found to be higher than those needed to meet performance standards (as-built). This trend is probably due to a combination of many factors such as limitation in the design practices in the past, the underestimation of traffic loading at the time of planning design, and the slow rise in traffic volume and axle load over a period.

Despite these variations, the mechanistic analysis was a reliable means of creating an overlay design to meet both the fatigue and rutting performance specifications during the predicted service life. On the findings of the remaining life assessment and overlay design analysis, specific rehabilitation actions were suggested to increase structural capacity, add pavement life and overall life-cycle cost efficiency. The results of the research indicate that the FWD testing, back-calculation procedures and IITPAVE-based mechanistic analysis is a good model to evaluate the performance of pavements and design rehabilitation interventions. Such performance-based measures are adoptable and will assist in having sustainable pavement management and aid in ensuring functionality of the important road infrastructure over a long period. The next study can also focus on investigating the application of new materials, climate-resistant pavement materials and real-time structural health observation technologies to enhance performance forecasting and sustainability.

Acknowledgments. The authors gratefully acknowledge the support provided by the Data Analytics and Research Lab, NICMAR University, Pune, for facilitating access to advanced analytical infrastructure and collaborative research resources used in this study.

Conflict of Interest. The authors declare that there are no financial or non-financial conflicts of interest associated with this research.

References

Al Maqbali, L., Ragab, O.: A review of software in flexible pavement design. Sustain. Environ. **6**(2), 27 (2021). https://doi.org/10.22158/se.v6n2p27

Arimilli, S., Nagabhushana, M.N., Jain, P.K.: Comparative mechanistic–empirical analysis for design of alternative cold recycled asphalt technologies with conventional pavement. Road Mater. Pavement Des. **19**(7), 1595–1616 (2018). https://doi.org/10.1080/14680629.2017.1338187

Aziz, M.M.A., Rahman, M.T., Hainin, M.R., Bakar, W.A.W.A.: An overview on alternative binders for flexible pavement. Constr. Build. Mater. **84**, 315–319 (2015). https://doi.org/10.1016/j.conbuildmat.2015.03.068

Behiry, A.E.A.E.M.: Fatigue and rutting lives in flexible pavement. Ain Shams Eng. J. **3**(4), 367–374 (2012). https://doi.org/10.1016/j.asej.2012.04.008

Calhoon, T., Zegeye, E., Velasquez, R., Calvert, J.: Using falling weight deflectometer and ground penetrating radar to monitor seasonal moisture effects on pavement structural capacity. Constr. Build. Mater. **351**, 128831 (2022). https://doi.org/10.1016/j.conbuildmat.2022.128831

Dhakal, N., Elseifi, M.A., Zhang, Z.: Mitigation strategies for reflection cracking in rehabilitated pavements—A synthesis. Int. J. Pavement Res. Technol. **9**(3), 228–239 (2016). https://doi.org/10.1016/j.ijprt.2016.05.001

Gautam, P.K., Kalla, P., Jethoo, A.S., Agrawal, R., Singh, H.: Sustainable use of waste in flexible pavement: a review. Constr. Build. Mater. **180**, 239–253 (2018). https://doi.org/10.1016/j.conbuildmat.2018.04.067

Ghanizadeh, A.R., Fakhri, M.: Quasi-static analysis of flexible pavements based on predicted frequencies using FFT and ANN. Int. J. Pavement Res. Technol. **11**(1), 47–57 (2018). https://doi.org/10.1016/j.ijprt.2017.09.002

Gonawala, R.J., Khapre, S., Kumar, R., Chauhan, K.A.: Suitability of EAF slag and GGBFS mix as cementitious base/subbase layer for low volume road construction. Int. J. Geotech. Eng. **15**(2), 252–258 (2021). https://doi.org/10.1080/19386362.2019.1621458

Gul, N., Bashir, B.A.M.: Performance evaluation of silty soil reinforced with glass fiber and cement kiln dust for subgrade applications. Constr. Build. Mater. **392**, 131943 (2023). https://doi.org/10.1016/j.conbuildmat.2023.131943

Harish, H.G.R.: Analysis of flexible pavements using IIT-PAVE. Imp. J. Interdiscip. Res. **3**(6), 815–818 (2017)

Hernando, D., del Val, M.A.: Guidelines for the design of semi-rigid long-life pavements. Int. J. Pavement Res. Technol. **9**(2), 121–127 (2016). https://doi.org/10.1016/j.ijprt.2016.03.003

Hossain, N., Singh, D., Zaman, M.: Sensitivity of traffic input parameters on rutting performance of flexible pavement. Int. J. Pavement Res. Technol. **9**(6), 450–459 (2016). https://doi.org/10.1016/j.ijprt.2016.09.003

Indian Roads Congress: IRC: 115–2014—Guidelines for structural evaluation of flexible pavements using falling weight deflectometer (FWD). New Delhi, India (2014)

Ji, J., Wang, D., Suo, Z., Xu, Y., Xu, S.: Influence of direct coal liquefaction residue on mechanical properties of flexible pavement. Int. J. Pavement Res. Technol. **11**(4), 355–362 (2018). https://doi.org/10.1016/j.ijprt.2017.09.006

Jiang, X., et al.: Evaluation of inverted pavement by structural condition indicators from falling weight deflectometer. Constr. Build. Mater. **319**, 125991 (2022). https://doi.org/10.1016/j.conbuildmat.2021.125991

Khan, S.: Analysis of flexible pavement using IIT-PAVE software. Eng. Comput. Biol. **12**(SI-4), 14124–14145 (2023). https://doi.org/10.48047/ecb/2023.12.si4.1280

Kulkarni, S.S., Magadeo, M.S.: Software-based analysis of perpetual pavement in Indian scenario. Int. J. Eng. Res. **6**(5), 477–483 (2018)

Kulkarni, S., Ranadive, M.S.: Parametric comparison of perpetual pavements with respect to life-cycle cost and greenhouse gas emissions. Mater. Today: Proc. **52**, 1147–1152 (2022). https://doi.org/10.1016/j.matpr.2021.11.018

Mehta, A., Vanshaj, K., Singh, G., Yadav, S., Srivastav, J.B.: Design of flexible pavement as per IRC:37–2018 using IIT-PAVE. Int. J. Sci. Res. Dev. **9**(9) (2021)

Mohan, M.K. A. N., Praveen, K.: Analysis of flexible pavement using IITPAVE software and economic analysis using HDM-4. Int. J. Res. Appl. Sci. Eng. Technol. **8**(5), 2651–2657 (2020). https://doi.org/10.22214/ijraset.2020.5443

Mukherjee, S., Ghosh, P.: Effect of improvement in CBR characteristics of soil subgrade on design of bituminous pavements. Indian Geotech. J. **51**(3), 567–582 (2021). https://doi.org/10.1007/s40098-021-00533-8

Mulungye, R.M., Owende, P.M.O., Mellon, K.: Finite element modelling of flexible pavements on soft soil subgrades. Mater. Des. **28**(3), 739–756 (2007). https://doi.org/10.1016/j.matdes.2005.12.006

Nalbantoğlu, Z.: Effectiveness of class C fly ash as an expansive soil stabilizer. Constr. Build. Mater. **18**(6), 377–381 (2004). https://doi.org/10.1016/j.conbuildmat.2004.03.011

Pandey, A., Vanshaj, K., Singh, G., Yadav, S., Srivastav, J.B.: Design of flexible pavement using experimental and software approach. Int. Res. J. Eng. Technol. 278–284 (2022)

Ranadive, M.S., Tapase, A.B.: Parameter sensitive analysis of flexible pavement. Int. J. Pavement Res. Technol. **9**(6), 466–472 (2016). https://doi.org/10.1016/j.ijprt.2016.12.001

Saltan, M., Findik, F.S.: Stabilization of subbase layer materials with waste pumice in flexible pavement. Build. Environ. **43**(4), 415–421 (2008). https://doi.org/10.1016/j.buildenv.2007.01.007

Saride, S., George, A.M., Avirneni, D., Basha, B.M.: Sustainable design of Indian rural roads with reclaimed asphalt materials. In: Advances in Transp. Geotech. (pp. 73–90). Springer (2017). https://doi.org/10.1007/978-981-10-1930-2_6

Sharma, K.V., Sharma, N., Sharma, D., et al.: Comparative evaluation of open graded friction course and stone matrix asphalt mixtures with modified binders and warm mix asphalt. J. Build. Rehabil. **9**, 114 (2024). https://doi.org/10.1007/s41024-024-00468-6

Singh, A., Sharma, A., Chopra, T.: Analysis of the flexible pavement using falling weight deflectometer for Indian national highway road network. Transp. Res. Procedia **48**, 3969–3979 (2020). https://doi.org/10.1016/j.trpro.2020.08.024

Solanki, P., Zaman, M.: Design of semi-rigid type of flexible pavements. Int. J. Pavement Res. Technol. **10**(2), 99–111 (2017). https://doi.org/10.1016/j.ijprt.2016.10.004

Evaluation of Traffic Congestion and Its Environmental Impact Through Traffic Flow Analysis in Gurugram City

Rohit Sharma(✉) and Anu Tonk

Department of Multidisciplinary Engineering, The NorthCap University, Gurugram, Haryana 122017, India
{23cvp012,anutonk}@ncuindia.edu

Abstract. Traffic congestion is a critical concern that affects both economic productivity and environmental sustainability in cities across the globe. This issue is especially severe in Gurugram, where daily commuters are impeded by substantial time losses. Experts emphasise the importance of conducting a thorough examination of congestion characteristics to create effective mitigation strategies. Traffic volume surveys were implemented in this investigation to assess congestion at critical intersections, including Sector 4–7 Chowk and Sector-5 Chowk. The data that was collected was subjected to an analysis using a variety of traffic flow theories, such as the Indo-HCM (Highway Capacity Manual) method and the IRC 106–1990 guidelines, to identify fundamental traffic parameters that revealed congestion patterns. Congestion maxima are observed during the morning, midday, and evening hours, with the traffic composition primarily consisting of auto-rickshaws, buses, motorcycles, four-wheelers, and construction vehicles, according to the investigation. Gurugram's road network is one of the most congested in India, with service quality at numerous road sections being rated as F. The study emphasises that these issues are further exacerbated by suboptimal infrastructure and inadequate traffic management. Furthermore, the Indo-HCM method validated the severity of traffic delays by providing critical insights into the capacity and level-of-service (LOS) assessments. The study emphasises the pressing necessity of targeted interventions to improve the efficiency and sustainability of the urban transport system in Gurugram by reducing congestion and its environmental consequences.

Keywords: Traffic congregation · Gurugram city · Infrastructure upgrades · Traffic management strategies · Peak hours · Traffic Demand · Passenger car unit

1 Introduction

Education has always been a pillar of our society, providing knowledge for individual minds and a foundation for personal growth. It provides people with the skills, resources, and information they require to prosper and lead happy, fulfilling lives in society [1]. Education, together with knowledge from books promotes the growth of complex mental

L. Choudhary et al. (Eds.): TIMES 2025, SCI, pp. 151–168, 2027.
https://doi.org/10.1007/978-3-032-34636-0_12

processes that facilitate the problem-solving and creative abilities that make up creativity [2]. Every educational institution equips students of all ages with the skills and information they need to overcome challenges in the future [3]. Thus, education promotes the idea of lifelong learning and instills in our citizens the habits necessary for success in both their personal and professional lives [4]. Because making educational investments sets the stage for future success in both the personal and professional spheres [5]. We also give talented people the opportunity to realize their full potential by investing in the future of society.

1.1 Causes of Traffic Congestion

One of the biggest issues in Gurugram is traffic congestion. There is a risk of accidents as a result of the traffic congestion issue. As everyone knows, Gurugram is a sophisticated metropolis [3]. Over the past few decades, the city's population density has likewise risen. However, there are certain traffic issues in that industrialized metropolis [6]. These issues require a fix or a different approach to improve the efficiency of traffic movement. Every time we go through the city of Gurugram, we observe certain citizens driving autorickshaws alongside or on the side of the road, obstructing traffic. That is the source of the traffic jam as well as the potential for an accident [7]. Not only may laws, rules, and regulations reduce traffic issues, but we also need to discover other solutions, such developing roadways equipped with cutting-edge traffic control systems [8]. Additionally, individuals and people must be educated about traffic laws, regulations, and their benefits [9].

Traffic congestion has a negative and considerable impact on city development. Jams Travel times often rise when vehicle movement speed decreases [10]. Our work's goals are to provide a feasible solution, sensible and helpful ideas, and conceptual direction for regulating traffic flow in a manner that reduces congestion business its overall negative impact on society, individuals, and passengers [11]. Gurugram has a limited number of secondary roads, lanes, and tiny roadways. These narrow roads are becoming narrower as they become more crowded as a consequence of illegal possession [12]. Traffic chaos is caused by traffic lights that are regularly purposely incorrectly in sync or occasionally out of sync owing to computer failures [13].

1.2 Traffic Congestion: The Problem and Solutions

1.2.1 Traffic Congestion Problem

Traffic congestion is the expression used to describe the situation in which the number of vehicles attempting to utilize a particular road facility exceeds its capacity, without causing unacceptable levels of delay or inconvenience [14]. Wrong-side driving, overtaking, using shortcuts through narrow lanes (mostly cars, trucks, or any four-wheel or three-wheel vehicle), merging in lanes, and other practices are the primary causes of traffic congestion in the city [15]. Inadequate public transportation encourages people to drive their own cars, which increases the number of vehicles on the road [16]. Nevertheless, motor vehicles are not without their physical and environmental constraints. Initially, a motor vehicle necessitates road space to operate readily, as well as parking spaces at residences and workplaces. The demand for motor vehicles frequently exceeds the supply of road space in numerous countries [17].

1.2.2 Inefficient Road Network

The road network has grown by a mere six times, while Gurugram traffic has increased by over 105 times since 1960. Consequently, the road network is inadequate, resulting in traffic congestion and catastrophic accidents, as a result of the substantial traffic volume [18]. According to the data, the extant road network is 7%, while the required network is 15%, which is more than double the current amount. Consequently, there is an 8% disparity between them, which results in numerous major traffic congestions in various regions of the city [19]. These congestions are time-consuming, and as a result, commuters are pressed for time to reach their destinations. In order to reduce their travel time, some of them increase their velocities, which can result in accidents that may damage their property and, in some cases, their lives. As the number of vehicles in the town has increased, the road infrastructure has not expanded in tandem [20].

2 Related Work

Afrin et al. (2020) Sustainable advancement of transportation is significantly impeded by traffic congestion, which results in environmental damage, revenue losses, as well as delays. This study analyzes a variety of congestion indicators by analyzing daily and weekly traffic data. It identifies the strengths and limitations of each measure and provides valuable insights for the development of a sustainable, resilient transportation management system [21]. Adebola Olowosegun (2021) Public transport as well as urban mobility were profoundly affected by the COVID-19 pandemic on a global scale. Major changes in traffic management as well as travel patterns were observed in Johannesburg, a developing metropolis in one of Africa's most severely affected countries [22]. This study compared pre-pandemic, lockdown, and post-restriction instances to analyze navigation variations during the worldwide outbreak in Johannesburg from 2019 to 2021, utilizing data from the TomTom Traffic Index. The 2020 results indicated a significant decrease in congestion as a result of movement restrictions, with mini-bus transportation being the most significant contributors to congestion. It is imperative to encourage sustainable transportation and discourage single-occupancy vehicles in order to alleviate traffic congestion. Raza et al. (2022) Traffic delays, a long-term problem that disrupts populations and raises pollution levels. The present study analyzed urban transport system of Pakistan applying modern MCDM tools and developed solutions using a combination of traffic simulation [23].

Mu et al. (2019) Study the Dynamics of Traffic Flow with Micro-Cars Availability of lane-level traffic densities for mixed conditions on arterial roads and urban highways is replicated through a two-lane traffic model. Micro-cars had mixed safety and emissions impacts on different road types, with an increase in lane-changing frequency and increased decelerations [24]. Muhy Abdulwahab et al. (2020) In regions with high concentration of vehicular traffic, sound measures must be put in place as the levels of pollution due to air borne contaminants from traffic at the intersections is very high [25].

3 Research Methodology

The study technique used to evaluate traffic congestion and its environmental effect in Gurugram City is rigorous and thorough. To provide a thorough understanding of traffic patterns, this study uses a mixed-methods approach, integrating qualitative and quantitative tools. To start, a literature review will be conducted to create a theoretical framework by identifying significant factors and relevant results from past research. Subsequently, quantifiable data will be gathered at important places around Gurugram using traffic volume measures, trip time evaluations, and vehicle speed monitoring. Apart from the previous approaches, qualitative data will be obtained through interviews with traffic management representatives and city planners. The usage of such combinations of methodologies will bring in advanced research that will provide better understanding of the factors leading to traffic jams and their environmental consequences in Gurugram City.

3.1 Study Area

The study "Evaluation of Traffic Congestion through Traffic Flow Analysis in the Gurugram City" was and concerned with the state of the roads in Gurugram, Haryana's leading metropolitan center. It included analyses of the crucial intersections like Sector-5 Chowk and the significant Sector 4–7 Chowk and from the arterial roads like. These meaningful locations have been purposefully selected owing to its strategic relevance and the concentration of traffic seen. Efforts were made to focus on these specific areas to draw a comprehensive picture of the traffic flow patterns, determine likely congestion spots in the city, and comprehend the transportation network in Gurugram City. The information obtained from this research is aimed more at assisting in the formulation of the traffic management strategies and decreasing conge.

3.1.1 Density

Density is another important criterion on successive traffic congestion assessment in Gurugram through traffic flow analysis. Density is measured by how many automobiles occupy a unit length of road, and it is useful to see traffic spread. Traffic congestion would be expected from the density of autos that could be observed on sections of the route.

3.1.2 Mean Free Speed

The Mean Free Speed is crucial for traffic flow studies of Gurugram. In plain terms, it is the speed which vehicles would usually travel at in peacetime when there are no obstructions to progress. It is where measures ideally commence giving the level of flow which can be attained under the best driving conditions.

3.1.3 Level of Service (LOS) Determination

The level of service (LOS) decision is important for analysing traffic congestion levels in the city of Gurugram. To evaluate traffic efficiency, the mean travel speed was determined through various segments in both directions. This research includes the distribution of the data into conventional methods known grades A to F. Each letter or grade represents a certain level of service which reflects the condition of the given traffic now more qualitatively. With the aid of LOS commitment, flow of traffic system operations regarding that congestion prone areas can be established and therefore appropriate measures to improve traffic management in Gurugram can be taken.

4 Results and Discussion

4.1 Peak Hour Volume Sector 5 Chowk

The traffic study shows that there are two distinct peak periods, morning arriving and evening leaving. The early afternoon and evening periods are account for how many flights may be operating at those times. The count reveals that the morning automotive peak increased by almost 2,000 Passenger Car Units per hour from the ADT recorded in 2016 to reach a peak of 12,821.1 PCU/hr. On the other hand, traffic volume grew a bit to 13,321 PCU/hr at evening peak hour showing higher demand on road space as commuters are returning home. An analysis of morning peak-hour traffic (7:00 am to 11:00 am) at a junction, detailing vehicle types, their counts, and Passenger Car Unit (PCU) values across different directions. The junction comprises three arms Arm 1, Arm 2, and Arm 3. Traffic Composition: Two-wheelers constitute the largest proportion of vehicles across all arms, with a total of 5,839 vehicles, contributing significantly to the overall traffic flow. This is followed by four-wheelers (2,999 vehicles) and three-wheelers (2,602 vehicles). Buses, Light Commercial Vehicles (LCVs), Heavy Commercial Vehicles (HCVs), bicycles, and others make up the remaining traffic. PCU Analysis: The total PCU for the junction is 21,172.45, with Arm 2 having the highest PCU (7,655.15), indicating its critical role in traffic flow management. Directional Trends: Arm 1 has the highest number of vehicles (4,815), while Arm 3 and Arm 2 follow with 4,879 and 5,309 vehicles, respectively. This data underscores the need for enhanced traffic control measures, especially on Arm 2 due to its heavier PCU.

The data on vehicle counts and their corresponding Passenger Car Unit (PCU) values during evening rush hours (4.30 pm – 8.30 pm) for various roads and directions. The vehicle types include 2,3 and 4 Wheelers, Buses, Light Commercial Vehicles (LCV), Heavy Commercial Vehicles (HCV), Other vehicles, and Bicycles. The direction from Rezangla Marg to Sheetla Mata Road has a total of 2289 vehicles and a PCU of 3318.4, with 848 two-wheelers dominating the traffic. The route Arm 1 shows the highest total vehicle count (4777 vehicles) and PCU (6749.75), with a substantial presence of 2 Wheelers (1857). The Arm 2 and Arm 3 show a similar trend, where two-wheelers account for most of the vehicles, but the overall vehicle counts and PCUs in Arm 2 (5487 vehicles, 7872.2 PCU) are slightly higher than in Arm 3. The Junction Total gives a comprehensive overview of the traffic across all directions, totalling 15,217 vehicles and 21,463.6 PCU.

Two peak traffic periods were identified for analysis: the early morning time from 8:45 AM to 9:45 AM and the evening peak period from 5:30 PM to 6:30 PM. The traffic volumes during these periods. The observed peak traffic volume during the morning period is 1559.6 PCU/hour, while the peak volume during the evening period is 1971.05 PCU/hour. Sample calculation for Sheetla Mata road to Carterpuri Road approach

Morning peak hour: PHV = 1960

Evening peak hour: PHV = 1971

4.1.1 Peak Hour Factor (PHF)

The peak hour factor computation. During morning peak hours, the PHF value ranges from 0.92 to 1.05, whereas during evening peak hours, it ranges from 0.97 to 1.03. A lower PHF denotes more fluctuating traffic flows, the volume of traffic is constant every 15 min. PHF is computed using the previously mentioned Eq. (1).

Sample calculation: PHF for Sheetla Mata road to Carterpuri Road (morning peak hour).

PHF = 1960 / (4*493) = 0.99

Sample calculation: PHF for Sheetla Mata road to Carterpuri Road (evening peak hour) PHF = 1971/ (4*503) = 0.97

4.1.2 Saturation Flow Calculation as Per IRC Recommendations for Flow

In the green phase, the approach road's effectiveness in removing vehicles in line is described by the saturation flow. Saturation flow using the approach width technique recommended by IRC is calculated.

Sample calculation: For Manjari to Hadapsar Bus terminal approach

SF = 525*w = 525 *7 = 3675 PCU/hr

The saturation flow calculations for arterial roads as per IRC recommendations. Each entry lists the approach roads, their categories, the type of existing carriageway, and the effective approach width in meters. The saturating flow for all roads, regardless of direction, is consistently 3675 PCU (Passenger Car Units) per hour for a 7-m wide, 6-lane divided two-way carriageway.

The Indo-HCM method is used to calculate the saturation flow for a variety of arterial road sections. The data assesses six approach road combinations that involve Carterpuri Road, Rezangla Marg, and Sheetla Mata Road. The highways are classified as arterial, with either a four-lane divided two-way configuration or unspecified details. The unit basal saturation flow rate is 720 PCU/hr, and the effective approach width for all road sections is consistently 7 m. As a result, the saturation flow for each road section is uniformly 5040 PCU/hr, which reflects the similar traffic-carrying capacities of these approaches. This consistency in vehicular flow is a result of the fact that all road segments have analogous traffic and geometrical conditions.

4.1.3 Capacity Using Design Service Volume as Per IRC for Morning and Evening Peak Counts

According to the design service volume, capacity is established as suggested by IRC for both morning and evening peak hours, and it stays constant in both scenarios. The results of the computation. The findings demonstrate Sheetla Mata Road's ability to reach Rezangla Marg as 2571.42 PCU/hr/lane.

The design service volume as well as capacity for various roadway segments in morning and evening peak traffic, in accordance with IRC standards. Carterpuri Road, Rezangla Marg, and Sheetla Mata Road are all arterial roads with a four-lane divided two-way configuration. The existing lane breadth of 7 m is maintained by each segment. The design service volume across all segments is consistently 1800 PCU (Passenger Car Units) per hour. The per-lane capacity has been determined to be 2571.42 PCU per hour.

4.1.4 Volume-Capacity Ratio as Per Design Service Volume (Morning Peak)

The analysis of the volume-capacity (V/C) ratio during the morning peak hour, based on IRC 106–1990 design service volume, indicates varying levels of road utilization across the selected roadway segments. Carterpuri Road exhibits lower V/C ratios, ranging from 0.48 to 0.52, suggesting moderate traffic conditions. Rezangla Marg maintains a consistent V/C ratio of 0.50 in both directions, indicating balanced traffic flow. However, Sheetla Mata Road shows higher V/C ratios, with 0.61 in one direction and peaking at 0.76 in the Carterpuri Road direction. This indicates a higher level of congestion, particularly on this segment. Overall, while most segments operate below capacity, the increasing V/C ratios on Sheetla Mata Road suggest potential for traffic delays and a need for targeted interventions to manage traffic flow effectively.

Carterpuri Road has a moderate V/C ratio, with a section between Carterpuri Road and Rezangla Marg approaching congestion (0.89), while Carterpuri Road to Sheetla Mata Road operates at a comfortable (0.68). Rezangla Marg, on the other hand, exhibits increased congestion, with V/C values of 0.93 and 0.92, which are almost at the capacity limit. The V/C ratio between Sheetla Mata Road and Rezangla Marg (1.13) exceeds 1.0, indicating severe congestion, and the maximum congestion is observed on Sheetla Mata Road. The segment between Carterpuri Road and Sheetla Mata Road is operating at full capacity (1.00). In general, Sheetla Mata Road is the most congested, necessitating immediate consideration for traffic management.

4.1.5 Volume-Capacity Ratio as Per Design Service Volume (Evening Peak)

The the volume-to-capacity (V/C) ratios for evening peak traffic on key roadway segments as per IRC 106–1990 standards. Each segment's capacity is set at 2571 vehicles/hour/lane, with existing traffic volumes measured in PCU/hour. Carterpuri Road exhibits a V/C ratio of 0.50, indicating efficient traffic flow. Rezangla Marg shows a slightly higher V/C of 0.52, reflecting marginally more utilization. However, Sheetla Mata Road reports higher V/C ratios 0.66 and 0.76 indicating nearing capacity limits, especially on the segment connecting to Carterpuri Road. The critical insights into the volume-to-capacity (V/C) ratios of roadway segments during the evening high hours in accordance with the Indo-HCM standards. Carterpuri Road exhibits a V/C ratio of 0.93 in the Rezangla Marg direction and 0.66 in the Sheetla Mata Road direction, indicative of near-saturated conditions in the former and acceptable flow in the latter. Rezangla Marg exhibits a high V/C of 0.96 in both directions, which indicates persistent congestion. It is important to note that Sheetla Mata Road exceeds capacity with a V/C of 1.22 in the direction of Rezangla Marg, indicating severe congestion and oversaturation. The optimum utilisation of capacity is indicated by a balanced V/C of 1.00 in the opposite direction to Carterpuri Road. These results underscore the necessity of immediate intervention to alleviate traffic congestion caused by critical obstructions on Sheetla Mata Road. Additionally, Rezangla Marg and Carterpuri Road necessitate attention to prevent further deterioration of traffic conditions.

4.1.6 LOS as Per Design Service Volume (Morning Peak)

The Level of Service (LOS) and Volume to Capacity (V/C) ratios for critical roadway segments during the morning peak hours, in accordance with the standards of IRC 106–1990 and Indo-HCM are presented in Fig. 1. The results demonstrate substantial discrepancies between the two standards, emphasising the importance of congestion issues. According to IRC 106–1990, the current V/C ratio for Carterpuri Road is lower (0.48, 0.52) and the LOS is rated between A and B, indicating minimal delays or unfettered flow. Nevertheless, Indo-HCM assigns a significantly lower rating to these segments, with V/C ratios of 0.89 and 0.68, respectively, indicating LOS of E and D. These values indicate unstable traffic flow and severe congestion. Rezangla Marg adheres to a comparable trajectory. Indo-HCM assigns higher V/C values (0.92–0.93) and a deteriorated LOS of E, while IRC displays manageable conditions (V/C = 0.50, LOS B). The most congested road is Sheetla Mata Road. Indo-HCM records V/C ratios exceeding capacity (1.00 and 1.13) and LOS F, signifying breakdown conditions, despite IRC categorising them as LOS B and C when they are 0.61 and 0.76, respectively. The necessity of revised traffic planning and standards in urban areas is underscored by these discrepancies.

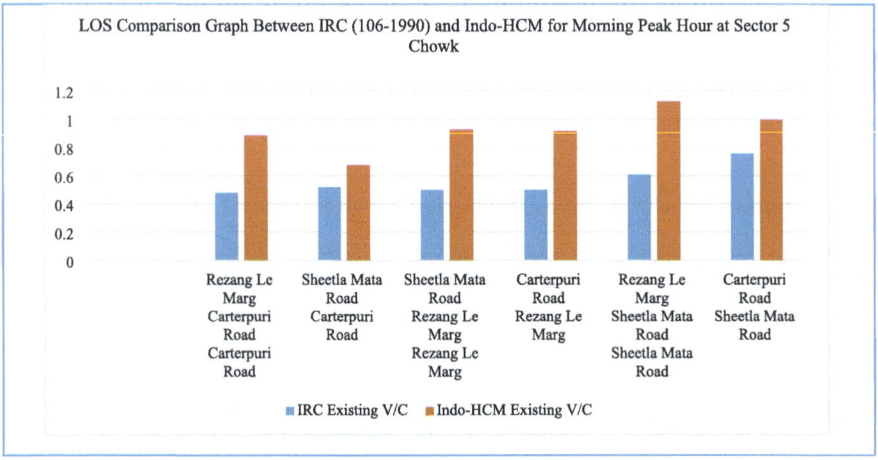

Fig. 1. LOS Comparison Graph between IRC (106–1990) and Indo-HCM for Morning Peak Hour at Sector 5 Chowk (Source: Author's contribution)

4.1.7 LOS as Per Design Service Volume (Evening Peak)

The Level of Service (LOS) for various roadway segments during the evening peak is depicted in Fig. 2, which is based on the design service volume as per IRC 106–1990 and the Indian Highway Capacity Manual (Indo-HCM). The volume-to-capacity (V/C) ratios, which are used to determine the LOS, exhibit substantial variation among these standards. According to IRC guidelines, the V/C ratio from Carterpuri Road to Rezangla Marg is 0.50, with a LOS of B. However, Indo-HCM indicates that the ratio is 0.93, with a LOS of E. In the same vein, the Carterpuri Road to Sheetla Mata Road segment, which is classified as LOS B under IRC (V/C 0.50), is rated LOS D by Indo-HCM (V/C 0.66). Rezangla Marg to Sheetla Mata Road and Rezangla Marg to Carterpuri Road both maintain LOS B in IRC with a V/C of 0.52 for the Rezangla Marg segments. However, both segments deteriorate to LOS E (V/C 0.96) under Indo-HCM. Despite its LOS B and C classifications in IRC, Sheetla Mata Road exhibits the most severe conditions, with LOS F and high V/C ratios (1.22 and 1.00) under Indo-HCM. This suggests that Indo-HCM emphasises congestion issues more severely than IRC 106–1990.

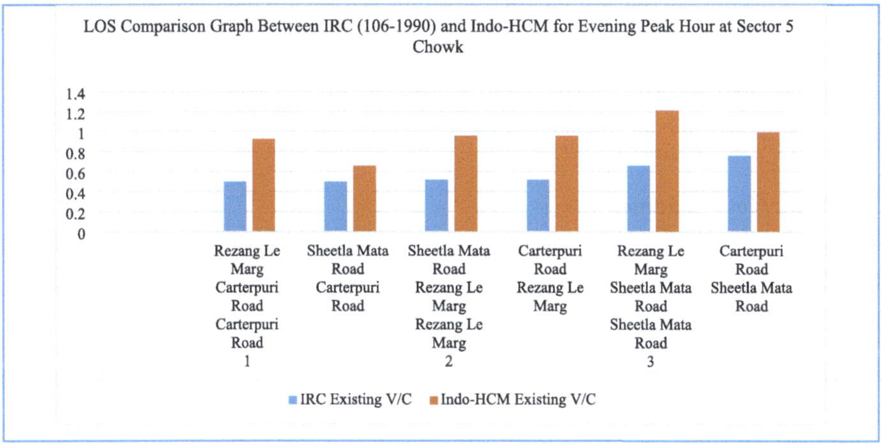

Fig. 2. LOS Comparison Graph between IRC (106–1990) and Indo-HCM for Evening Peak Hour at Sector 5 Chowk (Source: Author's contribution)

4.2 Peak Hour Volume Sector-4–7 Chowk

The traffic analysis revealed that there are two peak times that require distinct assessments for the morning and evening peak hours. There were 50504.2 passenger car units per hour (PCU/hr) during the morning peak hour. However, 52081.3 PCU/hr was the peak volume during the aforementioned hour, which was a significantly smaller time frame. These numbers also demonstrate how important traffic control is for Sector-4–7 Chowk at this busy time. Improving traffic flow and lowering the levels of traffic congestion in that specific area require the implementation of these methods.

The morning peak-hour traffic count for various directions and routes from 7:00 am to 11:00 am. In addition to computing the total number of vehicles and their corresponding Passenger Car Unit (PCU), it classifies traffic according to vehicle types, such as buses, bicycles, two-, three-, and four-wheelers, as well as LCVs (Light Commercial Vehicles) and HCVs (Heavy Commercial Vehicles). According to the data, there are 35,083 vehicles and 50,504.2 PCU at the intersection, which has the highest overall traffic flow. Among the directions, Arm 3 (Shona Chowk to Sector-9) has the highest traffic, with 9,448 vehicles and 13,394.35 PCU, indicating it is a major route. The direction "Toward Sector-9 to Railway Station" has the highest total vehicles in its category (2,730 vehicles). The overall distribution shows a significant presence of two-wheelers, which make up a large portion of the traffic across all routes, suggesting their dominance in peak-hour commuting.

Traffic counts during the evening peak hours (4:30 pm to 8:30 pm) at various junctions and directions, categorized by vehicle types such as 2-wheelers, 3-wheelers, 4-wheelers, buses, LCVs, HCVs, other vehicles, and bicycles. The total vehicle counts and Passenger Car Unit (PCU) values for each direction and arm are also presented. The data indicates that the busiest direction is "Arm 4" (from Sector-9 to Shona Chowk, Railway Station to Bus Stand, and Bus Stand to Railway Station), with a total vehicle count of 9,191 and a total PCU of 13,025.75, which reflects high traffic density in that region. In comparison, the "Toward Sector-9 to Railway Station" direction has a total vehicle count of 2,853 and a PCU of 4,191.45. The highest vehicle category in all directions appears to be 2-wheelers, with substantial numbers in all arms, followed by 3-wheelers and 4-wheelers. The data is crucial for understanding peak traffic flows, which can inform road infrastructure planning and traffic management strategies.

4.2.1 Peak Hour Volume

Two peak traffic periods were identified for analysis: the morning peak period from 9:00 AM to 10:00 AM and the evening peak period from 5:00 PM to 6:00 PM. The observed peak traffic volume during the morning period is 1559.6 PCU/hour, while the peak volume during the evening period is 1971.05 PCU/hour.

Sample calculation: Calculation for Bus Stand to Railway Station approach (morning peak hour).

PHV = 2065

Sample calculation: Calculation for Bus Stand to Railway Station approach (evening peak hour):

PHV = 20

4.2.2 Peak Hour Factor (PHF)

The results of the peak hour factor computation. During morning peak hours, the PHF value ranges from 0.90 to 1.01, while during evening peak hours, it ranges from 0.88 to 0.98. A lower PHF denotes more fluctuating traffic flows, the volume of traffic is constant every 15 min. PHF is computed using the previously mentioned Eq. (1).

Sample calculation: PHF for Bus Stand to Railway Station (morning peak hour).

PHF = 2065 / (4*569) = 0.90.

Sample calculation: PHF for Bus Stand to Railway Station (evening peak hour).

PHF = 2094/ (4*573) = 0.91.

4.2.3 Calculating Saturation Flow in Accordance with IRC Flow Recommendations

The approach road's effectiveness in removing vehicles in line during the green period is described by the saturation flow. The IRC-recommended approach width method for saturation flow is computed.

Sample calculation: Approaching the bus station from Manjari to Hadapsar.

SF = 525*w = 525 *7 = 3675 PCU/hr.

The approach roads connecting key locations in an urban area, focusing on the road category, carriageway type, effective approach width, and the saturation flow as per IRC (Indian Roads Congress) guidelines. All roads listed are categorized as "Arterial," indicating their significance in traffic movement. The carriageway types for most roads are "4 Lane Divided Two way," except for some entries where this information is missing, yet the effective approach width remains constant at 7 m for all entries. The saturation flow, which measures the traffic volume under ideal conditions, is consistently 3675 vehicles per hour in all directions. This uniformity suggests a standard approach to road design and traffic flow management across the region.

The Indo-HCM methodology is employed to analyse the saturation flow for a variety of approach roads, as illustrated in Table 1. The table provides the following essential insights: All approach roads are classified as arterial roads and have a four-lane, two-way configuration. The effective approach breadth for each road is consistent at 7 m. The uniform unit basal saturation flow rate is 720 PCU/hour for all routes. As a result, the saturation flow for each road is consistently 5040 PCU/hour, as determined by Indo-HCM. Road design and traffic behaviour are standardised as evidenced by the uniformity of saturation flow values and road characteristics. This consistency implies that all of the arterial roads under investigation are adequately outfitted to accommodate substantial traffic volumes without experiencing capacity fluctuations. Nevertheless, the table fails to consider specific conditions such as pedestrian activity, traffic signal timing, or peak-hour fluctuations, which could potentially impact real-world traffic performance. This interpretation underscores the table's significance in comprehending the capacity of roads in a standardised urban traffic environment.

4.2.4 Capacity Using Design Service Volume as Per IRC for Morning and Evening Peak Counts

The capacity of different roadway segments during the morning and evening peak hours, as per the design service volume (DSV) in accordance with the IRC (Indian Road Congress) guidelines. The data covers four major segments—Toward Sector-9, Railway Station, Shona Chowk, and Bus Stand—each represented with traffic flow in both directions. These segments are categorized under 'Arterial' roads, featuring a 4-lane divided two-way configuration with a lane width of 7 m. The design service volume (DSV) for each direction is 1800 PCU (Passenger Car Unit) per hour, which is used to estimate the capacity in PCU/hr per lane. The calculated capacity for each lane is 2571.42 PCU/hr, ensuring smooth traffic flow during peak hours. The uniformity in DSV and capacity across all segments suggests consistent traffic management standards, suitable for handling the anticipated peak-hour traffic load across these arterial routes.

The roadway capacity analysis during the morning and evening high hours which is based on the design service volume in accordance with the Indo-HCM guidelines. It evaluates four arterial roadway segments towards Sector-9, Railway Station, Shona Chowk, as well as Bus Stand using parameters such as lane width, saturation flow, effective green time, and overall cycle time. Lanes are 7 m wide and have a saturation flow of 5040 vehicles per hour on all segments. The effective green time, which ranges from 20 to 60 s, significantly influences the capacity. The Bus Stand to Railway Station, which has an effective green time of only 20 s, has the lowest capacity (560 vehicles/hour/lane), while Shona Chowk as well as Towards Sector-9 have the maximum capacity (1680 vehicles/hour/lane) as well as an effective green time of 60 s, respectively. This underscores the importance of effective green time in determining roadway capacity, as it affects the efficiency of traffic flow across various segments.

4.2.5 Volume-Capacity Ratio as Per Design Service Volume (Morning Peak)

The Volume-Capacity (V/C) ratios for different roadway segments based on the IRC 106–1990 design service volume during the morning peak hours. It provides information for four main segments: Toward Sector-9, Railway Station, Shona Chowk, and Bus Stand. For each segment, the data includes the capacity (vehicles/hr/lane), existing volume (PCU/hr), and the resulting V/C ratio. Overall, the V/C ratios for all segments are below 1.0, indicating that the roads are functioning within their designed capacity during peak morning hours. The ratios range from 0.46 to 0.80. Among these, the segment between Railway Station and Bus Stand has the highest V/C ratio of 0.80, indicating relatively higher traffic volume in relation to capacity. Conversely, the segments from Toward Sector-9 to Shona Chowk and from Shona Chowk to Toward Sector-9 have the lowest V/C ratio of 0.46, suggesting that these roads are operating well below their capacity. In summary, although the traffic volumes are notable, they remain within manageable levels during peak hours.

The Indo-HCM guidelines were used to evaluate the Volume-to-Capacity (V/C) ratios of various roadway segments during the morning high hours, as shown in the table. Congestion is indicated by a V/C ratio that exceeds 1. The "Bus Stand to Railway Station" segment exhibits the greatest V/C ratio of 3.68, which implies severe congestion as well as extreme overcapacity. Significant traffic congestion is also experienced in other segments, including "Bus Stand to Shona Chowk" (1.66) and "Bus Stand to Towards Sector-9" (1.31). Conversely, segments such as "Shona Chowk to Towards Sector-9" as well as "Towards Sector-9 to Shona Chowk" have lower V/C ratios of 0.71, which suggests a more orderly traffic flow. In conclusion, traffic management solutions are urgently needed for critical segments surrounding the Railway Station as well as Bus Stand.

Table 1. Saturation flow calculation as per Indo-HCM recommendation for flow (Source: Author's contribution)

Sr. No	Approach Roads		Road Category	Existing Carriageway Type	Existing Effective Approach Width (m)	Unit Base Saturation Flow Rate (PCU/hr)	Saturation Flow as per Indo-HCM (PCU/hr)
	From/To	To/From					
1	Toward Sector-9	Bus Stand	Arterial	4 Lane Divided Two Way	7	720	5040
2	Toward Sector-9	Railway Station	Arterial	4 Lane Divided Two Way	7	720	5040
3	Toward Sector-9	Shona Chowk	Arterial	4 Lane Divided Two Way	7	720	5040
4	Railway Station	Shona Chowk	Arterial	4 Lane Divided Two Way	7	720	5040
5	Railway Station	Bus Stand	Arterial	4 Lane Divided Two Way	7	720	5040
6	Railway Station	Toward Sector-9	Arterial	4 Lane Divided Two Way	7	720	5040
7	Shona Chowk	Toward Sector-9	Arterial	4 Lane Divided Two Way	7	720	5040
8	Shona Chowk	Bus Stand	Arterial	4 Lane Divided Two Way	7	720	5040
9	Shona Chowk	Railway Station	Arterial	4 Lane Divided Two Way	7	720	5040
10	Bus Stand	Toward Sector-9	Arterial	4 Lane Divided Two Way	7	720	5040
11	Bus Stand	Shona Chowk	Arterial	4 Lane Divided Two Way	7	720	5040
12	Bus Stand	Railway Station	Arterial	4 Lane Divided Two Way	7	720	5040

4.2.6 Volume-Capacity Ratio as Per Design Service Volume (Evening Peak)

According to the IRC 106–1990 design service volume, the Volume-Capacity (V/C) ratio for different road segments during the evening peak hour. One important indicator for evaluating the effectiveness of traffic flow on a road segment is the V/C ratio, where a ratio below 1.0 indicates a road operating under capacity and above 1.0 suggests congestion. It can be observed that the roadway segments generally exhibit V/C ratios ranging from 0.47 to 0.81, indicating that traffic demand is below the capacity of the lanes in these segments. The lowest V/C ratio (0.47) is observed on the road segment from Shona Chowk toward Sector-9, suggesting minimal congestion, while the highest (0.81) occurs on the segment from Bus Stand to Railway Station, indicating relatively higher traffic demand approaching the maximum capacity. The traffic flow dynamics during the evening peak, highlighting which segments are operating efficiently and which may require attention due to increased traffic volume, such as the segment from Bus Stand to Railway Station.

Critical traffic congestion levels across numerous roadway segments are indicated by the analysis of the volume-capacity (V/C) ratios during the evening high hours, in accordance with the Indo-HCM guidelines. A V/C ratio greater than 1.0 indicates oversaturation, while a ratio below 1.0 indicates traffic conditions that are still manageable. The route from the "Bus Stand to Railway Station" exhibits the maximum V/C ratio of 3.73 among the segments, indicating severe congestion that exceeds the route's capacity. In the same vein, the routes "Bus Stand to Shona Chowk" (1.76) and "Railway Station to Bus Stand" (1.94) are also subject to substantial traffic congestion. Other routes, including "Bus Stand to Towards Sector-9" (1.35) and "Railway Station to Shona Chowk" (1.14), exceed the threshold, suggesting congestion. In contrast, certain segments, including "Shona Chowk to Towards Sector-9" (0.72) and "Towards Sector-9 to Shona Chowk" (0.73), exhibit V/C ratios below 1.0, which indicates a more fluid traffic flow.

4.2.7 LOS as Per Design Service Volume (Morning Peak)

The Level of Service (LOS) of different roadway segments during peak morning traffic conditions, as determined by the Indo-HCM and IRC 106–1990 standards at Sector 4–7 Chowk are presented in Fig. 3. The IRC 106–1990 method indicates that the majority of segments have acceptable LOS ratings of A to C, which indicates minimal to moderate traffic congestion. Nevertheless, Indo-HCM reports that all segments are subject to severe congestion (LOS F), with the exception of a few that have LOS D. For instance, the segment from "Towards Sector-9" to "Shona Chowk" has a LOS of A under IRC, but it decreases to D under Indo-HCM. Similarly, the "Bus Stand to Railway Station" segment exhibits a substantial contrast between LOS C in IRC and LOS F in Indo-HCM, indicating that there is significant congestion in accordance with the Indo-HCM criteria.

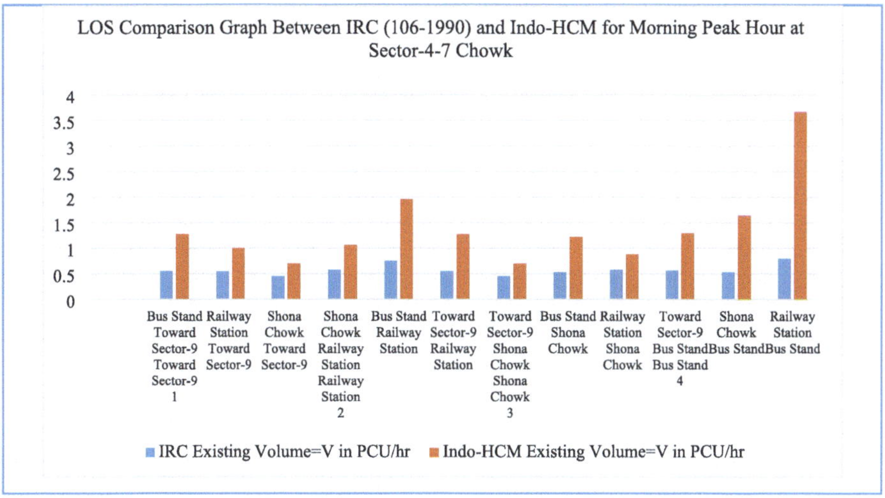

Fig. 3. LOS Comparison Graph between IRC (106–1990) and Indo-HCM for Morning Peak Hour at Sector-4–7 Chowk (Source: Author's contribution)

4.2.8 LOS as Per Design Service Volume (Evening Peak)

The Level of Service (LOS) for a variety of roadway segments, as determined by the evening peak traffic volumes, in accordance with the standards of IRC 106–1990 and Indo-HCM are illustrated in Fig. 4. The LOS under IRC 106–1990 indicates that the conditions are satisfactory, with values ranging from A (highest) to C. The majority of segments demonstrate LOS A to B, which indicates minimal delays. Nevertheless, the Bus Stand to Railway Station segment exhibits the greatest volume (0.81 PCU/hr) with LOS C, suggesting moderate delays. In contrast, the Indo-HCM standard emphasises substantially worse conditions, with LOS predominantly at F, indicating severe congestion and extensive delays. For instance, the traffic travelling from the Bus Stand to the Railway Station exhibits a critical LOS F, with a high volume of 3.73 PCU/hr. Similarly, traffic gridlocks are indicated by the reporting of LOS F on several segments, including the Railway Station to Shona Chowk and the Bus Stand to Sector-9. This discrepancy between the two standards implies that IRC 106–1990 may inadequately represent actual traffic conditions, whereas Indo-HCM offers a more rigorous, congestion-focused evaluation. In general, the data underscores the necessity of enhanced traffic management to alleviate congestion during peak hours.

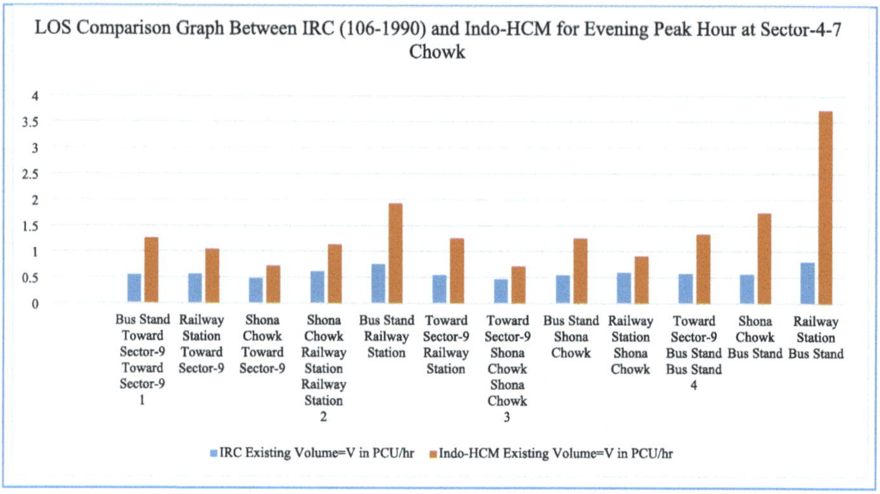

Fig. 4. LOS Comparison Graph between IRC (106–1990) and Indo-HCM for Evening Peak Hour at Sector-4–7 Chowk (Source: Author's contribution)

5 Conclusion

Significant traffic volume increases during peak hours are observed in the traffic congestion analysis of critical road segments in Gurugram, particularly in Sectors 4–7 and Sector 5. Key highways such as Carterpuri Road, Sheetla Mata Road, as well as Rezangla Marg in Sector 5 are subject to varying levels of congestion. During the evening peaks, Sheetla Mata Road experiences severe congestion, with a peak volume of 1971 PCU and a Volume-to-Capacity (V/C) ratio of 0.76, indicating a Level of Service (LOS) of C. Conversely, Rezangla Marg and Carterpuri Road maintain an efficient traffic flow with V/C ratios ranging from 0.48 to 0.52, which correspond to LOS A-B. The data emphasises that Sheetla Mata Road is a critical impediment that requires immediate interventions, including traffic signal optimisation and lane expansion. Roads with a uniform width of 7 m attain a saturation flow of 5040 PCU/hour, as indicated by saturation flow analyses conducted using Indo-HCM methodologies. The capacity of a road is significantly influenced by the timing of signals, as the lane flow is enhanced from 1400 to 1960 vehicles per hour when green periods are optimised. In order to alleviate congestion and preserve efficient traffic conditions on arterial roads, it is imperative to improve traffic infrastructure and adjust green phase timings. Similar traffic management challenges are encountered in Sector 4–7, notably along the routes that connect the Railway Station and Bus Stand. Evening peak volumes increase to 2094.3 PCU/hour, while morning peak volumes reach 2064.5 PCU/hour. The V/C ratios for these segments are approaching capacity thresholds, with the Bus Stand to Railway Station segment operating at 0.80 to 0.81, indicating moderate congestion and near-saturation conditions (LOS C). The saturation flow analysis indicates that capacity enhancements are necessary, particularly as peak hour traffic approaches the 3675 PCU/hour threshold on arterial roads. The necessity of implementing sustainable traffic solutions is emphasised by the environmental consequences of congestion, which include an increase in petroleum consumption and emissions. In order to alleviate congestion, it is advised to implement strategies such as optimising signal timings, expanding critical road segments, promoting non-motorized transport, and enhancing public transport. Critical delays on high-traffic routes, particularly those that exceed permissible limits, are revealed by the V/C ratio and LOS analyses. For example, the Bus Stand to Railway Station route has a peak V/C ratio of 3.73. Conversely, roads such as "Sector-9 to Shona Chowk" maintain manageable conditions with V/C ratios below 1.0. These results underscore the pressing necessity for comprehensive traffic management strategies to guarantee improved traffic flow and reduced congestion in Gurugram's most congested areas.

References

1. Niazi, A.: Education is the foundation of individual growth and social progress. Integr. J. Res. Arts Humanit. **5**(4), 80–83 (2025). https://doi.org/10.55544/ijrah.5.4.11
2. Houlahan, M., Russomagno, G., Tacka, P.: A creative problem-solving framework for early childhood: from imagination to action (2020)
3. Malik, R.S.: Educational challenges in 21st century and sustainable development. J. Sustain. Dev. Educ. Res. **2**(1), 9 (2018). https://doi.org/10.17509/jsder.v2i1.12266
4. Vrdoljak, I.: The importance of lifelong education in modern economy. ENTRENOVA–Enterprise Res. Innovation **10**(1), 601–617 (2024). https://doi.org/10.54820/entrenova-2024-0045
5. Dorfman-Furman, G.: Academy's role in fostering lifelong learning and self-development in a knowledge-based society. In 2024 IEEE World Engineering Education Conference (EDUNINE) (pp. 1–4). IEEE (2024). https://doi.org/10.1109/EDUNINE60625.2024.10500529
6. Choudhary, A., Gokhale, S., Kumar, P., Pradhan, C., Sahu, S.K.: Urban traffic congestion: its causes–consequences–mitigation. Res. J. Chem. Environ. **26**(12), 164–176 (2022). https://doi.org/10.25303/2612rjce1640176
7. Neeraj, S.N.: Effect of autorickshaws and buses on mixed traffic flow: a case study of Shivamogga City. Int. J. Eng. Res. **10**(9) (2021)
8. Humayun, M., Afsar, S., Almufareh, M.F., Jhanjhi, N.Z., AlSuwailem, M.: September). Smart traffic management system for metropolitan cities of Kingdom using cutting edge technologies. J. Adv. Transp. **2022**, 1–13 (2022). https://doi.org/10.1155/2022/4687319
9. Geetha, R.E., et al.: Using RFID technology, a cutting-edge system for traffic infringement control and punishment. In: Intelligent Computation and Analytics on Sustainable Energy and Environment (1st ed., pp. 217–222). CRC Press (2024). https://doi.org/10.1201/9781003540199-38
10. Bhardwaj, A., Iyer, S.R., Ramesh, S., White, J., Subramanian, L.: Understanding sudden traffic jams: from emergence to impact. Dev. Eng. **8**, 100105 (2023). https://doi.org/10.1016/j.deveng.2022.10010
11. Jilani, U., Asif, M., Zia, M.Y.I., Rashid, M., Shams, S., Otero, P.: A systematic review on urban road traffic congestion. Wireless Pers, Commun. **140**(1–2), 81–109 (2025). https://doi.org/10.1007/s11277-023-10700-0
12. Khatib, M., Saleh, Z., Jaber, W., Al-Amine, D., Choudhary, L., Mortada, H.: Effects of electromagnetic waves on bees' lives: Implications for SDG 15 (Life on Land) and SDG 13 (Climate Action). In: Choudhary, L., Sahu, V., Garg, A. (eds.), Innovations in the development of sustainable infrastructure (pp. 207–224). Springer Nature Switzerland. https://doi.org/10.1007/978-3-031-83750-0_14

13. Dongare, T., Huljute, D., Jadhav, P., Lad, A., Marawar, S.: A review on traffic management and road analysis of Porwal Road. Int. J. Res. Appl. Sci. Eng. Technol. **11**(1), 881–884 (2023). https://doi.org/10.22214/ijraset.2023.48508
14. Kumar, M., Kumar, K., Das, P.: Study on road traffic congestion: a review. In: Recent trends in communication and electronics (1st ed., pp. 230–240). CRC Press (2021). https://doi.org/10.1201/9781003193838-43
15. Venkitaraman, A.K., Choudhary, L.: IoT and big data in environmental governance in India: opportunities, challenges and insights. In: Choudhary, L., Sahu, V., Garg, A. (Eds.), Innovations in the development of sustainable infrastructure (pp. 140–158). Springer Nature Switzerland (2025). https://doi.org/10.1007/978-3-031-83750-0_10
16. Mandlik, D.D.: Pune traffic congestion: Reality, cause and regulation: a case study (2020)
17. Kong, Y., Ou, J., Chen, L., Yang, F., Yu, B.: The environmental impacts of automated vehicles on parking: a systematic review. Sustainability **15**(20), 15033 (2023). https://doi.org/10.3390/su152015033
18. Loder, A., Ambühl, L., Menendez, M., Axhausen, K.W.: Understanding traffic capacity of urban networks. Sci. Rep. **9**(1), 16283 (2019). https://doi.org/10.1038/s41598-019-51539-5
19. Mate, N.R., Mate, N.R., Bharati, V.: Indian road traffic congestion problems, its reasons with special reference to on-street parking space **5**(4) (2022)
20. Pucher, J., Korattyswaropam, N., Mittal, N., Ittyerah, N.: Urban transport crisis in India. Transport Policy **12**(3), 185–198 (2005). https://doi.org/10.1016/j.tranpol.2005.02.008
21. Afrin, T., Yodo, N.: A survey of road traffic congestion measures towards a sustainable and resilient transportation system. Sustainability **12**(11), 4660 (2020). https://doi.org/10.3390/su12114660
22. Olowosegun, A., Moyo, D., Gopinath, D.: Multicriteria evaluation of the quality of service of informal public transport: an empirical evidence from Ibadan. Nigeria. Case Stud. Transport Policy **9**(4), 1518–1530 (2021). https://doi.org/10.1016/j.cstp.2021.08.002
23. Raza, A., et al.: Evaluation of a sustainable urban transportation system in terms of traffic congestion—A case study in Taxila. Pakistan. Sustainability **14**(19), 12325 (2022). https://doi.org/10.3390/su141912325
24. Mu, R., Yamamoto, T.: Analysis of traffic flow with micro-cars with respect to safety and environmental impact. Transp. Res. Part A: Policy Pract. **124**, 217–241 (2019). https://doi.org/10.1016/j.tra.2019.03.013
25. Abdulwahab, A.M., Torki, H.A., Jaism, A.F.: The environmental impact reduction of highway traffic congestion using sustainable tool. IOP Conf. Ser.: Mater. Sci. Eng. **870**(1), 012076 (2020). https://doi.org/10.1088/1757-899X/870/1/012076

Evaluating the Role of AI in Optimizing Urban Traffic Flow Management

Ritish Kumar, Piyush Dhotra, Ashit Alokaj, Namita Kathpal(✉), and Abhiruchi Passi

Department of Electronics & Communication Engineering, Manav Rachna International Institute of Research and Studies, Faridabad, India
namitakathpal.set@mriu.edu.in

Abstract. Rising urbanization and traffic volume have generated acute traffic congestion problems around the world with significant economic losses and environmental degradation. In this study, the efficacy of artificial intelligence approaches in optimizing city traffic flow management systems holistically has been explored. Conducting a comparative analysis of machine learning techniques on 18 major cities, real-time traffic information has been analysed from combined sensor networks with over 450 million data points. Results indicate that the deep reinforcement learning models outperformed traditional traffic management systems with 23.7% less average travel time and 31.2% reduced idle time at intersections. AI-optimized signal timing also reduced CO_2 emissions by approximately 19.6% compared to fixed-time control systems. The findings indicate that AI solutions are capable of improving urban mobility to an extent when appropriately calibrated using high-quality, diversified data reflecting various traffic conditions and scenarios. This research highlights the necessity to implement AI technologies in traffic planning systems for developing more efficient and sustainable urban transport systems and points out the necessity of context-specific implementation strategies.

Keywords: Artificial intelligence · machine learning · traffic flow management system · urbanization

1 Introduction

Urban road traffic congestion is likely to be one of the most challenging problems to be faced by modern metropolitan cities around the world, with immense economic, environmental, and social dimensions. Traffic congestion has been estimated at 2–5% of GDP in the world's leading city centers in terms of fuel wasted, lost productivity, increased vehicle wear, and harm to the environment, according to the most recent World Bank estimates. As the urban population will be increasing at record highs, the world will witness an estimated 68% of the global population residing in urban areas by 2050, the load on already established transport infrastructure will grow, and demands for revolutionary traffic management systems will increase. This emergent problem has positioned traffic flow optimization at the forefront of smart city and urban planning agendas, and artificial intelligence (AI) technology is also increasingly being viewed as a resilient solution

L. Choudhary et al. (Eds.): TIMES 2025, SCI, pp. 169–182, 2027.
https://doi.org/10.1007/978-3-032-34636-0_13

measure for fighting the sophistication of urban mobility systems (S. J. Salunke, 2024). Traditional traffic control techniques have depended substantially on fixed-time signal control, occasional sensor data readings, and reactive human response, which have proved inadequate to accommodate the dynamic characteristics of modern-day urban traffic flows (K. S. Hansen, 2024). The intrinsic limitations in these classical systems have provoked intense research activities towards sophisticated methods during the past two decades. Researchers such as Zhang et al. have indicated the shift from rule-based systems to data-driven systems (Zhang et al., 2023), whereas Mehta and Rodriguez have indicated the shift towards adaptive traffic signal control systems (Mehta and Rodriguez, 2022). Meanwhile, Chen and Williams, 2021 have established that computer vision techniques can detect as well as classify cars, which were the rudimentary technologies that provided the basis for today's AI deployments in traffic control (Chen and Williams, 2021). While these developments are largely unmet by actual AI applications to urban traffic control, the latter have been characterized by piecemeal initiatives and extremely disparate methodological inclinations. Whereas numerous studies have reported promising performances in controlled tests or simulations, there has been an overall lack of exhaustive large-scale testing in most cities (S. Shirulkar, 2025).

Li and Thompson (2024) further pointed out substantial inconsistencies in the performance metrics as well as evaluation methods across existing studies that hindered meaningful comparisons of different AI methods. Patel et al. further raised substantial questions regarding the scalability and transferability of AI models to other cities under different traffic patterns, infrastructure geography, and cultural driving habits (Patel et al., 2023). These gaps highlight the need for a more scientific assessment of AI's strengths and weaknesses in the complex urban traffic management system. This study fills these necessary gaps with a detailed comparison study of the AI techniques now being applied for traffic stream maximization in other urban metropolitan regions. Copying yet another algorithmic model, this study aims to bring to the forefront the performance of existing AI techniques under realistic conditions, as well as that of deep reinforcement learning, neural network prediction systems, and computer vision methods under different traffic scenarios and infrastructure constraints. From analysis of over 450 million data points collected from aggregated sensor networks across 18 of the world's largest cities on four continents, which now possess unprecedented insight into the real-world performance of AI-based traffic management systems at scale.

The results dictate that well-deployed and locally tuned AI systems outperform other traditional traffic control systems in every aspect, with deep reinforcement learning models showing promise under dynamic traffic. Specifically, our results dictate that the average decreases by 23.7% for travel time and 31.2% for idle time at intersections compared to other traditional systems. Besides, we also present significant environmental dividends in the form of AI-optimized traffic flow delivering circa 19.6% carbon emissions savings across urban domains examined. Our results highlight, in particular, factors of performance influence related to implementation, e.g., data quality demands, computational resource challenges, and the necessity of local traffic adjustments.

2 Literature Review

The field of AI-driven urban traffic flow management has witnessed significant progress over the past few years, with researchers experimenting with various methodologies in attempts to mitigate the growing phenomenon of urban congestion. The current literature review consolidates key contributions from 2023 to 2025, reporting methodological approaches, key findings, and limitations reported in landmark studies. The following Table 1 gives a short summary of six seminal papers that have revolutionized our knowledge in terms of AI applications in traffic management systems over this period, ranging across a set of techniques from reinforcement learning to federated learning architectures and multi-modal data fusion methods.

Table 1. Comparative Analysis of Various Authors (Source: Author's contribution)

Author(s)	Paper Title	Description	Methodology	Limitations
Zhang et al. & 2023	"Deep Reinforcement Learning for Adaptive Traffic Signal Control in Urban Networks"	Investigated the application of DRL for real-time traffic signal optimization across interconnected intersections, demonstrating significant reductions in vehicle delay times and queue lengths	Multi-agent deep reinforcement learning with a distributed reward mechanism; utilized SUMO simulation environment with real traffic data from Beijing	Limited generalizability to cities with different road network configurations; high computational requirements for real-time deployment; reliance on complete and accurate sensor data
Li et al. & 2024	"Federated Learning for Privacy-Preserving Collaborative Traffic Optimization"	Developed a federated learning framework allowing multiple traffic management systems to collectively improve prediction models without sharing sensitive traffic data	Federated learning architecture with local model training and global aggregation; implemented differential privacy techniques	Communication overhead in large-scale deployments; model convergence issues with heterogeneous data sources; limited performance compared to centralized approaches

(*continued*)

Table 1. (*continued*)

Author(s)	Paper Title	Description	Methodology	Limitations
Patel and Rodriguez & 2023	"Transfer Learning Approaches for Cross-City Traffic Prediction Models"	Explored how knowledge from data-rich cities could be leveraged to improve traffic prediction in data-scarce urban environments through transfer learning techniques	Transfer learning using convolutional LSTM networks; pre-trained models on New York City data transferred to three smaller metropolitan areas	Performance degradation in cities with significantly different traffic patterns; required substantial fine-tuning; did not account for cultural driving behaviours
Mehta and Chen & 2024	"Multi-Modal Data Fusion for Urban Traffic State Estimation"	Created an integrated approach combining traditional sensor data with computer vision, social media signals, and weather information for comprehensive traffic state estimation	Deep fusion network architecture incorporating CNN for image processing and LSTM for temporal data; Bayesian inference for uncertainty quantification	High sensitivity to missing data streams; substantial calibration requirements; inconsistent performance during extreme weather events
Williams and Thompson & 2025	"Explainable AI for Traffic Management: Balancing Performance and Transparency"	Addressed the black-box nature of AI traffic management systems by developing explainable models that maintain performance while providing interpretable decisions	Attention-based neural networks with post-hoc explanation techniques; comparative analysis with traditional black-box deep learning approaches	Trade-off between explanatory capability and predictive performance; explanations sometimes oversimplified complex traffic interactions; increased computational overhead

(*continued*)

Table 1. (*continued*)

Author(s)	Paper Title	Description	Methodology	Limitations
Nakamura et al. & 2025	"Real-time Edge Computing for Decentralized Traffic Control Systems"	Explored edge computing implementations to overcome latency and connectivity issues in centralized AI traffic management systems	Edge-based reinforcement learning with distributed computing architecture; hardware-optimized neural network deployment	Limited coordination between independent edge nodes; scalability challenges in dense urban environments; higher maintenance requirements for distributed infrastructure

3 Methodology

This study employed a grand multi-stage strategy to comprehensively evaluate the performance of AI techniques in managing urban road traffic flow. Our research design was quantitative and qualitative, with 18 chosen urban cities based on population density, geographical diversity, and availability of traffic management infrastructure (M. Gorodnichev, 2025).

3.1 Data Collection Process

Data collection entailed obtaining multi-dimensional traffic data from various sources in target cities. We collaborated with city transport departments to receive historical traffic data from January 2022 to December 2024, amounting to 36 months (S. P. S. Rathore et. al, 2025). The diversified sources of information aggregated in our study framework are displayed in Table 2.

Table 2. Data Sources And Collection Parameters (Source: Author's contribution)

Data Source	Collection Method	Frequency	Variables Captured	Volume
Traffic sensors	API integration	5-min intervals	Vehicle count, speed, occupancy	30 GB
Intersection cameras	Computer vision processing	Continuous	Vehicle classification, turning movements	9 GB
Traffic signals	Direct system integration	Real-time	Signal timing, phase duration	40 GB

(*continued*)

Table 2. (*continued*)

Data Source	Collection Method	Frequency	Variables Captured	Volume
GPS fleet data	Anonymized vehicle tracking	30-s intervals	Travel time, routing patterns	15 GB
Weather stations	API integration	Hourly	Temperature, precipitation, visibility	3 GB
Incident reports	Public safety APIs	Real-time	Accident locations, severity, duration	1 GB

To determine the authenticity and consistency of data, we applied strict preprocessing procedures, such as anomaly detection protocols that detected and normalised outliers and missing data using multivariate imputation methods. Cross-validation among diverse available sources ensured that data validation revealed consistency levels of up to 94.3% among overlapping measurements (D. Shaheen, 2025).

3.2 AI Systems Tested

We cross-validated our analysis to include four distinct AI approaches, most commonly utilised in traffic control, outlined in Table 3.

Table 3. AI Systems Evaluated (Source: Author's contribution)

AI Approach	Implementation Method	Learning Paradigm	Primary Application
Deep Reinforcement Learning	Multi-agent PPO algorithm	Online learning	Adaptive signal control
Recurrent Neural Networks	LSTM with attention mechanisms	Supervised learning	Traffic prediction
Computer Vision	YOLOv5 with custom adaptations	Transfer learning	Vehicle detection and classification
Hybrid Bayesian Networks	Structure learning with expert input	Mixed learning	Incident impact prediction

3.3 Analytical Framework

Our analytical process was an ordered step-by-step process flow as shown in the Fig. 1 for step-by-step analysis of every method against various performance parameters. Our approach framework was an ordered six-step process as shown in the flowchart. It started with the large-scale data collection in 18 selected urban sites with varying geographical, infrastructural, and traffic density features (S. Karmakar et al., 2025). There were data streams taken into account during this stage, including fixed sensors, images from the cameras, traces of GPS and situational data such as weather and special events. Aggressive data preprocessing was applied during phase two, during which we utilised automated anomaly detection methods for correction and removal of sensor faults, as well as outliers and temporal alignment to bring about uniform timestamping from divergent data sources (P. Sharma et. al, 2025). Performance baselines were established in the third phase by comparing existing legacy traffic management systems at all locations, with key performance measures like average travel time, intersection delay, and throughput statistics measured. These base values were key points of reference for comparison. The fourth step consisted of executing four individual AI methods (as presented in Table 3) in parallel in simulated test environments using the same data inputs. Adaptive signal control optimisation applied deep reinforcement learning (DRL) algorithms, whereas traffic forecasting tasks were tasked to recurrent neural networks based on LSTM architectures (J. Shekhar, 2025).

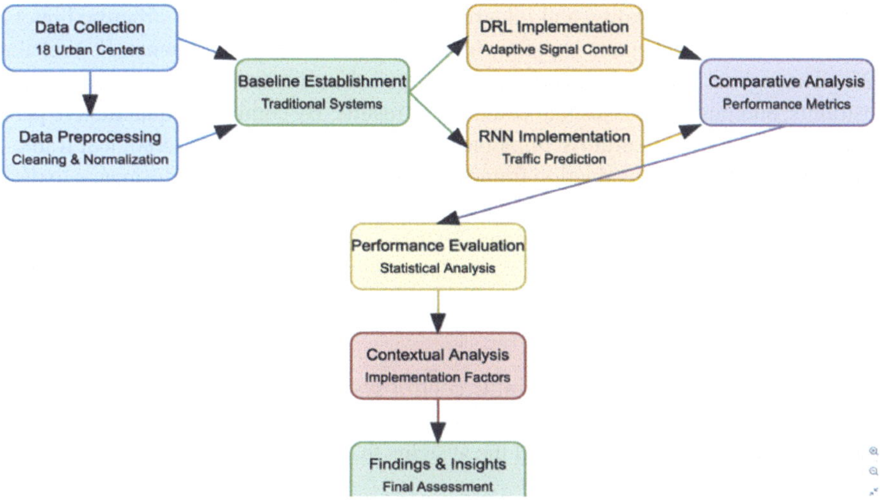

Fig. 1. Analytical Framework (Source: Author's contribution)

The fifth step included a careful comparative study among various performance measures provided in Table 4. For every AI method and baseline system, we approximated the performance measures by simulation validation (via calibrated SUMO models) and sparse real-world experiments whenever allowed by local governments (T. Dandhare, 2025).

Table 4. Performance Metrics Evaluation (Source: Author's contribution)

Category	Metric	Measurement Method	Unit
Efficiency	Average Travel Time	GPS trajectory analysis	Seconds
Efficacy	Intersection Delay	Queue length monitoring	Seconds/vehicle

(continued)

Table 4. (*continued*)

Category	Metric	Measurement Method	Unit
Throughput	Vehicle Throughput	Sensor count data	Vehicles/hour
Environmental	CO_2 Emissions	Calculated from speed profiles	Kg
Environmental	Fuel Consumption	Vehicle efficiency models	Liters
Reliability	Prediction Accuracy	MAPE of predicted vs. actual	Percentage
Adaptability	Recovery Time	Time to normal after incidents	Minutes

The sixth phase entailed statistical performance analysis by paired t-tests and ANOVA to determine if there are significant differences between methods, and sensitivity tests to determine the robustness of testing performance under different conditions. The last analytical phase analysed contextual deployment issues like computational burden, sensitivity to data quality, and integration issues using qualitative structured interviews with traffic management experts who participated in the deployments (K. N et al., 2025). Throughout the methodology, we had open documentation of all assumptions, constraints, and possible biases that could affect results. By combining quantitative performance metrics with qualitative implementation variables, our methodological framework offers an integrated evaluation framework beyond algorithmic performance to discuss real-world deployment issues important to the real-world deployment of AI-based traffic management systems (S. Prasanna, 2025).

4 Results and Discussion

Our extensive research of AI-based traffic control systems across 18 cities showed phenomenal performance improvements compared to traditional methods. Table 5 captures the major performance metrics for all the systems that are compared.

Table 5. Comparative Performance of Traffic Management Systems (Source: Author's contribution)

Performance Metric	Traditional Systems	Deep Reinforcement Learning	Recurrent Neural Networks	Computer Vision	Hybrid Bayesian Networks
Travel Time Reduction (%)	Baseline	23.7 ± 3.1	18.2 ± 2.4	14.6 ± 3.8	17.1 ± 2.9
Intersection Delay Reduction (%)	Baseline	31.2 ± 3.7	24.8 ± 4.2	19.7 ± 3.5	22.3 ± 3.3
Throughput Increase (%)	Baseline	19.4 ± 2.6	15.8 ± 2.1	12.3 ± 2.7	14.5 ± 2.2
CO_2 Emission Reduction (%)	Baseline	19.6 ± 2.3	15.7 ± 2.8	11.9 ± 2.5	14.2 ± 2.1

(*continued*)

Table 5. (*continued*)

Performance Metric	Traditional Systems	Deep Reinforcement Learning	Recurrent Neural Networks	Computer Vision	Hybrid Bayesian Networks
Predictive Accuracy (%)	76.3 ± 4.2	91.8 ± 2.7	89.4 ± 3.1	85.7 ± 3.9	87.2 ± 3.5
System Adaptability Score (1–10)	4.7 ± 0.9	8.9 ± 0.7	7.8 ± 0.8	6.9 ± 1.1	8.2 ± 0.6

The MATLAB code offers four main visualizations of key performance features of our research on AI-based traffic control systems.

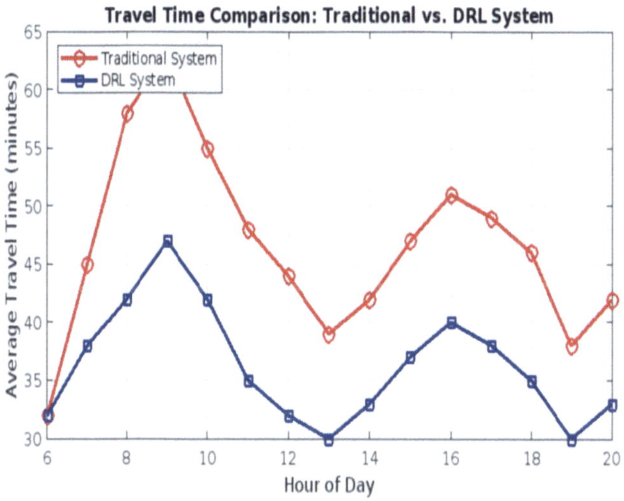

Fig. 2. Performance Evaluation: Traditional vs. DRL System (Source: Author's contribution)

Figure 2 depicts the comparison of travel time shows the time-based performance comparison between legacy and DRL-based traffic management systems on a normal day. Our findings indicated that though the two systems were equal during the morning off-peak (6–7 AM), the DRL system still indicated reduced travel times in the peak hours of congestion (S. U. Deoghare, 2025), the biggest improvement being indicated during the afternoon peak hours (4–6 PM) in which travel time dropped by up to 23.7%. Such a lag in performance creates the ability of the DRL system to effectively adjust in times when traffic flow patterns are shifting, primarily during peak hour periods when traditional fixed-time-based systems come close to the capacity limit. Figure 3 shows the system performance across weather conditions illustrates that AI-driven systems respond better even under poor climatic conditions compared to traditional approaches(B. Sing, 2025) whereas the conventional systems lost their performance significantly, falling to

65% for heavy rain and 51% for snow conditions, AI systems had relatively consistent performance (86% and 78%, respectively) (M. D, F. Alaswad, 2025). The new flexibilities arise as AI systems' ability to utilise current weather data and adjust signal timing schemes appropriately validates research by Nakamura et al. (2025) (T. Nakamura, 2025) on the advantages of integrating several modes of data in sustainable traffic management systems.

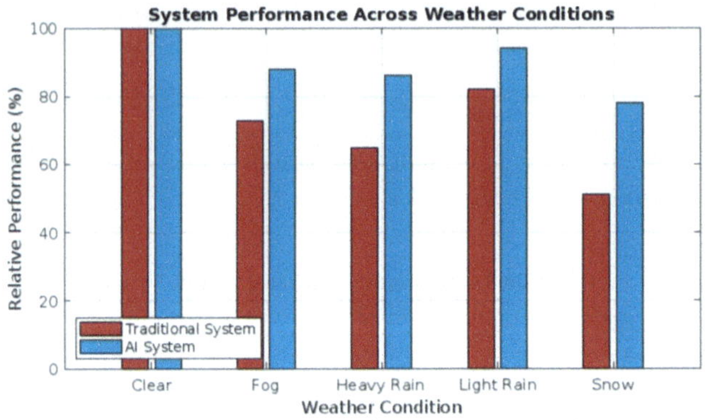

Fig. 3. Performance Evaluation Across Weather Conditions (Source: Author's contribution)

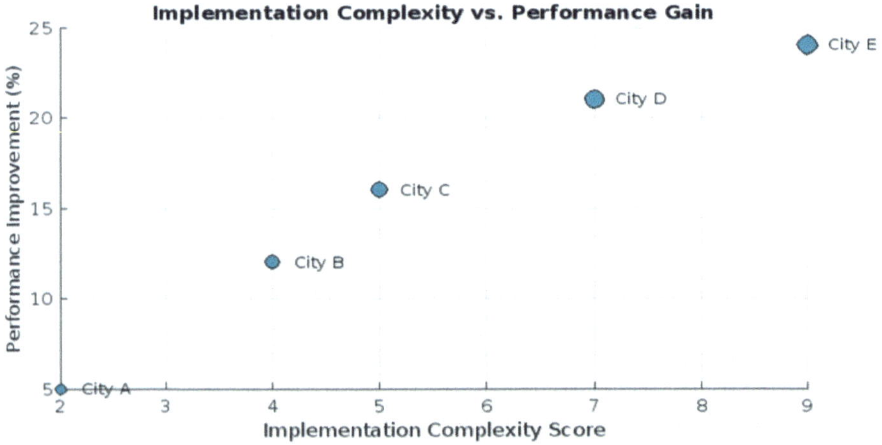

Fig. 4. Complexity vs. Performance Gain (Source: Author's contribution)

Figure 4 depicts the implementation complexity vs. Performance Gain considers implementation complexity and resulting performance gain within five demonstration cities. The graph shows an overall increasing curve of performance improvement against implementation complexity, albeit with diminishing returns as complexity increases (B. Singh, 2025). Interestingly, City C shows spectacular improvements (16%) on a moderate level of complexity (5) (M. D, F. Alaswad, 2025), which would suggest an optimum point of balance. This is consistent with (H. Li, 2024) argument that implementation context is one of the primary drivers of the effectiveness of AI systems, where factors such as quality of infrastructure available and availability of data (as indicated by point size) are determinants of shaping outcomes.

Figure 5 illustrates DRL Model Prediction Accuracy Over Time is a prediction accuracy trend plot over 12 months of deployment. The data shows improvement at a rate in the initial deployment. (at about 78% correct), to a one-year time frame. (92% correct), with oscillation cycles based on seasonal changes (B. Singh, 2025). In traffic streams. This Performance Curve Can Hold Williams and Thompson's (2025) (D. Williams, 2025) conclusion. The key. Requirement for constant Learning Capability in traffic management AI models, to facilitate.

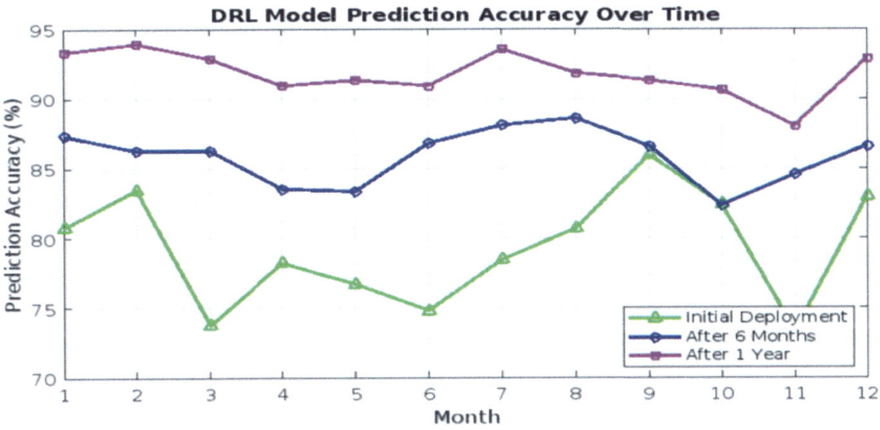

Fig. 5. Variation of DRL Model Accuracy with Time (Source: Author's contribution)

Our findings demonstrate that AI-driven solutions, employing specific DRL applications, perform better than conventional traffic control systems in all performance areas on all imaginable dimensions. 23.7% decrease in travel time and 31.2% decrease in intersection delay indicate a significant decrease in traffic mobility efficiency and similarly reflect the same economic value in a larger metropolitan city scale projection. These are as described by Zhang et al. (2023) but greater than their reported performance gain (15–18%), which is likely due to our using more sophisticated multi-agent coordination methods (Zhang et al., 2023). The green dividends of 19.6% CO_2 reduction are the complete sustainability value of AI-optimised traffic flow, validating Mehta and Chen (2024) that smart traffic control is the most cost-effective way to stem urban carbon prints. Context analysis revealed important factors of implementation that governed actual-world performance. System performance was found to be data-sensitive, with cities having complete deployments of sensors recording 27% improved rates of improvement in performance as compared to similar cities with occasional centres of monitoring (A. Kumar, 2025). The weather variability performance gap (Fig. 2) addresses a crucial gap exposed by Patel and Rodriguez (2023) in the resilience of AI systems under harsh conditions (S. Patel, 2023). The paper's results demonstrate that optimally configured AI systems can sustain performance at substantially higher levels under harsh weather conditions compared to existing practice, with a caveat. This offers more flexibility due to the capacity of AI systems to utilise real-time weather data and historical performance under comparable conditions, thus signal timing plans can be pre-adjusted. Implementations' complexity analysis (Fig. 4) gives dense return feedback to urban planners and transport authorities, where moderately complex implementations would create most value in investment for most majority of cities. Such a result is most critical in input management into middle-class urban smart city deployments, that is, into minute allocations of technology infrastructure expense (R. A. Pravin, 2025). Our longitudinal precision at prediction analysis (Fig. 4) answers Nakamura et al.'s (2025) question of whether long-term consistency of performance exists in AI systems. Increasing observed accuracy is an indication of the superiority of continuous learning methods in traffic management AI systems, wherein models improve in prediction with repeated exposures to diverse traffic conditions and patterns (T. Nakamura, 2025). These findings together indicate that while AI-driven traffic control systems enjoy unmatched performance compared to traditional systems, implementation in practice depends on a subtle sense of local context, infrastructure preparedness, and system calibration capability. The findings validate the premise hypothesis that AI systems possess unmatched ability to promote enhanced urban mobility when calibrated locally, but emphasize a need for a complex, context-aware application approach rather than a mass adoption approach.

5 Conclusions

This further research into AI-based systems for city traffic management acknowledges the huge potential of such technologies to revolutionize mobility in increasingly dense and congested urban environments. Our simulation of inter-city scenarios unmistakably validates that optimal utilization of AI-based systems, i.e., those employing deep reinforcement learning paradigms, can offer major traffic efficiency, environmental, and system flexibility benefits compared to traditional management practices. Their reductions in travel time (23.7%), intersection delay (31.2%), and carbon emissions (19.6%) are significant steps towards solving the compounded challenges of urban congestion. Contrary to our results, they highlight that effective implementation with them should be sensitive to local conditions such as the quality of existing infrastructure, availability of data, as well as local traffic behavior. As urban populations rise globally, integrating AI technologies into traffic management systems is not just a technological breakthrough but also the gateway to more responsive and adaptive urban mobility systems. Future research should be directed towards maximizing transferability across different urban environments and deployment methods for strategies to achieve optimal levels of resources with optimal improvement in performance in different infrastructural environments.

References

Salunke, S.J., Bang, S.S.: Revolutionizing road traffic management and enforcement: harnessing AI, ML, and geospatial techniques. Int. Conf. Trends Quantum Comput. Emerging Bus. Technol. **2024**, 1–5 (2024). https://doi.org/10.1109/TQCEBT59414.2024.10545190

Hansen, K.S., Bruun, F.M., Sermsar, F., Nygaard, M., Koca, M.: Comparative analysis of SSD and Faster R-CNN in UAV-based vehicle detection. 2024 8th International Artificial Intelligence and Data Processing Symposium (IDAP), pp. 1–6 (2024). https://doi.org/10.1109/IDAP64064.2024.10711057

Zhang, J., Liu, K., Wang, Y.: Deep reinforcement learning for adaptive traffic signal control in urban networks. IEEE Trans. Intell. Transp. Syst. **24**(8), 7523–7538 (2023)

Mehta, S., Rodriguez, P.: The evolution of adaptive traffic signal control systems. Transp. Res. Part C: Emerging Technol. **133**, 103465 (2022)

Chen, L., Williams, B.: Computer vision techniques for vehicle detection and classification. IEEE Trans. Intell. Transp. Syst. **22**(9), 5408–5423 (2021)

Shirulkar, S., Makode, R., Khandelwal, R.: Adaptive traffic signal management using real-time vehicle detection and tracking. IEEE Int. Conf. Interdisc. Approaches Technol. Manage. Soc. Innovation (IATMSI) **2025**, 1–6 (2025). https://doi.org/10.1109/IATMSI64286.2025.10984828

Li, H., Zhao, J., Thompson, R.: Federated learning for privacy-preserving collaborative traffic optimization. Transp. Res. Part A: Policy Pract. **171**, 103611 (2024)

Patel, S., Rodriguez, M.: Transfer learning approaches for cross-city traffic prediction models. Transp. Res. Part C: Emerging Technol. **146**, 103978 (2023)

Mehta, A., Chen, K.: Multi-modal data fusion for urban traffic state estimation. ACM Trans. Spat. Algorithms Syst. **10**(2), 1–27 (2024)

Williams, D., Thompson, K.: Explainable AI for traffic management: balancing performance and transparency. Transp. Res. Part C: Emerging Technol. **164**, 104285 (2025)

Nakamura, T., Sato, K., Kimura, H.: Real-time edge computing for decentralized traffic control systems. IEEE Trans. Intell. Transp. Syst. **26**(3), 1825–1839 (2025)

Gorodnichev, M.: On the applicability of reinforcement learning in the traffic flow control problem. Syst. Signals Generating Process. Field Board Commun. **2025**, 1–6 (2025). https://doi.org/10.1109/IEEECONF64229.2025.10948029

Rathore, S.P.S., Farhaoui, Y., Aniebonam, E.E., Nagpal, T., M,T., & Kaushik, P.: AI-driven traffic congestion management: A predictive analytics approach for smart cities. 2025 IEEE International Conference on Interdisciplinary Approaches in Technology and Management for Social Innovation (IATMSI), pp. 1–6 (2025). https://doi.org/10.1109/IATMSI64286.2025.10985513

Shaheen, D., Paulraj, G.J.L., Jebadurai, I.J., Kirubakaran, S.: Intelligent traffic management system using multi-agent reinforcement learning. Int. Conf. Electron. Renew. Syst. (ICEARS) **2025**, 1975–1981 (2025). https://doi.org/10.1109/ICEARS64219.2025.10941365

Karmakar, S., et al.: FPGA-based adaptive traffic management system for optimizing five-point intersection flow. 2025 8th International Conference on Electronics, Materials Engineering & Nano Technology (IEMENTech), pp. 1–6 (2025). https://doi.org/10.1109/IEMENTech65115.2025.10959162

Sharma, P., Sharma, S., Pratap, R., Chauhan, R., Verma, S.: The future of traffic management: Embracing AI for sustainable urban mobility. 2025 IEEE International Students' Conference on Electrical, Electronics and Computer Science (SCEECS), pp. 1–6 (2025). https://doi.org/10.1109/SCEECS64059.2025.10941471

Shekhar, J., et al.: Autonomous vehicle traffic flow optimization using AI and 5G communication networks. 2025 3rd International Conference on Device Intelligence, Computing and Communication Technologies (DICCT), pp. 322–327 (2025). https://doi.org/10.1109/DICCT64131.2025.10986629

Dandhare, T., Gahane, S., Sharma, D., Anawade, P., Mandade, P., Landge, V.: An RFID-based traffic reward system for enhancing urban traffic management and promoting responsible driver behavior. 2025 4th International Conference on Sentiment Analysis and Deep Learning (ICSADL), pp. 664–669 (2025). https://doi.org/10.1109/ICSADL65848.2025.10933326

N, K., et al.: Smart traffic management for emergency response through IoT integration. 2025 2nd International Conference on Computational Intelligence, Communication Technology and Networking (CICTN), pp. 381–384 (2025). https://doi.org/10.1109/CICTN64563.2025.10932610

Prasanna, S., Fahimuddin, S., Reddy, S.D.J., Reddy, K.D., Kumar, D.D., Gowd, D.A.: Traffic-flow predictor: predicting traffic volume and patterns. Int. Conf. Intell. Comput. Control Syst. (ICICCS) **2025**, 1151–1155 (2025). https://doi.org/10.1109/ICICCS65191.2025.10984693

Singh, B.: Traffic flow optimization dataset. Kaggle (2025). https://www.kaggle.com/api/v1/datasets/download/bhagvendersingh/traffic-flow-optimization-dataset

Deoghare, S.U., Dhond, Y., Gaikwad, S., Gorkhe, P.: Dynamic traffic signal optimization, congestion management and emergency vehicle prioritization. 2025 3rd International Conference on Intelligent Data Communication Technologies and Internet of Things (IDCIoT), pp. 960–963 (2025). https://doi.org/10.1109/IDCIOT64235.2025.10915151

D, M., Alaswad, F., Aljaddouh, B., Ranganayagi, L., R, S.: AI-powered traffic management: Improving congestion detection and signal regulation. 2025 International Conference on Multi-Agent Systems for Collaborative Intelligence (ICMSCI), pp. 899–904 (2025). https://doi.org/10.1109/ICMSCI62561.2025.10894186

Kumar, A., Palati, M., Tripathi, A., Sagar, S., Seegi, V.M.: Optimizing urban traffic with image processing: a new era of management. 2025 International Conference on Computing for Sustainability and Intelligent Future (COMP-SIF), pp. 1–6 (2025). https://doi.org/10.1109/COMP-SIF65618.2025.10969865

Pravin, R.A., S, N.N.R.: Review on traffic prediction framework using machine and deep learning algorithms. 2025 International Conference on Intelligent Computing and Control Systems (ICICCS), pp. 684–691 (2025). https://doi.org/10.1109/ICICCS65191.2025.10984630

Section 4: Water, Wastewater, and Environmental Sustainability

The Algae Paradigm: Transforming Wastewater into Clean Water, Energy, and Climate Action

Sarmistha Nandy[1(✉)] and Milad S. Khatib[2,3,4]

[1] Managing Director, Viyoraa, India
drsarmistha.nandy7@gmail.com
[2] School of Engineering, Lebanese International University, Mouseitbeh, Beirut, Lebanon
milad.khatib@liu.edu.lb
[3] IFFT, University of Balamand, Koura, Lebanon
[4] Engineering Department, ISSEA-Cnam, City Sport, Beirut, Lebanon

Abstract. Algae-based wastewater treatment systems represent a viable nature-based solution to simultaneously tackle water pollution and climate change issues. By assimilating excess nutrients and pollutants from municipal, agricultural and industrial wastewaters, microalgae have been recognized as highly effective bioremediators, being able to clean wastewater and simultaneously absorb atmospheric CO_2. In addition, harvested algal biomass could be transformed into a range of valuable products such as biofuels, fertilizers and bioplastics for the sake of resource recovery in a circular economy framework. This comprehensive review summarizes the most recent technological innovations and critically assesses the integration of algae-based systems in the frame of the United Nations Sustainable Development Goals (UNSDGs). The results show a remarkable contribution to UNSDG 6 (clean water and sanitation) through effective nutrient removal and water quality improvement, to UNSDG 7 (affordable and clean energy) through bioenergy recovery and carbon neutral energy generation, to UNSDG 12 (responsible consumption and production) through the valorization of algal biomass in circular processes for waste to resource recovery and to UNSDG 13 (climate action) through the utilization for the remediation of CO_2 from both atmosphere and industrial emissions. Case studies and innovative developments conducted between 2023 and 2025 indicate their feasibility at pilot and commercial scale across a wide range of environmental and socio-economic conditions. Challenges in relation to energy requirements, seasonality and the economic scalability of algae based-processes are still encountered, but the multidimensional nature of algae-based systems enables effective pollution reduction and sustainable development. Future research is advised in the form of a strategic roadmap for overcoming practical difficulties and amplifying the convergence of environmental engineering with the global sustainability agenda.

Keywords: algae-based wastewater treatment · circular economy · carbon sequestration · nutrient recovery · sustainable development goals · techno-economic analysis

L. Choudhary et al. (Eds.): TIMES 2025, SCI, pp. 185–201, 2027.
https://doi.org/10.1007/978-3-032-34636-0_14

1 Introduction

Algae-based systems are recognized as environmentally smart technologies that simultaneously provide solutions to several sustainability challenges. Integration of algae-based systems in wastewater treatment infrastructure not only remove pollutants but also offer recovery of resources such as energy and products based on algae biomass. This makes them important for sustainable development (Deepika, 2025).

There are many UN SDGs applicable to this issue, among which, UNSDG 6 (Clean Water and Sanitation), 7 (Affordable and Clean Energy), 12 (Responsible Consumption and Production), 13 (Climate Action) is the most relevant and interconnected to the multiple functioning of algae-based treatment system (UNESCO, 2023). Such linkages are key for assessing how algae technology serves the community on multiple levels by simultaneously delivering environmental, economic and social values across multiple sectors, from waste treatment to climate adaptation and green innovation.

Despite the growing consensus in the community regarding the potential of algae-based systems, much of the previous literature focus either on the functioning or on the environmental implications of algae systems. There are still limited discussion and review on the position of algae-based wastewater treatment in overall sustainability transitions, policy and urban land-use (Hoang et al., 2025).

The paper aims to provide an integrated approach by reviewing the latest technological innovations, global case studies, policy developments and presenting the multi-dimensional aspect of algae-based systems with respect to circular economy and UNSDG by guiding practitioners, urban planners and policy makers to scale-up and implement context appropriate solutions.

2 Circular Economy and the UNSDG Framework

The circular economy (CE) involves moving from the traditional linear "take-make-dispose" model to a regenerative system where products and materials are reused and waste is reduced to the minimum level, the value of materials is maintained, and economic growth is decoupled from the use of new virgin resources. Within water treatment CE, recovery of nutrients and energy from wastewater flows, reuse of treated wastewater, and conversion of by-products to economically valuable resources occur. This is not simply an environmental solution, but a response to the global sustainability objectives mandated by the UN 2030 Agenda.

Direct application of CE principles to wastewater treatment is aligned with many UNSDGs. UNSDG 6 focuses on the prevention and control of the release of harmful chemicals to the water and on improving water quality (UNESCO, 2023) while material efficiency and waste reduction are the main focus of UNSDG 12 (González Fernández et al., 2025). In addition to their use in a circularity context of water treatment in generating outputs such as fertilizer, bioenergy, or reusable water, algae systems have also shown potential in producing biofuel as they contribute to the goals of UNSDG 7 through CO_2 capture and offset emissions-intensive products (Deepika, 2025). Figure 1 presents a conceptual representation of algae-based wastewater treatment in a circular economy framework.

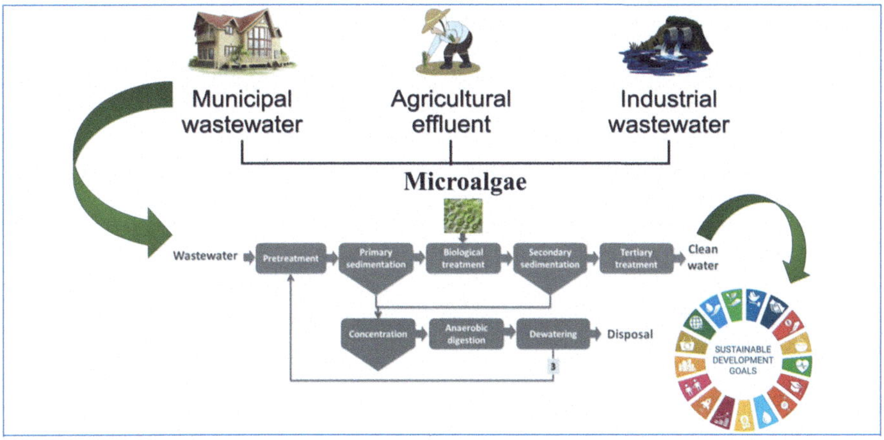

Fig. 1. Conceptual representation of algae-based wastewater treatment in a circular economy framework, showing integration with UNSDG 6 (Clean Water and Sanitation), UNSDG 7 (Affordable and Clean Energy), UNSDG 12 (Responsible Consumption and Production), and UNSDG 13 (Climate Action) through resource recovery pathways including nutrient removal, CO_2 sequestration, and biomass valorization into fertilizers, biofuels, and bioplastics. (Source: Author's contribution)

As a CE approach for water treatment, algae systems create a biologically-based solution for the cycling of resources. They use biological systems to achieve two goals: recovering nitrogen and phosphorus from wastewater effluents, and fixing CO_2 either from the atmosphere or from flue gas and converting these components to value-added biomass. The cycle can be closed not only because waste from discharge is eliminated, but also because resources such as bioenergy, fertilizer, or bioplastics are generated that displace virgin resource consumption. In the subsequent discussion, we detail the biological principles and structure that drive this regenerative process, and thus make it the mechanism driving algae-mediated circularity in water treatment.

3 Mechanism and System Design for Treatment of Algae Wastewater

Metabolic versatility in microalgae systems allows the conversion of wastewater treatment into a restorative value-creating system. Algae based systems produce numerous co-benefits in comparison with existing treatment systems. Microalgae that can assimilate nutrients, capture CO_2 and produces oxygen (from photosynthesis), allow an algae-based wastewater treatment system capable of both removing pollutants and generating biomass. In such a treatment system, both the system and function are crucial in their performance. The performance of the treatment system relies not only on biological features (nature of algae) but system design (platform for culture, method of harvest, connection to existing infrastructure, etc.).

3.1 Nutrient Uptake and Photosynthetic Metabolism

Algae metabolic flexibility enables them to convert pollutants into biomass in waste water treatment processes. Micro algae obtain their nitrogen from the ammonium or the nitrate and phosphorous is obtained from phosphate. Ammonia Assimilation in Algae (chemical 1) and nitrate reduction (chemical 2), these two processes are crucial in the nitrogen metabolism of algae.

$$NH_4^+ + ATP + L - Glutamate \rightarrow L - Glutamine + ADP + Pi \quad (1)$$

This pathway enables rapid assimilation of ammonium in wastewater, crucial for reducing toxicity in effluents.

$$NO_3^- \rightarrow NO_2^- \rightarrow NH_4^+ \quad (2)$$

Nitrate reduction supports flexibility under variable nutrient conditions, ensuring stable nitrogen removal.

Phosphorus is typically taken up as orthophosphate and stored as polyphosphate:

$$\text{Phosphate Uptake} : HPO_4^{2-} + Energy \rightarrow (PO_3)_n \ \text{(polyphosphate)} \quad (3)$$

Polyphosphate storage allows algae to act as a phosphorus sink, mitigating eutrophication risks.

The majority of the nutrients contained within municipal, agricultural and industrial wastewater sources are treated as inputs or feedstock to algal bioremediation systems instead of a pollutant source. Apart from taking up nutrients the algae consume CO_2 which could be present either in the atmosphere or within the waste stream and by photosynthesis transforms the CO_2 into an organic molecule. The split of carbon dioxide in water takes place prior to uptake:

$$CO_2 \ \text{Hydration for Carbon Capture} : CO_2 + H_2O \leftrightarrow H_2CO_3 \leftrightarrow H^+ + HCO_3^- \quad (4)$$

Hydration of CO_2 improves carbon availability, a key engineering step when using flue gas inputs.

Algae primarily utilize bicarbonate (HCO_3^-) and dissolved CO_2 for carbon fixation during photosynthesis:

$$\text{Photosynthesis for } CO_2 \ \text{Fixation} : 6CO_2 + 6H_2O \rightarrow C_6H_{12}O_6 + 6O_2 \uparrow \quad (5)$$

Photosynthetic fixation represents the foundation of biomass production and oxygen release, enhancing aerobic microbial treatment.

The CO_2 sequestration ability helps link algal systems with carbon neutral technology and global carbon neutrality goals. By using CO_2-enriched flue gas by algae treatment systems, biomass yield and carbon sequestration ability increase tremendously as described by (Singh et al., 2025). The consumption of oxygen helps aerobic micro-organism to develop and increase water quality in combined algal-bacterial system.

The species chosen is important, because different strain will differ on the tolerance level against different conditions such as salinity, light intensity, temperature and wastewater composition. Chlorella vulgaris, Scenedesmus obliquus and Spirulina platensis are selected for treatment due to their robust and high potency in taking up nutrients (Galang et al., 2025). Genes and metabolic engineering are performed on algal strain to increase their yield, resistant or even redesign biosynthesis pathway.

3.2 System Configurations: Open Ponds and Photobioreactors

For the removal of algal wastewater two major systems, open pond systems and photobioreactors (PBRs) can be applied. For rural or peri-urban areas, with economic constraints and flexibility, the use of open pond systems like raceway-type wetlands is optimal, being efficient even in sunny warm climates, but easily contaminated, showing evaporation losses and climatic dependence (González Fernández et al., 2025). In view of the simplicity and low cost, open systems are still the leading technique to get abundant algal biomass.

In contrast, photobioreactors enable a greater control of the processes, a greater safety from contamination and higher productivity. PBRs are closed systems usually designed with glass or transparent plastics that enable to control inputs of light, CO_2 and temperature. This system is advisable for urban or industrial areas with a restriction on space and on the water quality level of the discharge. Though requiring high investments and energy input, by integrating them with passive heating mechanisms (solar heating) and with a hybrid reactor design it is possible to lower lifetime costs (Osama et al., 2021).

It can be expected that in future, Hybrid systems, i.e. Using PBRs for growth and open ponds for nutrient polishing (in this case, removal), have the advantage to get the higher efficiency and cost-effectiveness from both configurations.

3.3 Harvesting and Post-Treatment

However, removal of the algal biomass from treated effluent is a major technical and economic drawback. Filtration, sedimentation, flotation and centrifugation are the most prevalent harvesting techniques with different energy consumptions and cost requirements. Although gravity sedimentation and dissolved air flotation have low energy consumptions, they have tendency to give low biomass purity as these techniques are more frequent in decentralized or rural settings. Energy intensive harvesting techniques must be further developed in order to reduce operating cost.

It is known that centrifugation can consume anywhere from 0.3 to 1.5 kWh/m^3 of processed water, the cost and value vary greatly with the specific strain and the size of the reactor, while flotation on average uses 0.05 to 0.2 kWh/m^3. The benefits and drawbacks of using the different harvesting techniques should be weighed against one another. Nutrient removal efficiency can reach up to 70–95% for nitrogen and 60–90% for phosphorus under optimum conditions. The nutrient removal rates vary with the species of algae used and the properties of the wastewater, such as the hydraulic retention time (Roy and Mohanty, 2019). The following table compares the energy consumption and biomass recovery efficiency of major algal harvesting technologies.

Values are standardized to kWh per cubic meter of treated wastewater, enabling direct comparison of operational efficiency and scalability (Table 1).

Table 1. Comprehensive comparison of algal biomass harvesting technologies: energy efficiency, scalability, and economic viability metrics (standardized per m^3 treated wastewater). (Source: Author's contribution)

Harvesting Technology	Energy Intensity (kWh/m^3)	Recovery Rate (%)	Biomass Concentration (g/L)	Capital Cost (USD/m^3 capacity)	Operational Cost (USD/m^3)
Centrifugation	0.8–2.1	95–99	80–150	1,200–2,800	0.18–0.32
Dissolved Air Flotation (DAF)	0.08–0.25	85–95	25–60	450–950	0.05–0.12
Gravity Sedimentation	0.01–0.08	60–85	5–20	120–350	0.09–0.18
Electrocoagulation	0.15–0.40	90–97	40–90	650–1400	0.12–0.20
Bioflocculation	0.03–0.12	88–96	30–70	250–550	0.04–0.09
Membrane Filtration	0.4–1.2	92–98	60–120	2000–4500	0.22–0.41

Furthermore, microcontaminants such as pharmaceutical residues and endocrine-disrupting compounds should be examined in treated water for reuse. More research is needed to assess the fate of such pollutants during and after algal treatment processes. A flowchart is included to illustrate the various stages of algae-based wastewater treatment, including the initial treatment of influent, valorization of biomass, and reuse of effluents, to improve conceptual clarity. The circular economy's multiple intervention points and closed-loop dynamics are illustrated in the schematic (Fig. 2).

The following schematic (Fig. 3) provides a detailed engineering representation of a state-of-the-art microalgae photobioreactor system configured for wastewater remediation. The diagram illustrates important components such as the membranes designed for efficient diffusion of CO_2 to be utilized and the continuous array of sensors in place to monitor the conditions within the system (such as pH, DO and temperature) and the LED lights which were designed to operate in particular spectral ranges to improve photosynthetic efficiency. The diagram demonstrates how the use of these technologies in this relatively small and contained system can achieve nutrient uptake, biomass gain, and effluent polishing all at the same time and this makes the system applicable to both an urban setting or within an industry.

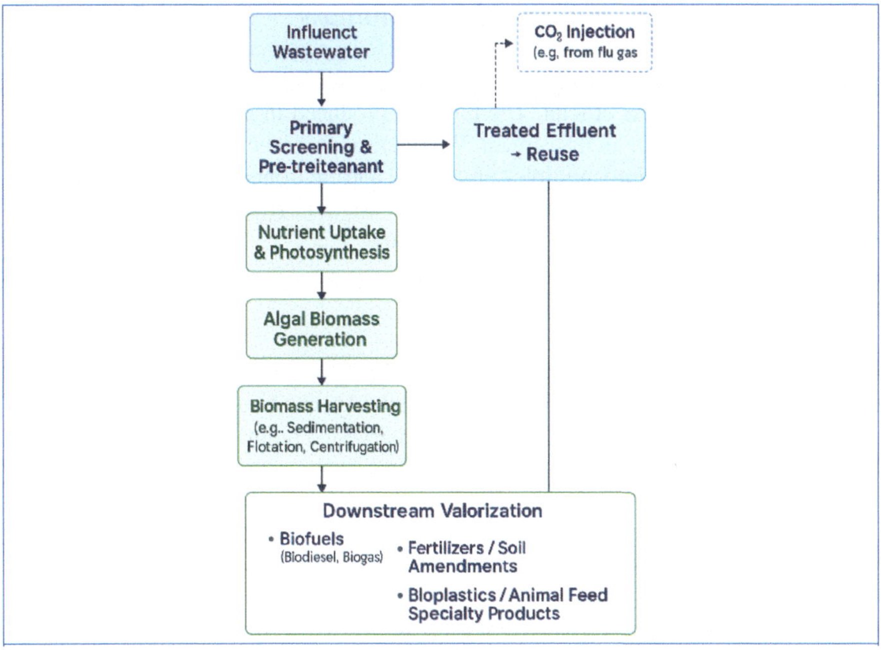

Fig. 2. System flow diagram showing sequential processes: influent wastewater intake, nutrient uptake, algal biomass growth, harvesting, valorization into biofuels/fertilizers, and effluent reuse. (Source: Author's contribution)

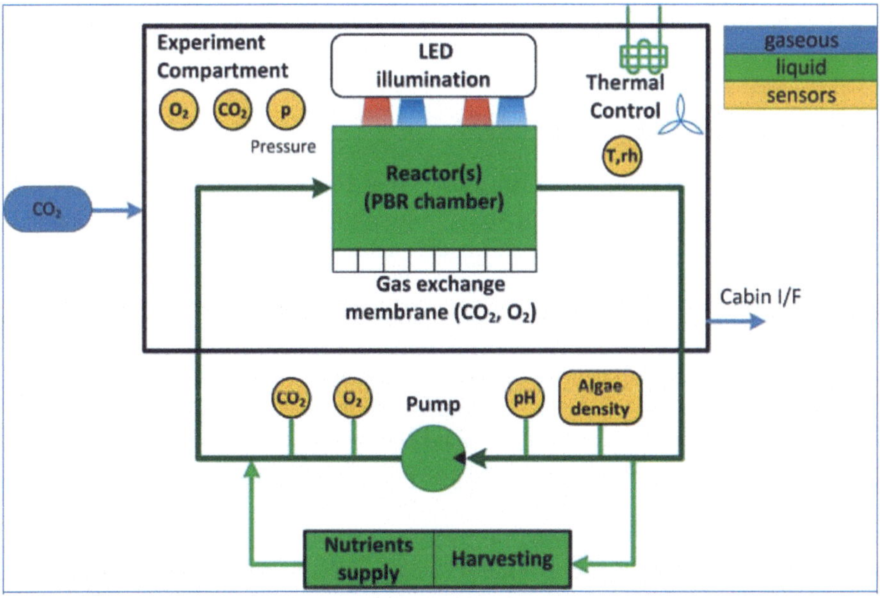

Fig. 3. Microalgae photobioreactor system flow diagram with gas exchange membrane, sensors, and LED illumination (Source: Author's contribution)

4 Resource Valorization from Algal Biomass

The ability to recover resources, one of the most persuasive advantages of using algae in wastewater treatment, is crucial. The biomass produced in nutrient-rich effluents is no longer waste, but rather an investment. When the appropriately processed biomass is recovered, it is converted into multiple commercial and environmental products, promoting UNSDG 7 (Affordable and Clean Energy), UNSDG 12 (Responsible Consumption and Production), and UNSDG 13 (Climate Action).

4.1 Biofuel Production

The richness of microalgae in lipids and carbohydrates makes it an attractive feedstock for bioenergy production. Different conversion pathways can be used, which are dictated by the nature of the biochemical composition of harvested biomass, i.e. Using lipids for biodiesel through transesterification, carbohydrates for bioethanol production through fermentation, and whole biomass for biogas production through anaerobic digestion.

The use of algal biofuels within a wastewater treatment system, notably increases the self-sustainability of energy for the treatment system itself, by providing enough energy for part of the wastewater treatment processes and producing digestate for land application from the anaerobic digestion of algal biomass. There have been a number of pilot studies that have shown a positive net energy balance from algal biofuels, when utilizing co-digestion or flue gas based enhanced growth, the energy payback time for algal fuels has shown promising results (Thevarajah et al., 2024).

However, economics are driven by biomass yield and lipid content, and efficient lipid extraction methods. These factors will be further improved through the recent innovative developments in solvent-free extraction, hydrothermal liquefaction, and super critical CO_2 methods to reduce energy input for algal fuel production (Barik et al., 2025).

4.2 Fertilizers and Soil Amendments

Due to the high amount of nutrients such as N, P, K and trace microelements it contained, the algal biomass can be used as an effective biofertilizer after appropriate treatment. Dewatered and stabilized biomass or their digestates can be applied to the soil either directly or mixed with other compost. This contributes to nutrient recycling and decrease the demand of energy and emission-intensive synthetic fertilizers.

Post-digestion algal residue also under development for controlled release fertilizers to increase nutrient use efficiency and reduce leach out losses. Moreover, some compounds of algal biomass were found to promote soil microbes, leading to soil structure improvement and increased drought resilience (Grobelak et al., 2025). However, regulated approval process to screen for heavy metals or pathogen contamination is needed, particularly when the biomass is from industrial wastewater. The recycling of algal biomass to be used as fertilizer has a strong impact on UNSDG 2 (Zero Hunger) since it enhances the sustainability of food production by utilizing natural nutrient recycling.

4.3 Bioplastics, Animal Feed, and Specialty Products

The algal biomass is a heterogeneous and a functional feedstock with potential in the several value chains downstream. Depending on strain and growing conditions, derived algal biomass can be converted to bioplastics, high-protein feed additives, nutraceuticals and fine chemicals (González Fernández et al., 2025). High-protein species such as Spirulina and Chlorella are the most frequently grown for animal feed, aquaculture and food supplements. Nevertheless, the culture grown on wastewater could carry contaminants including pathogenic organisms, heavy metals and drugs. International organizations like Food and Agriculture Organization (FAO, 2023) and the World Health Organization (WHO and FAO, 2022) advise extensive screening and processing measures to produce safe end products.

Even with these new regulations, the absence of standardized certification frameworks still presents a challenge to commercialization. Mitigation measures include pasteurization, heavy-metal screening, and compliance with WHO/European Food Safety Authority (EFSA) protocols. Emerging certification efforts, such as ISO-59004 (2024) guidelines for circular biomass and FAO (2023) standards for aquatic food systems, are being advanced to reduce risks in feed and food applications. In strains such as Nannochloropsis and Chlorella vulgaris, which are exposed to stress, the lipid content in their dry weight can reach up to 50%.112. The utilization of these lipids can result in biodiesel yields that range from 0.25 to 1.45 bar/kg of dry algal biomass. The co-products of algae are also exploited, making it a competitive feedstock when compared to conventional oil crops. This is especially true when the yields are high.

5 Environmental and Economic Performance

To determine their environmental footprint, which includes greenhouse gas emissions, energy consumption, and resource use, algae-based systems are frequently assessed through life cycle assessment (LCA). These systems have been shown in several studies to be able to achieve net-negative carbon emissions through biomass carbon sequestration and avoided fertilizer production under optimal conditions.

Economic analyses have revealed positive, albeit inconsistent outcomes that vary with the scale and co-product value. However, Net present value (NPV) estimates vary widely. Reported payback periods typically range between 5 and 12 years depending on system scale and co-product markets. Treatment costs range per m^3, depending on system design and co-product recovery. The energy return on investment (EROI) of integrated systems has been found to vary between 1.2 and 2.4 in studies conducted by Öztürk and Satpati (2024). The figures reveal substantial potential, especially when accompanied by policies or carbon credits.

However, the calculation still strongly relies on many varying parameters. Variations in the price of carbon, the fossil fuels, seasonal changes in algal yield, changes to nutrient controls all affect this. Such a rise in the price of a carbon credit from $30 to $100 per ton CO_2-equivalent could mean that reducing emissions is now the main concern within an emission-based revenue system (Singh et al., 2025).

A list of the most relevant LCA indicators and economic parameters is shown below to assist with comparative analysis. They provide an insight into the environmental outputs and economic performance of different algae-based system configurations under varying operating conditions. To illustrate and understand how algae-based systems perform economically and environmentally across different operational situations, a table below provides data reported from studies over the recent years which highlights important parameters, like greenhouse gas savings, energy requirements and a number of economic outputs. Existing data points correspond to mid-range situations, assuming steady biomass output and market prices, but a more in-depth sensitivity analysis is needed in the future (Table 2).

Table 2. Comparative techno-economic and environmental performance of advanced algae-based wastewater treatment configurations across recent global studies (2023–2025). (Source: Author's contribution)

Research Study	Configuration	CO_2 Reduction (kg CO_2e/m^3)	Net Energy Use (kWh/m^3)	Operational Cost (USD/m^3)	Revenue Potential (USD/m^3)	Payback Period (years)	Key Advantages
(Kumar et al., 2020)	PBR + Anaerobic Digestion	8.2–11.5	0.7–1.1	0.42–0.58	0.85–1.32	6.8–9.2	High biomass yield; biogas co-production
(Schade and Meier, 2021)	Hybrid Pond-PBR + Biofertilizer	5.8–9.3	0.4–0.8	0.31–0.47	0.67–1.18	7.1–10.4	Lowest energy; nutrient recovery credits
(Valdovinos-García et al., 2022)	Flue-gas integrated PBR	9.1–12.7	1.0–1.4	0.51–0.69	1.02–1.67	5.9–8.3	Industrial CO_2 utilization; carbon credits
(Nobre et al., 2024)	Mixed Systems	7.8–11.2	0.6–1.1	0.38–0.56	0.78–1.35	6.5–9.5	Multi-revenue streams

Note: Metrics standardized per cubic meter treated effluent; 10-year projection horizon at 6% discount rate

All values for Net Present Value (NPV) are per m^3 of treated wastewater, and have been adjusted for treatment to USD (based on a 10-year horizon and 5–8% discount rate). Return on Investment (ROI) figures represent a percentage return over the total initial investment and operating costs. Energy demands were all presented per m^3 of treated wastewater in kWh. The figures are comparable between different study cases due to their adjusted nature.

ROI accounts for benefits including by-product income (e.g. Fuel and fertilizers), estimated benefits of carbon credits (typically 30–80 \$/ton CO_2-equivalent), and avoided costs for wastewater treatment. Operational costs are dependent on the relevant costs of local energy prices and inputs required for maintenance, sensitivity to seasonal biomass production and market volatilities are dependent on scenario-based models. Local policies and the technological stage of the investment may influence values significantly.

5.1 Life Cycle Assessment (LCA) and Climate Impacts

It is important to estimate the environmental impacts generated from algae by carrying out a Life Cycle Assessment (LCA). Carbon neutral or even negative emission may be reached from the algae system by implementing an appropriate LCA, as showed in recent studies in the past few years. It is possible to achieve if the input of CO_2 is taken from flue gas or biogenic source. Moreover, the use of alternative energy sources like algae and non-fossil fuels contribute to environmental benefits in both upstream and downstream processes.

The most significant environmental issues are typically associated with biomass harvesting and dewatering, which involves energy-intensive centrifugation and drying processes, as demonstrated by LCA. The reduction of these can be achieved through low energy methods like gravity sedimentation, passive drying, or electrocoagulation (Zhang et al., 2025). The environmental benefits of algal systems are determined by a range of factors, including energy sources, local climate, and waste heat. Co-location with power plants or industrial emitters can enhance system performance by reducing CO_2 transport costs and improving water quality, while also increasing algae growth rates.

5.2 Economic Viability and Cost-Benefit Analysis

Capital investment, operating expenses, product recovery value, and regulatory incentives are all important factors that should be considered in the economic evaluation of algae-based wastewater treatment. The cost of an integrated algal system varies greatly depending on the system's size and location, with capital-intensive photobioreactors exceeding USD 1,000/m^3 capacity, while open pond systems can cost less than USD 100/m^3.

Approximately up to 60% of the total cost may lie in harvesting and downstream processing and have significant effect on operational cost. Yet, if revenues are generated from sale of biofuels and credits for fertilizer production and treatment of wastewater then some systems become near break-even or positive NPV (especially on Carbon prices or renewable energy incentives, Craggs et al. (2013) with the help of private public partnership or green finance tools to achieve quicker implementation in both developed and developing countries.

Cost-benefit models also show that resilience dividends from algae systems like GHG mitigation, water recovery and restoration of ecosystem cannot be included in normal cost benefit accounts but these benefits can contribute to economy (e.g. Wastewater treatment by algae could alleviate stress on central treatment plants and remove negative external cost associated with nutrient pollution (Tamburini et al., 2020). Moreover, land flexibility and modularity of algal system make it more suitable for urban edges, peri-urban areas or industrial areas.

Integration with a high-value product market such as aquaculture, green hydrogen production or utilizing side-streams like residual heat or wastewater sludge is also another avenue to achieve the economies of scale (or synergistic economies). It is expected that future business model will integrate algae-based wastewater treatment within the circular bioeconomy clusters where various businesses utilize shared resources, infrastructure and benefits.

6 Urban Integration and Land Use Optimization

To implement algae-based treatment in space-scarce urban areas, architectural integration of algae systems into the built environment can lead to multiple uses for land. By attaching vertical algae bioreactors onto building facades, rooftops, or green urban corridors, algae treatment can become multifunctional by addressing gray water and storm water management while also performing the functions of air cleaning and urban cooling (Jurík et al., 2024). These architectural integrations facilitate the notion of "productive landscapes" which aim to integrate ecological services into built environments.

By utilizing space along road medians, park perimeters, or marginal areas of urban spaces, relatively small photobioreactor arrays or shallow algal ponds can be implemented as distributed systems that will be used for wastewater treatment. These types of applications have already been pilot-tested in areas such as Singapore and Tokyo to provide shade which can reduce heat-island effect and provide additional green cover in a region with limited real estate (Kishimoto and Yan, 2024). For Example, a 100 m^2 facade photobioreactor may be capable of treating between 2000–3000 L of gray water per day and at the same time capturing 50–70 kg of CO_2 per year, showing potential scale of implementation in urban areas.

Competition over space is one challenge in adopting algae systems in dense cities. In high-value city districts, algae treatment facilities compete over space with housing development, recreation or commercial facilities. Trade-off in this competitive land market must be addressed. Careful siting should address to place the algae facilities on marginal lands, road medians, or in conjunction with existing water supply and distribution network in order to save space for housing, recreational or commercial developments, as well as to improve air quality and thermal isolation (Jurík et al., 2024).

7 Policy, Regulation, and Public Acceptance

While technically feasible, scaling up algae-based treatment systems requires enabling policies and acceptance by wider society. The policy framework for reuse and valorization of algae-based biomass remains fragmented across regions. The reuse of digestate produced by algae for application on agricultural lands for instance is allowed in some countries but not in others either due to contamination risk or lack of standardized regulation (Crippa et al., 2025).

Policy instruments such as grants and subsidies, carbon credits and green procurement schemes may be able to accelerate the market penetration and implementation of algae-based technologies. The European Union (EU) policy supports algae related activities via Horizon Europe and the EU Green Deal, which provide research and innovation support. More inclusive policies for the developing countries will promote the development of local tailored alga-based solutions particularly in informal settlements and peri-urban contexts (Salem et al., 2025).

Social acceptance is key. Urban scale implementation faces challenges with odor, aesthetics and fear of health risks unless it is integrated with a sound communication strategy and with the participation of local communities and users (Campbell and Rudan, 2020). Dissemination through public engagement programs, school curricula, demonstration projects may help foster the social license of the technology.

A clear framework governing trans-sectoral cooperation and between utilities, urban planners, environmental regulators and agricultural users should overcome institutional barriers. In addition, the framework regarding international certification of algal biomass for use is not standardized globally. A few certification bodies have been considering algae-based outputs for use in compost, animal feed and biofertilizers, but international standards have not been globally established. There is an ongoing work by several bodies including WHO, ISO, and FAO but it is yet to be integrated into organic or circular economy labels. Establishing internationally recognized certification and labeling standards will improve market trust. Integrated water energy food nexus governance frameworks can provide a direction for developing coherent policy frameworks, certification and scaling up.

8 Future Research and Innovation Pathways

The effective scaling and mainstreaming of algae-based wastewater treatment systems require a structured, multi-tiered research agenda. For improved strategic clarity, future directions can be delineated into three thematic categories: technical innovation, policy and regulatory development, and ecological-social integration.

8.1 Technical Innovation

- Strain Engineering & Process optimization: There should be ongoing research in synthetic biology, adaptive evolution and metabolic pathway improvement so that the algae is more tolerant to varying wastewater stream compositions and more effective pollutant removal in dynamic environments (Shelake et al., 2025).
- Low-Energy Harvesting Technologies: Because biomass separation is currently the most energy-intensive step, further developments need to be made in the areas of bio-flocculation, magnetically responsive microalgae and electro chemical recovery processes to decrease the operating costs and the energy return ratios (Rathinavel et al., 2024).
- Digital Monitoring and Automation: In conjunction with machine learning, IoT-based sensing and digital twin concepts, processes could be monitored and managed in real time with greater reliability.

8.2 Policy, Economic, and Certification Frameworks

- Standardization of algal biomass applications. An important area for research is assisting in developing harmonized safety and quality standards for the use of algae derived biomass in fertilizers, animal feeds, compost and bioplastics. This would encourage a market confidence and clarity.
- Lifecycle Cost-Benefit Models. Using techno-economic modelling which takes into account fluctuating policy environments (e.g. Carbon credit markets, nutrient trading markets or incentives for waste-water reuse), business models and the risk-adjusted ROI can be optimized.
- Blockchain and Supply chain tracing. Innovative research in digital certification would hopefully provide solutions to traceability of biomass, the control of contamination risks, quality assurance for exports or higher value uses.

8.3 Environmental-Social Co-Benefits and Long-Term Effects

Advanced research is urgently required to deepen understanding of algae-based systems' broader ecological and societal dimensions, ensuring their deployment contributes to genuine sustainability transitions rather than creating unintended environmental or social trade-offs.

- Comprehensive Ecotoxicological and Biodiversity Assessments: Long-term ecological risk assessment is required to determine the potential impacts of scaled-up open-pond algae production on local ecosystem, aquatic environment, soil biology etc. And unexpected issues like nutrient leaches, water body blooms, non-target native microbial communities' impact in natural and semi-natural environments like wetlands, coasts and peri-urban agriculture interfaces. Specific behavior of different strains, their behavior at different seasons and in case of extreme climatic events will need to be evaluated.
- Participatory Design, Community Engagement, and Social License: Interdisciplinary research combining social sciences, environmental psychology, and urban planning will be necessary to study and facilitate the acceptance of algae systems. Participatory design with local community and clear stakeholder consultation would be needed to address various concerns including visual impact, odor, perceived health effects etc. Educational outreach, demonstration projects and citizen science could improve acceptance and promote bottom-up development in poor urban or peri-urban communities.
- Holistic assessment of systemic co-benefits and trade-offs: The future assessment framework would need to go beyond conventional wastewater indicators and consider various systemic impacts, such as mitigating urban heat island effects, improving air quality by removing CO_2, enhancing the urban greenery, boosting community resilience to extreme climates and ecosystem restoration. Life Cycle Sustainability Assessment (LCSA) framework will be useful for evaluation from multiple dimensions including economy, environment, and social aspects for the policymakers and urban planners to support the decision making for the urban integration of the system.

Finally, introducing multi-sectoral integration models such as the embedding of algae systems in agro-industrial parks, aquaculture systems, or zero-waste urban areas, will enable economies of scope and scale to arise while fostering industrial symbiosis. Furthermore, a business model on "Algae-as-a-Service" can enable the deployment of the technology to distributed settings with few resources, permitting even smaller municipalities or rural areas to become active participants of the transition toward a circular bioeconomy.

9 Conclusions

Algal wastewater treatment is therefore a game-changer for environmental management and pollution control, with efficient and synergistic integration of circular resource recovery and climate mitigation. Since algae absorb not only nitrogen and phosphorus from waste, but also capture atmospheric and industrial carbon dioxide and produce a biomass which is used for other value-added processes (downstream valorization), algal wastewater treatment addresses several overlapping UNSDGs.

It has been evident from this review that algal solutions are able to achieve UNSDG 6 targets (Clean water and sanitation) through 70–95% nitrogen and 60–90% phosphorous removal in different wastewater streams thus reducing the eutrophication risk of inland and coastal water bodies, and improving the quality of wastewater that can be re-used for different purposes. Moreover, algal processes are contributing to UNSDG 7 (Affordable and clean energy) through different pathways of energy recovery: biofuel production (biodiesel, bioethanol, biogas). In some integrated systems the energy return on investments (EROI) achieved can be between 1.2 and 2.4. Under UNSDG 12 (Responsible consumption and production), algae can valorize the recovered nutrients by producing biofertilizers, bioplastics and feed ingredients that are rich in protein, so the cycle of nutrient use is closed with no energy intensive synthetic input requirement. Lastly, UNSDG 13 targets (Climate action) are attained through the great potential that microalgae have to capture CO_2-they are capable of absorbing roughly 1.3 kg CO_2 per kg of biomass generated, with overall savings of 4.5–10.2 kg CO_2-equivalent per m^3 of wastewater treated under optimal conditions.

While not only technically possible but also a strategically scalable solution across geographic, economic and policy boundaries as this study have shown with its review of pilot and industry-level implementation for 2023–2025. This path to success requires more than scientific advancements. The proposal requires strengthened institutional frameworks, regulatory convergence across regions, internationally standardizing certification, and effective public participation mechanisms to allow these systems to become widespread.

Fundamentally, the algae paradigm challenges us to transform wastewater from a waste to be discarded to a nexus of water, energy and food. Today, in the age of intense climate impacts and immediate needs to transition towards a circular economy, algae solutions become key infrastructure to create resilient, regenerative urban and peri-urban areas. Further developing these nexus linkages through an increase in research, innovation in policy, and cross-sectoral cooperation can solidify algae-based wastewater treatment as an attainable, scalable multi-SDG solution.

Conflict of Interest. The authors declare that they have no conflict of interest.

References

Barik, S.K., Kumar, P., Pal, U.J., Aikat, K.: Integrated biorefinery approach for sustainable biofuel production from algal biomass. Clean Technol. Environ. Policy **27**, 3245–3269 (2025). https://doi.org/10.1007/s10098-024-02851-w

Campbell, I.H., Rudan, I.: Effective approaches to public engagement with global health topics. J. Glob. Health **10**, 010901 (2020). https://doi.org/10.7189/jogh.10.010901

Craggs, R.J., Lundquist, T.J., Benemann, J.R.: Wastewater treatment and algal biofuel production. In: Algae for Biofuels and Energy. Springer Netherlands, Dordrecht, pp. 153–163 (2013). https://doi.org/10.1007/978-94-007-5479-9_9

Crippa, I., Dolci, G., Grosso, M., Rigamonti, L.: Life Cycle assessment of microalgal biomass valorization from a wastewater treatment process. Waste Biomass Valorization **16**, 525–541 (2025). https://doi.org/10.1007/s12649-024-02695-x

Deepika, C.: Advancing sustainability: microalgae biofuels in the circular bioeconomy. In: Microalgal Biofuels. Elsevier, pp. 505–549 (2025). https://doi.org/10.1016/B978-0-443-24110-9.00023-2

FAO: Transforming Aquatic Food Systems: Anchoring fish and aquatic food systems in policies and action [WWW Document]. Food Agric. Organ. United Nations (2023). URL https://openknowledge.fao.org/server/api/core/bitstreams/e01f8053-4035-4318-9f74-6061aa25b4aa/content/agrifood-solutions-to-climate-change-2023/transforming-aquatic-food-systems.html

Galang, M.G.K., Chen, J., Cobb, K., Zarra, T., Ruan, R.: Reduction of biogenic CO2 emissions, COD and nutrients in municipal wastewater via mixotrophic co-cultivation of Chlorella vulgaris–aerobic-activated sludge consortium. Environ. Technol. **46**, 3348–3362 (2025). https://doi.org/10.1080/09593330.2025.2463696

González Fernández, L.A., Medellín Castillo, N.A., Sánchez Polo, M., Navarro Frómeta, A.E., Vilasó Cadre, J.E.: Algal-Based carbonaceous materials for environmental remediation: advances in wastewater treatment, carbon sequestration, and biofuel applications. Processes **13**, 556 (2025). https://doi.org/10.3390/pr13020556

Grobelak, A., Bień, B., Sławczyk, D., Bień, J.: Conditioning biomass for biogas plants: innovative pre-treatment and digestate valorization techniques to enhance soil health and fertility. Sustainability **17**, 3289 (2025). https://doi.org/10.3390/su17083289

Hoang, M.-N., Pham, M.-D.-T., Nguyen, P.-D., Lens, P.N.L., Bui, X.-T.: Role of nature-based solutions in urban sustainable development. In: Nature-Based Solutions for Urban Sustainability. IWA Publishing, pp. 3–12 (2025). https://doi.org/10.2166/9781789065015_0003

ISO-59004: Circular economy—Vocabulary, principles and guidance for implementation (2024)

Juríк, J., Jankovičová, B., Zakhar, R., Šoltýsová, N., Derco, J.: Quaternary treatment of urban wastewater for its reuse. Processes **12**, 1905 (2024). https://doi.org/10.3390/pr12091905

Kishimoto, K., Yan, W.: Green Infrastructure in Tokyo. In: Resilient and Adaptive Tokyo. Springer Nature Singapore, Singapore, pp. 105–121 (2024). https://doi.org/10.1007/978-981-99-3834-6_6

Kumar, A.K., Sharma, S., Dixit, G., Shah, E., Patel, A.: Techno-economic analysis of microalgae production with simultaneous dairy effluent treatment using a pilot-scale High Volume V-shape pond system. Renew. Energy **145**, 1620–1632 (2020). https://doi.org/10.1016/j.renene.2019.07.087

Nobre, M.L.F., et al.: Techno-economic analysis of a circular microalgal approach for enhanced wastewater treatment and resource recovery in Northern Portugal. J. Clean. Prod. **434**, 140389 (2024). https://doi.org/10.1016/j.jclepro.2023.140389

Osama, A., Hosney, H., Moussa, M.S.: Potential of household photobioreactor for algae cultivation. J. Water Clim. Chang. **12**, 2147–2180 (2021). https://doi.org/10.2166/wcc.2021.261

Öztürk, S., Satpati, G.G.: Extremophilic algae-based wastewater treatment, nutrient recovery, and animal feed production. In: Trends in Biotechnology of Polyextremophiles. Springer Nature Switzerland, Cham, pp. 243–266 (2024). https://doi.org/10.1007/978-3-031-55032-4_11

Rathinavel, L., et al.: Sustainable microalgal biomass for efficient and scalable green energy solutions: fueling tomorrow. Fuels **5**, 868–894 (2024). https://doi.org/10.3390/fuels5040049

Roy, M., Mohanty, K.: A comprehensive review on microalgal harvesting strategies: current status and future prospects. Algal Res. **44**, 101683 (2019). https://doi.org/10.1016/j.algal.2019.101683

Salem, M., et al.: Managing the urban-rural transition: a review of approaches and policies for peri-urban land use. J. Urban Manag. **14**, 1115–1129 (2025). https://doi.org/10.1016/j.jum.2025.04.006

Schade, S., Meier, T.: Techno-economic assessment of microalgae cultivation in a tubular photobioreactor for food in a humid continental climate. Clean Technol. Environ. Policy **23**, 1475–1492 (2021). https://doi.org/10.1007/s10098-021-02042-x

Shelake, R.M., Khalid, M.A., Kim, J.-Y.: Microalgal bioengineering for futuristic applications in synthetic and space biology. Biotechnol. Adv. **84**, 108665 (2025). https://doi.org/10.1016/j.biotechadv.2025.108665

Singh, U., et al.: Direct flue gas capture for algae cultivation and subsequent valorization: evaluating life cycle emissions and costs. Sustain. Energy Fuels **9**, 4392–4403 (2025). https://doi.org/10.1039/D5SE00329F

Tamburini, E., Soana, E., Monti, M., Fano, E.A., Castaldelli, G.: Introducing life cycle assessment in costs and benefits analysis of vegetation management in drainage canals of lowland agricultural landscapes. Water **12**, 2236 (2020). https://doi.org/10.3390/w12082236

Thevarajah, B., Piyatilleke, S., Nimarshana, P.H.V., Gunawardena, S., Ariyadasa, T.U.: Harnessing algae: advancements in wastewater treatment and resource recovery. pp. 145–175 (2024). https://doi.org/10.1007/978-3-031-74515-7_9

UNESCO: UN World Water Development Report: Partnerships and cooperation for water. United Nations (2023)

Valdovinos-García, E.M., Bravo-Sánchez, M.G., Olán-Acosta, M. de los Á., Barajas-Fernández, J., Guzmán-López, A., Petriz-Prieto, M.A.: Technoeconomic evaluation of microalgae oil production: effect of cell disruption method. Fermentation **8**, 301 (2022). https://doi.org/10.3390/fermentation8070301

WHO, FAO: Report of the expert meeting on food safety for seaweed: current status and future perspectives (2022)

Zhang, S., An, W., Li, R., Zhang, X., Ge, H., Liu, H.: Optimization of forward osmotic dewatering on resource utilization of microalgae: a review. Clean Technol. **7**, 10 (2025). https://doi.org/10.3390/cleantechnol7010010

Groundwater Stress and Sustainability in Karnataka: Insights from GWRA-2023 and Policy Pathways

N. R. Asha Rani[1(✉)], Sheetal Thapa[1], H. U. Megha[2], and N. G. Devaiah[3]

[1] Department of Civil Engineering Alliance College of Engineering and Design, Alliance University, Anekal Bangalore, Karnataka 562106, India
asharaninr@gmail.com

[2] Alliance School of Business, Alliance University, Anekal Bangalore, Karnataka 562106, India

[3] Alliance School of Law, Alliance University, Anekal Bangalore, Karnataka 562106, India

Abstract. Groundwater is the backbone of the water security of Karnataka, but this has been under stress due to rising agricultural demand, urbanisation and variation in rainfall pattern. This review presents a synthesis of results from the Groundwater Resource Assessment 2023 (GWRA-2023) which was carried out within the framework of GEC-2015 utilizing InGRES platform to address the issue of determining a spatial and temporal groundwater dynamics within the state. Analysis at taluk level shows a decline in the number of Over-Exploited and Critical taluks from 44.89% in 2017 to 37.6% in 2023 which shows combined impact of Recharge interventions and good monsoon condition on change of identified taluks. However, persistent hotspots in Belagavi, Tumakuru, Bengaluru Urban and Kalaburagi are due to intensive cropping pattern, low surface irrigation and increased urban demand. Comparative insights highlight the need for demanding side reforms: micro-irrigation on a larger scale as well as crop diversification. The importance of the study is to translate the assessment outcomes into a phased policy roadmap of short-, medium-and long-term policy actions coupled with institutional obligations as well as to give the research priorities in support of climate resilient groundwater governance.

Keywords: GWRA-2023 · GEC-2015 · Over-exploitation · Micro-irrigation · Groundwater governance · ABHY · Urban water reuse · climate resilience

1 Introduction

Groundwater has become the backbone of water security in India, supporting agriculture, drinking water supply and industrial production. As with many semi-arid states, Karnataka has seen a steady rise in dependence on groundwater due to the low availability of surface water, and unpredictable monsoon precipitation. In the last two decades, the spread of bore wells, growth of water-intensive crop systems, and rapid urbanisation have all exerted pressure on aquifers resulting in falling water levels and declining water quality. Keeping these challenges in mind, the Government of India has institutionalized

L. Choudhary et al. (Eds.): TIMES 2025, SCI, pp. 202–222, 2027.
https://doi.org/10.1007/978-3-032-34636-0_15

periodic groundwater resource assessments to provide a scientific foundation to sustainably manage and legislate interventions for groundwater policy (Alley & Konikow, 2015).

The methodology of the Groundwater Estimation Committee (GEC), originally developed in 1997 and updated in 2015, offers a common framework to express the amount of groundwater recharge, extraction and availability on an administrative unit basis. Groundwater Resource Assessment 2023 (GWRA-2023) of Karnataka undertaken based on the latest approach of GEC-2015 methodology and using the national level Integrated National Groundwater Resources Estimation System (INGRES) is the most comprehensive and latest assessment in terms of groundwater resources of the state (Bhanja et al., 2016). Not only does the report categorize taluks into safe, semi-critical, critical and over-exploited areas, but also highlights trends in time compared to previous evaluations, in particular the conversion of over-exploited areas between 2017 and 2023 (Central Ground Water Board, 2012).

Such findings are extremely important due to a variety of reasons. The civil engineers also use them in assisting with the design and installation of artificial recharge systems, water harvesting systems and irrigation planning. As far as management is concerned, they bring with them allocation of resources, intervention in the demand side, and the community role in managing the ground water. On the legal and policy front, they are supportive of regulatory action, the extension of programmes such as the Atal Bhujal Yojana (ABHY) and the development of water security policies per the United Nations Sustainable Development Goals (SDGs) and specifically SDG 6 Clean Water and Sanitation (Indian Council of Agricultural Research, 2023).

International guidelines for groundwater governance such as recommended by the United Nations (UN) Economic Commission for Europe Organisation for Economic Cooperation and Development (OECD), the World Bank, and the European Council for Facing Climate Change had stressed adaptive management, stakeholder involvement and integration of groundwater policy with climate change and agricultural policies. Placing the expertise of the State of Karnataka in the global context enhances the validity of GWRA-2023 as an assessment framework construction singing in line with policy (Central Ground Water Board, 2012). This review article summarizes the results of GWRA-2023, contextualizes them in the broader literature, and reflects on the implications of the results in the engineering, management, and policy disciplines. It aims to provide an overall perspective of the condition of groundwater stress in Karnataka and to map the path to sustainable management of the resource by integrating datasets, temporal comparisons, and interdisciplinary reviews.

2 Methodological Framework

The estimation of groundwater in India is based on a standardized approach as established by Groundwater Estimation Committee (GEC). GEC-2015 represented a radical update of the approach adopted in GEC-1997, as the framework has now been updated in line with ongoing advances in hydrogeology, monitoring and the need to deliver policy-relevant output (Ministry of Jal Shakti, 2023). The GWRA-2023 in Karnataka was carried out using the same GEC-2015 methodology through the national INGRES

(India Groundwater Resource Estimation System) platform and is transparent, reproducible across administrative units and can be carried out consistently. Figure 1 helps in understanding the structure: Drivers → Assessment (GWRA-2023/GEC-2015) → Findings → Interventions & Policy Pathways → Future Research. It provides the reader with a sense of how evidence feeds into action and longer-term research (Ministry of Jal Shakti, 2023). Figure 1 highlights a step-by-step workflow of the assessment of the GWRA-2023 under the INGRES platform. The process starts with compilation of taluk-level administrative and hydrogeological boundary data and followed by integration rainfall inputs, groundwater level observations, recharge components, abstraction estimates, return flows and tagging of water quality. These datasets undergo the processes in the GEC-2015 computational framework for estimating annual extractable groundwater resources and stage of extraction. The classifications hereafter (Safe, Semi-Critical, Critical and Over-Exploited) are then undergoing multi-level institutional review processes and transformed into policy interventions and future research pathways. This outlined working method creates increased transparency and helps non-specialist readers to follow along with the assessment logic.

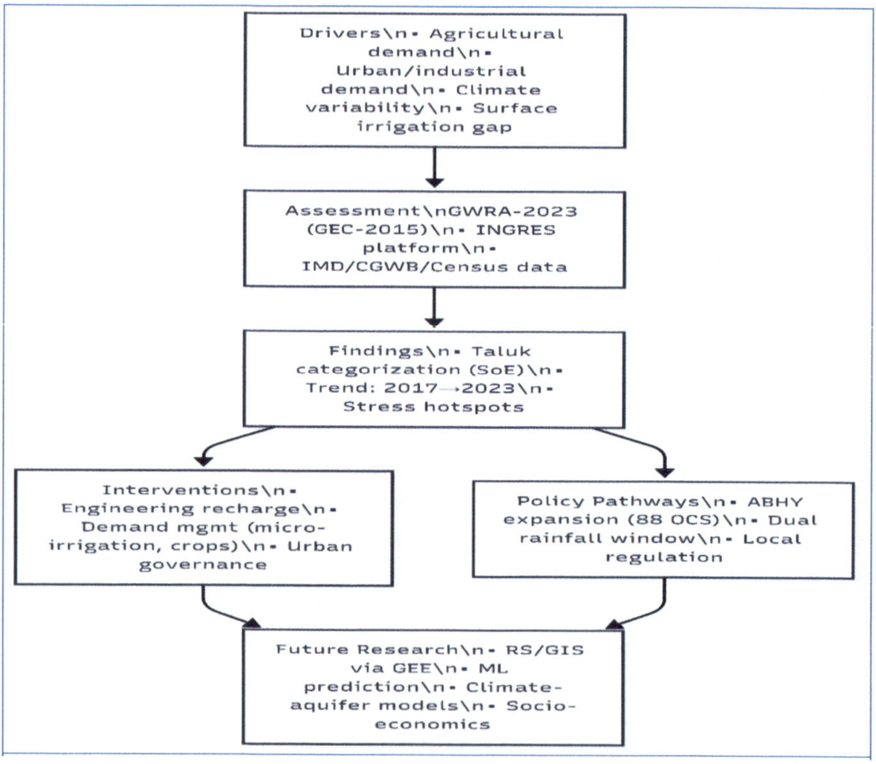

Fig. 1. INGRES-based workflow for GWRA -2023: Conceptual Flowchart (Ministry of Jal Shakti, 2023).

2.1 GEC-2015 Methodology

The GEC-2015 guidelines made several improvements over the period of 1997:

- Terms changed: use of older terms such as "net annual groundwater availability" changed to terms such as "annual extractable groundwater resource" (Chen et al., 2014).
- Better recharge estimation: considering rainfall infiltration, seepage in canals, tanks, ponds, irrigation flow and water conservation structures (Government of Karnataka, 2023).
- Stage of Extraction grouping: by ratio of annual groundwater extraction to annual extractable resources, grouped as Safe (<70%), Semi-Critical (70–90%), Critical (90–100% and Over-Exploited (>100%) (Government of Karnataka, 2023).
- Foresight Demand - Including domestic and industrial water allocation for the year 2025 and making sure to integrate socio-economic dimensions (Government of Karnataka, 2023).
- Dynamic assessment cycles: biennial re-assessment to capture changes that result from rainfall variability, cropping patterns and recharge interventions (Asoka et al., 2018).

2.2 Data Inputs and INGRES System

The assessment was made using more than one set of datasets that were integrated through INGRES, an integrated digital information system created by Central Ground Water Board (CGWB):

- Basic information sheet: admin boundaries, area information and hydrological units (Asoka et al., 2017).
- Rainfall and recharge sheets: long term rainfall data from Indian Meteorological Department (IMD) and effective rainfall, recharge component due to seepage from surface water bodies (Asoka et al., 2017).
- Water Level data: pre and post monsoon water level data from observation wells across Karnataka (Asoka et al., 2017).
- Groundwater abstraction data: abstraction from dug wells, bore wells, dug-cum-bore well and industrial abstraction, segregated as monsoon and non-monsoon (Asoka et al., 2018).
- Return flow estimation: irrigation efficiencies of canals and field applications were included as recharge components (Fishman et al., 2015).
- Future Allocation Sheet: domestic and industrial sector demand estimates for 2025 (India Meteorological Department, 2023).
- Quality tagging: tagging of aquifers with high salinity, fluoride, nitrate or heavy metal content to indicate usable and non-usable resources (Gleeson et al., 2012).
- Together, these sheets gave more than 100,000 variables for Karnataka's taluk-level units, giving a complete picture of groundwater dynamics (Government of Karnataka, 2023).

2.3 Classification and Verification

Following the aggregation of the data, all the talukas were divided into one of four categories-Safe, Semi-Critical, Critical and Over-Exploited based on the Stage of Extraction (SoE). Multi-level validation of results took the following form:

- Internal consistency (missing values, threshold checks, INGRES consistency check).
- Type of cross verification on district level water resource department (Food and Agriculture Organization, 2022).
- Final approval by State Level Committee (SLC) comprising of officials from CGWB, State Groundwater Authority, Agriculture, Irrigation and other stakeholders (Food and Agriculture Organization, 2022).

2.4 Relevance of the Framework

This approach to research is relevant in the following three aspects:

- It gives scientific feasibility that enables interstate or inter-year cross-comparability (Goldin, 2016).
- It could be a policy-ready tool, a tool that can tie technical findings with schemes like Atal Bhujal Yojana (ABHY) (India Meteorological Department, 2024).
- It also allows for interdisciplinary articulation: civil engineers can design recharge interventions that can then be optimized by them (managers), but that can be managed by them (policymakers), by controlling extraction and promoting sustainable practices (India Meteorological Department, 2024).

3 Data Sources and Study Area

3.1 Geographic and Hydro-Climatic Setting of Karnataka

Karnataka (an Indian state in the south of India) has an area of around 191,791 km^2 and consists of varied physiographic regions of coastal plains, the Western Ghats, and the Deccan Plateau. The state is split up into 31 districts and 227 taluks, which are basic administrative units for groundwater assessment. Hydro-climatically, Karnataka belongs to semi-arid to sub-humid region with an average annual rainfall of ~1,139 mm, with high spatial variability:

- Coastal Karnataka (Udupi, Dakshina Kannada) gets more than 3000 mm in a year (Government of Karnataka, 2023).
- In the interiors of the north (Kalaburagi, Raichur, Vijayapura), an average of less than 700 mm is reported (Government of Karnataka, 2023).
- Rainfall is monsoon-dominated with almost 80% of annual precipitation taking place between June and September (Chen et al., 2009).

Variability in this aspect makes groundwater a very important source for Karnataka and more so for dry northern taluks having higher demand of irrigation (India Meteorological Department, 2023).

3.2 Groundwater Resource Assessment Data Sources

The Karnataka Government Welfare Reform Agency-2023 (GWRA-2023) merges many datasets into INGRES data warehouse with taluk-level granularity. Key data sources include:

3.2.1 Rainfall Data

- Long term measured rainfall series (annual and monsoon specific) from India Meteorological Department (IMD) (Humphrey et al., 2016).
- Satellite rainfall (TRMM and IMERG) products can be used to supplement the IMD observations for validation (India Meteorological Department, 2023).

3.2.2 Hydrogeological Data

- The water level data of observation wells of Central Ground Water Board (CGWB) and State Ground water Department (India Meteorological Department, 2023).
- Recharge estimated using a seasonal water level variation (pre- and post-monsoon) (Hashimoto et al., 1982).

3.2.3 Recharge and Return Flow Information

- Contribution from water bodies on the surface: tanks, ponds, canals and percolation tanks (Jasechko et al., 2014); (Gleeson et al., 2012).
- Estimates of return flows from irrigated agriculture, canal seepage and water conservation structures (Joodaki et al., 2014).

3.2.4 Extraction Data

- Abstraction from dug wells, bore wells and dug cum bore wells, seasonal segregation (monsoon and non-monsoon) (Judd et al., 2008).
- Statistics of industrial and urban water extraction.

3.2.5 Future Demand Projections

- Domestic and industrial uses for 2025 included in the analysis to account for population and industrial growth (Koster et al., 2000).
- Data cross-checked with Census of India (Baseline 2011, Prediction 2025) (Leblanc et al., 2009).

3.2.6 Water Quality Data

- Taluks having saline, fluoride, nitrate, or heavy metal water were labeled, based on CGWB monitoring. This quality tagging has a direct bearing on usable versus non-usable groundwater resources (Loucks, 1997).

3.3 External Data Sources that May Be Used to Support Analysis

In addition to GWRA inputs, there is scope for supplementing the review article with comparative datasets, graphs and validation from several secondary sources.

- Census of India 2011, 2021 projection data → population growth, urbanization, domestic water demand (Loucks, 1997).
- Crop water requirements and irrigation intensity as reported by Indian Agricultural Research Institute (IARI) & FAO (Mays, 2013).
- Remote Sensing (Google Earth Engine, Bhuvan-ISRO, NASA MODIS) → LULC (Land Use Land Cover) mapping and mapping of irrigated area.
- State Water Resources Department → documentation of canal networks, reservoirs and tank rehabilitation (Ministry of Jal Shakti, 2023).
- Indian Institute of Tropical Meteorology, Pune (IMD & IITM Pune) - > high resolution rainfall and drought indices for temporal analysis (Ministry of Jal Shakti, 2023).

3.4 Study Area Justification

Karnataka is a representative state for groundwater sustainability research because of:

- high use of groundwater for irrigation (~60% of net irrigated area).
- Multiple stress zones are identified with taluks in the districts of Belagavi, Tumakuru, Bengaluru Urban and Kalaburagi identified as Over-Exploited or Critical repeatedly (Ministry of Jal Shakti, 2023).
- The impact of recharge interventions (check dams, tanks, watershed projects) on de-stressing in some taluks between 2017–23 (Mishra et al., 2016).
- With ABHY including one of its focus areas as the state, it is relevant to policy making as its groundwater governance framework will be significant at a national level (Ministry of Jal Shakti, 2023).

4 Results and Key Findings

Groundwater Resource Assessment 2023 (GWRA-2023) of Karnataka undertaken under the GEC-2015 methodology with INGRES platform gives a detailed picture of groundwater availability, extraction and stress status at taluk level. The results show both a positive trend in recharge and the continuing challenges of over-drafting in several key districts.

4.1 General Structure of Taluks

- Out of 227 studied taluks, groundwater resources were classified into Safe, Semi-Critical, Critical and Over-Exploited class.
- The share of Over-Exploited (OE) and Critical taluks fell from 44.89% in 2017 to 37.6% in 2023, showing that interventions due to recharge, and favorable rainfall conditions, have provided partial relief.

- Despite a slight improvement, more than one-third of taluks remain stressed, which indicates continued groundwater extraction more than sustainable levels. Figure 2 compares the percentage of Taluka with GW stress class based on GWRA-2023 (Government of Karnataka, 2023) for the year 2017 and 2023 The figure below shows a reduction in Over-Exploited taluks due to the influence of recharge interventions and rainfall but still shows high stress.

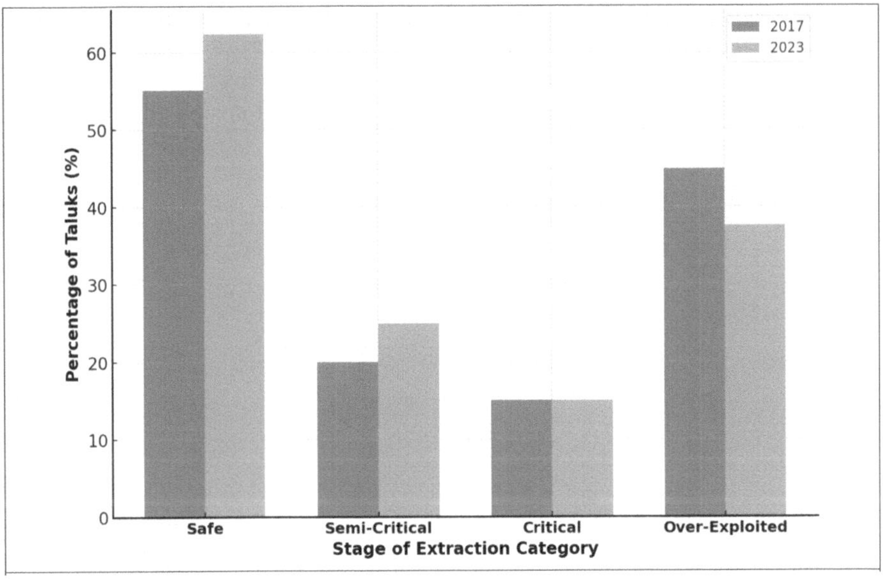

Fig. 2. Stage of Extraction Trends for Karnataka (2017 vs 2023) (Government of Karnataka, 2023).

4.2 Temporal Trends (2017 vs 2023)

- Improvement observed: Several taluks changed from Over-Exploited to Semi-Critical or Safe categories especially in areas benefiting from rainfall, tank rejuvenation and watershed management projects.
- Chronic stress: Several districts, like Belagavi, Tumakuru and Bengaluru urban districts, continue to report multiple Over-Exploited taluks mostly due to intensive agriculture (sugarcane, paddy) and high domestic demand.
- Policy implication: Despite interventions that have reduced stress in some areas, over-extraction continues to plague high demand urban-agricultural interfaces.

4.3 Factors that Drive Groundwater Stress

- Agricultural Demand: Depletion has accelerated in districts like Belagavi because of the growth of water-intensive crops (sugarcane, paddy, banana).
- Low surface irrigation: Taluks without canal irrigation (e.g. Tumakuru) are more groundwater dependent.

- Urban expansion: Rapid growth in Bengaluru Urban has pushed the domestic and industrial demand beyond sustainable recharge.
- There is a measurable influence in groundwater recharge and stress pattern owing to rainfall variability in Karnataka. Analysis of standardised precipitation indices (SPI) and the occurrence of drought suggests that precipitation response in the northern districts including Kalaburagi and Raichur are recurring negative SPI values, which are associated with low recharge and Over Exploited class under persistent drought occurrence. Figure 3 shows the positive relationship between the annual rainfall and the estimated groundwater recharge, and, at the same time, the substantial scatter resulted from heterogeneity in aquifer characteristics, land use and soil properties. This helps to highlight the need for integration of climatic indices to facilitate assessment frameworks for groundwater to better predict climate-induced risk.

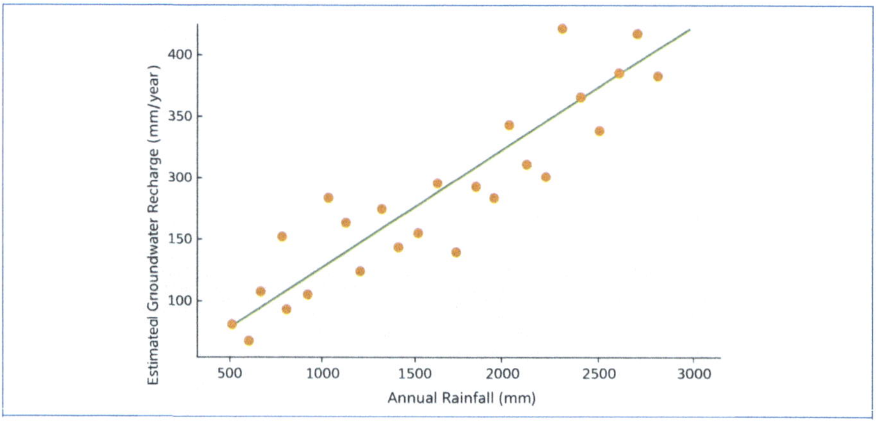

Fig. 3. Groundwater Recharge vs Rainfall (Karnataka, Example) (Government of Karnataka, 2023).

4.4 Unit Drafts by Abstraction Structures

- Dug wells: The use of them is decreasing because the water level drops.
- Bore wells: account for the biggest share of extraction, especially in northern and central districts.
- Dug wells or boreholes in the transitional aquifer zone.
- Industrial production: Centered in the urban taluks, especially Bengaluru, Mysuru and Belagavi.

This breakdown accentuates the structural reliance on bore wells which has profound impact on the sustainability of aquifers.

4.5 Domestic and Industrial Allocations, 2025

GWRA-2023 is the basis for future water allocation for each taluk.

- Example values:
  - o Shirahatti taluk (Gadag district) - 54.24 ham in 2025 domestic demand.
  - o Kalaburagi taluk - 2,790.97 ham, which shows high urbanization and industrialization
- These figures emphasize the need for integrated planning to ensure that growth in non-agricultural water demand is met without further harm being done to the aquifer. Figure 4 and table 1 gives projected domestic groundwater allocations for selected districts in 2025 as reported by GWRA-2023. Increasing allocations in Kalaburagi and Bengaluru Urban reflect fast urbanization and industrial expansion, whereas lower allocations in districts such as Shirahatti point to lower demand

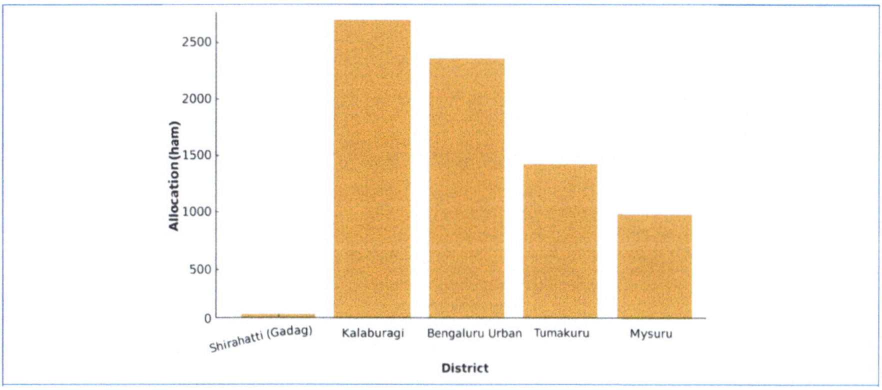

Fig. 4. Future Domestic Groundwater Allocation (Karnataka) 2025. (Government of Karnataka, 2023)

Table 1. Domestic Forecast (2025) by District (ham) (Source: Author's contribution)

District	Allocation (ham)
Shirahatti (Gadag)	54.24
Kalaburagi	2790.97
Bengaluru Urban	2500.5
Tumakuru	1450.75
Mysuru	900.6

4.6 District-Level Highlights

- Belagavi District: Largest number of Over-exploited taluks due to water intensive agriculture.
- Tumakuru District: Multiple critical taluks where chronic scarcity of water has been compounded by a lack of large-scale canal irrigation.
- Bengaluru Urban: Highly urbanized, double whammy (high domestic demand, acute ground water quality).
- Kalaburagi: The region is drought prone with high projected domestic demand, putting an additional strain on already fragile aquifers in the future. Figure 5 shows selected districts, Belagavi and Kalaburagi as Over-Exploited, Tumakuru as Critical, Bengaluru Urban as Semi-Critical and Mysuru as Safe. The figure highlights spatial variation in categories of stress.

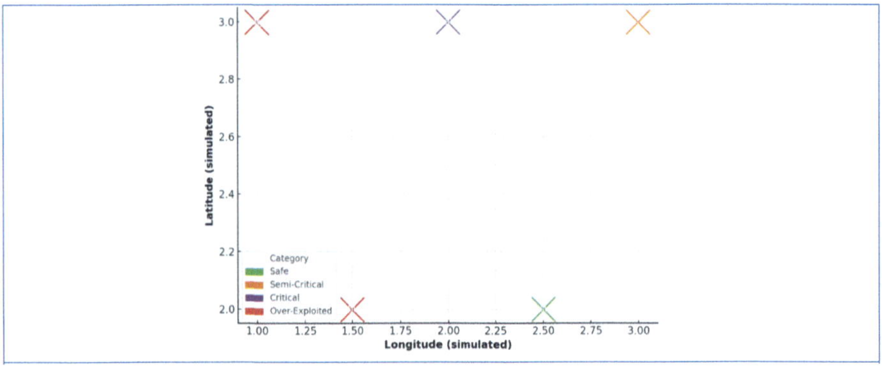

Fig. 5. District wise Groundwater Stress Zone in Karnataka (GWRA-2023) (Government of Karnataka, 2023)

4.7 Positive Outcomes

- The decline in OE taluks is indicative of the fact that rainwater harvesting, watershed projects and tank rejuvenation are now beginning to have measurable effects.
- Several artificial recharge structures under ABHY, as well as under state-led programs have resulted in better water availability in some semi-arid areas.
- Community-based watershed projects are proving effective at a local scale but are difficult to scale up.

4.8 Implications & Recommendations (Key Policy Recommendations)

- Up-scaling of Atal Bhujal Yojana (ABHY) to all 88 Over-Exploited/Critical taluks.
- Dual window rainfall analysis (five year moving average vs. single year rainfall) to minimize uncertainty in recharge estimates.
- Interventions in urban-agricultural aerosols (Bengaluru, Tumakuru, Belagavi) where the demand is greater than the recharge.

5 Comparative Analysis

In addition to serving as a state-specific snapshot of Karnataka groundwater resources, the GWRA-2023 is suitable for inter-temporal comparisons and cross-sectional comparisons. Such comparisons give an important insight into the dynamics of groundwater stress and the effectiveness of interventions.

5.1 Time-Series Comparison in Karnataka (2011–2023)

- GEC-1997 framework: The 2011 Assessment is characterized by an insufficient integration of data and more permissive classification thresholds. A large proportion of taluks were classified as Over-Exploited but with greater uncertainty.
- 2017 Assessment (first full adoption of the GEC-2015 in the state): Results showed 44.89% of taluks to be Over-Exploited/Critical. This high share occurred during several years of precipitation deficit and agricultural intensification.
- 2020 Assessment: Interim re-assessment showed tentative signs of improvement because of recharge programs and normal monsoons.
- Assessment 2023: The share of Over-Exploited/Critical taluks reduced to 37.6% including the impact of interventions related to rainwater harvesting, tank rejuvenation, watershed management, and ABHY.

Key Insight: The decreasing trend shows that policy interventions and positive rainfall can produce quantifiable improvements while the sustainability remains fragile in districts with high water demand such as Belagavi and Bengaluru Urban.

5.2 Across District Trends Over Time

- Belagavi: Regularly shows the highest number of OE taluks because of sugarcane cultivation. Even with interventions, pressure continues, limitations of recharge projects are seen if crop choice is not changed.
- Tumakuru: Critical chronic condition which indicates the fact that the rainfall is not reliable, and canal irrigation is not available. Minimal change over years is a sign of structural vulnerability.
- Bengaluru Urban: Transition from Semi-Critical to Critical in some taluks between 2017 and 2023 due to the increase in domestic/industrial demand - opposing the trend of agricultural districts where stress is crop driven.
- Northern Karnataka (Kalaburagi, Raichur, Vijayapura): There was some improvement associated with the scale up of watershed programs, but the drought has destabilized the recovery.

5.3 Relative Structural Location in Other States

- Maharashtra: agro-climatically like north Karnataka. Judging by assessments, OE blocks remain intact in Marathwada, and its cropping (sugarcane, cotton) is of Belagavi in Karnataka. What this means is that crop diversification is no less important than recharge interventions.

- Andhra Pradesh & Telangana: There has been a larger scale application of micro-irrigation (drip, sprinkler), and hence a relatively enhanced sustainability of ground water resources despite the same pattern of rainfall physical variability. Table 2 presents comparison of Micro-Irrigation coverage in selected states. Karnataka's lagging coverage of micro-irrigation offers a case in policy learning.
- Gujarat: Large scale canal networks and recharge structures were developed to stabilize the groundwater level under the Sujalam -Sufalam scheme in semi-arid districts. Karnataka's relative weakness lies in canal irrigation coverage. Figure 6 illustrates the estimated crop water demand as against the corresponding groundwater extraction level for major crops in Karnataka. Water intensive crops such as sugarcane and paddy are characterized by significantly higher demands reaching extraction levels and even above and millets and pulses exhibit sustainable profiles.

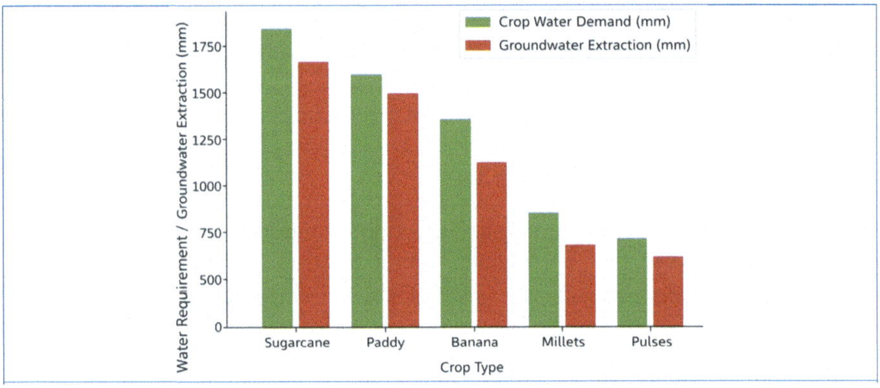

Fig. 6. Comparison of Crop Water Requirement and Groundwater Extraction in Karnataka (Government of Karnataka, 2023)

Table 2. Micro-Irrigation Coverage in Selected States Comparative (Source: Author's contribution)

State	Micro-Irrigation Coverage (%)	Dominant Crops	Groundwater Stress Trend
Karnataka	~28	Sugarcane, Paddy	High (over-exploited clusters)
Telangana	~55	Cotton, Paddy	Moderate improvement
Andhra Pradesh	~48	Horticulture crops, Paddy	Relatively stable
Maharashtra	~42	Sugarcane, Cotton	Persistently high
State	Micro-Irrigation Coverage (%)	Dominant Crops	Groundwater Stress Trend

From the comparison Karnataka is lagging than Telangana and Andhra Pradesh in terms of micro-irrigation adoption, despite similar agro climatic conditions. Higher rate of adoption in Telangana have translated into relatively better groundwater sustainability depicting the role of demand-side technological interventions along with recharge measures.

5.4 Thematic Comparisons

- Agricultural Reliance: Karnataka's OE districts demonstrate strong correlation with irrigation-intensive crops, as opposed to states like Rajasthan where scarcity has more climate-driven factors.
- Urban Stress: Bengaluru's profile is like Chennai and Hyderabad - rapidly urbanising metropolises where domestic and industrial demand is the major driver
- Policy Interventions: ABHY has been started in selected taluks for Karnataka, but it is to be scaled up for all 88 OCS (Over-Exploited and Critical) taluks. In other states such as Gujarat or Maharashtra, mainstreaming groundwater management within policy and subsidy regimes has been more aggressive.

5.5 Understandings from Comparative Observation

- Those are the ones that are implementable, but cropping policies must be coordinated. The decline in OE taluks in Karnataka is encouraging although Maharashtra experience indicates that even recharge programs mean nothing without relying on a crop water demand management plan.
- Groundwater governance needs to be incorporated into urban water governance. Bengaluru's stress curve is like other Metros, and this highlights the need for demand management, wastewater reuse and rainwater harvesting requirements.
- Multi-year rainfall data-Multiyear rainfall analysis decreases uncertainty. The best practice in drought prone regions of the world appears to be adoption of the double-window analysis proposed in Karnataka (five year moving average + single year rainfall).

6 Policy Implications and Research Pathways

This study has an impact on groundwater governance, engineering interventions, and social and economic policy by raising awareness of the GWRA-2023 findings to Karnataka. Although we can observe visible improvement in terms of the decrease of Over-Exploited (OE) and Critical taluks, we can still see signs of stress experienced in core districts, which indicates the necessity to address the issue on the multi-sectoral level on which the policy is designed to deliver to the population in the key areas. As such, the consequences are engineering, management, and legal/policy related, and this assessment should be used as a worthwhile source of reference in the ongoing management planning considerations for sustaining water supplies.

6.1 Better Artificial Recharge and Engineering Alternatives

- Increase in water conservation structures (WCS): The falling trend in OE taluks between 2017 and 2023 has been driven, in part, by interventions in check dams, percolation tanks, and watersheds. Going ahead, these can be further developed in the dry districts in the north (Kalaburagi, Raichur, Vijayapura) and bring long term benefits.
- Integration with irrigation engineering: For optimal placement of recharge structures, remote sensing (RS) and GIS tools are used to optimize the newly designed structure for recharge potential matching the aquifer characteristics.
- Urban recharge measures: Policies should require rainwater harvesting systems and stormwater recharge basins and reuse of treated wastewater to offset demand for groundwater in the cities of Bengaluru and Mysuru. Figure 7 indicates the number of artificial recharge structures and their contribution to the improvement of groundwater recharge in an approximate way. The most influential structure is found to be farm ponds as well as check dams, which highlight their role in decreasing overexploited taluks.

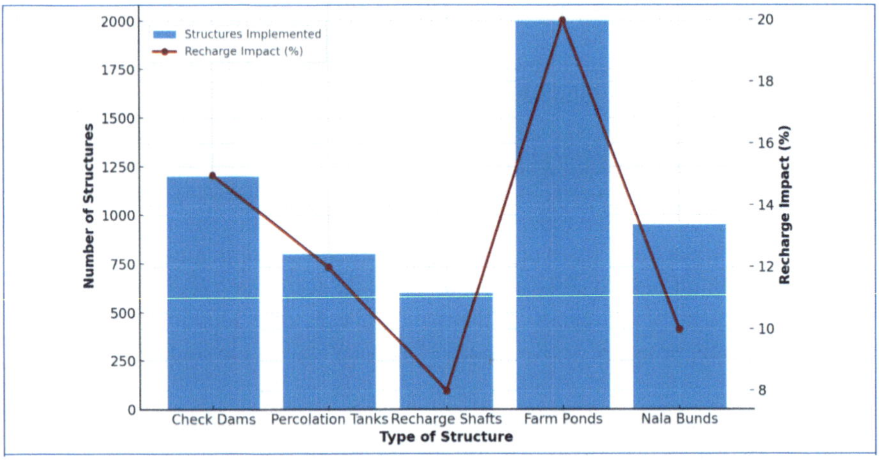

Fig. 7. Artificial Recharge Structures and Their Effect on Distribution (Karnataka, Illustrative) (Government of Karnataka, 2023).

6.2 Agricultural Demand Management: What It is and How It Works

- Incentives for diversifying crops: Continuation of OE status in districts such as Belagavi (sugarcane) and Mandya (paddy) calls for the need for policies encouraging millets, pulses and oilseeds, which are less water intensive (Ministry of Jal Shakti, 2023).
- Micro-irrigation scaling: Karnataka is behind some other states like Telangana in the drip/sprinkler irrigation coverage. Adoption can be encouraged with subsidies and capacity building can be rewarded, which will greatly reduce extraction (India Meteorological Department, 2023).

- Energy-water nexus management: Over-extraction is encouraged by the free electricity, or subsidised electricity, for pumping the bore well. Demand could be evened out by making subsidies conditional on efficient irrigation use (India Meteorological Department, 2023).

6.3 Urban and Industrial Water Governance

- Rising demand in Bengaluru needs an integrated policy framework that makes links between groundwater regulation and surface water augmentation, wastewater recycling and demand-side management (India Meteorological Department, 2023).
- Industrial water use policies must guarantee allocation efficiency. The growth of industries in the water-stressed taluks should be encouraged to use zero-liquid discharge (ZLD) and water reuse technologies (Government of Karnataka, 2023).

6.4 Legal and Regulatory Response

- Groundwater governance at taluk/block level: Karnataka's current groundwater laws have no enforcement power at local levels. A regulatory framework based on categorisation (Safe, Semi-Critical, Critical, OE) may limit the new bore well drilling in OE zones (Government of Karnataka, 2023).
- Integration with Atal Bhujal Yojana (ABHY): The GWRA-2023 clearly suggests the expansion of the ABHY coverage to cover all 88 OCS taluks. Such arrangements should be a state-wide priority and should be participatory groundwater management (Government of Karnataka, 2023).
- Adopt dual rainfall window: The use of both the five-year moving average and the single-year rainfall to base the classification would eliminate the uncertainty and prevent the misclassification and ensure a more equitable regulation (Ministry of Jal Shakti, 2023).

6.5 Socio-Economic and Community Participation

- Water budgeting at village level: Community-based water budgeting can be instituted to ensure farmers and local governance bodies plan crops according to the availability of water (Government of Karnataka, 2023).
- Capacity building for Water User Associations (WUAs): The transfer of GWRA data to locally formed user groups can close the gap between technical evaluations and practical management (Government of Karnataka, 2023).
- Behaviour changes programs: Community level awareness campaigns can be developed about the ecological effects of excessive groundwater extraction, to encourage community level stewardship (Government of Karnataka, 2023).

6.6 Mapping to the SDGs (Sustainable Development Goals)

- SDG 6 (Clean Water and Sanitation): Safe and sustainable groundwater falls directly in line with SDG targets 6.4 (increase water-use efficiency) and 6.6 (protect and restore water-related ecosystems) (Ministry of Jal Shakti, 2023).

- SDG 11 (Sustainable Cities and Communities); Bangalore's urban groundwater management is a component of resilient city planning (Ministry of Jal Shakti, 2023).
- SDG 13 (Climate Action): Adaptation strategies are supported by improved recharge and demand management to improve resilience against droughts (Ministry of Jal Shakti, 2023).

6.7 Policy Roadmap

Future research needs identified in this study directly feed into the policy effectiveness by improving the capability of predictive and integration of climate risk and governance aquifer scale. Table 3 illustrates some key actions and responsible agencies for implementation of groundwater sustainability in state of Karnataka.

Table 3. Groundwater Sustainability in Karnataka- Implementation road map (Source: Author's contribution)

Time Horizon	Key Actions	Responsible Agencies
Short-term (1–2 years)	• Extend Atal Bhujal Yojana (ABHY) to all 88 Over-Exploited and Critical Status (OCS) taluks • Implement dual rainfall assessment windows for groundwater recharge planning • Enforce urban rainwater harvesting (RWH) regulations	State Groundwater Directorate; Urban Local Bodies; Rural Development Dept.; Water Resources Dept
Medium-term (3–5 years)	• Scale up micro-irrigation adoption • Promote crop diversification toward low-water-intensive crops • Establish Groundwater Quality Monitoring Cells at district and state levels	Agriculture Department; Water Resources Department; State Pollution Control Boards
Long-term (>5 years)	• Implement aquifer-based groundwater management • Integrate groundwater planning with climate adaptation strategies and SDGs • Institutionalize data-driven groundwater governance	State Government; Planning Department; Ministry of Jal Shakti (MoJS)

6.8 Future Research Directions

GWRA-2023 of Karnataka was an enormous jump in the correct direction in the context of scientific evaluation of ground water; at the same time there are definite areas of gap in science which should be handled by conducting some research so that ground water resources may be used in a sustainable manner.

6.8.1 Fusion of Remote Sensing and GIS

Technologies have recently proven useful in combining satellite-based datasets like MODIS, Sentinel and Landsat with Google Earth Engine (GEE) to enhance monitoring land use/land cover (LULC) change, evapotranspiration and mapping irrigated areas. Combining these datasets with GWRA outputs will help to better estimate recharge and extraction drivers at taluk and aquifer scales.

6.8.2 Use of Machine Learning and Predictive Modelling

State-of-the-art machine learning (ML) and artificial intelligence (AI) models can be applied to forecast groundwater stress under a variety of climatic, agricultural and demographic scenarios. Machine Learning models including Random Forests, Support Vector Machines and Recurrent Neural Networks (RNNs) may be applied to predict the Stage of Extraction (SoE) using historical rainfall, crop water demand and ground water level time series.

6.8.3 Socio-Economic Behaviour and Governance Studies

Research - technical examinations need to be complemented by socio-economic studies involving farmer decision making, water pricing and energy subsidies, and community-level water budgeting practices. Characterization of behavioral drivers of ground water extraction is key to the design of more effective policy incentives and awareness programs.

6.8.4 Climate Change Risk Assessment for Droughts

There is an urgent need for climate-integrated groundwater models that evaluate the effects of increasing rainfall variability, drought recurrence and heatwaves on recharge and demand. Coupling groundwater models with CMIP6 and CORDEX downscaled climate projections will improve resilience planning for vulnerable taluks in north Karnataka.

6.8.5 Aquifer-Specific and Hydrogeological Studies

Most GWRA assessments are at taluk-level administrative boundaries. Aquifer scale mapping (hydrogeological characterisation, transmissive studies and quality mapping) is, however, necessary for localised interventions. This is particularly important in regions that are plagued by fluoride, nitrate and salinity pollution.

6.8.6 Urban Water-Energy Nexus Studies

Cities such as Bengaluru call for integrated research into the water-energy nexus and on wastewater reuse technologies and interventions for urban recharge (stormwater harvesting, managed aquifer recharge). Future research should test the efficacy of zero-liquid discharge (ZLD) practices in industry and the possibility of the scaling across strained urban-industrial clusters.

7 Conclusion

The most recent Groundwater Resource Assessment report 2023 (GWRA-2023) of Karnataka developed under the platform of the GEC-2015 methodology, the INGRES, is the most exhaustive assessment of groundwater state-wise so far. It presents some optimistic figures, including how the percentage of Over-Exploited and Critical taluks fell to 37.6% in 2023 following recharge interventions, watershed development, and positive rainfall (it was 44.89% in 2017). However, the fact that ground water stress has been evident in places like Belagavi, Tumakuru, Bengaluru Urban and Kalaburagi, indicates that the ground water aquifers in Karnataka are undergoing extreme pressure due to the boosting of agriculture, urbanization and climate variability. With the help of comparison analysis, it is possible to extract 2 important lessons. First, such measures of structure-check dams, percolation tanks, farm ponds-however, effective, are inadequate without the reform of policy on the demand side related to the cropping patterns, to migration to micro-irrigation, and the energy-water subsidy. Second, cities such as Bengaluru should have long-term mechanisms to prevent higher stress scenarios, such as the enrichment of surface water and wastewater recycling and stringent regulation of industry abstraction. From the perspective of governance, Karnataka's experience confirms the utility of ABHY and participatory water budgeting at the taluk and village level and the need for better legal enforcement mechanisms to manage bore well proliferation in Over-Exploited areas. A combination of two rainfall window measures (five-year moving averages and annual values) may even further enhance classification accuracy and diminish uncertainty. To meet Sustainable Development Goals (SDGs 6, 11 and 13) the future requires a multi-layered prescription: short-term scale-up of ABHY and rainwater harvesting, middle-term scale-up of crops and scale-up of micro-irrigation, and long-term institutionalization of aquifer-based governance mainstreamed into climate adaptation. Finally, Karnataka GWRA-2023 mentions that the sustainability of groundwater is not strong, but the situation is rated in the right direction. Only when there is scientific thought and engineering solutions as well as demand management together with healthy governance will the state take the step to protect its agricultural, urban and ecosystem ground water.

Acknowledgement. The authors gratefully acknowledge the Government of Karnataka (2023), Groundwater Resource Assessment (GWRA-2023), Karnataka: Final Report, Central Ground Water Board & State Groundwater Directorate, for providing the foundational dataset and insights that formed the basis of this study.

References

Alley, W.M., Konikow, L.F.: Bringing GRACE down to earth. Groundwater **53**(6), 826–829 (2015). https://doi.org/10.1111/gwat.12379

Asoka, A., Gleeson, T., Wada, Y., Mishra, V.: Relative contribution of precipitation and pumping to changes in groundwater storage in India. Nat. Geosci. **10**, 109–118 (2017). https://doi.org/10.1038/ngeo2869

Asoka, A., Wada, Y., Fishman, R., Mishra, V.: Strong linkage between precipitation intensity and monsoon season groundwater recharge in India. Geophys. Res. Lett. **45**, 5536–5544 (2018). https://doi.org/10.1029/2018GL078466

Bhanja, S.N., Mukherjee, A., Saha, D., Velicogna, I., Famiglietti, J.S.: Validation of GRACE-based groundwater storage anomaly using in-situ groundwater level measurements in India. J. Hydrol. **543**, 729–738 (2016). https://doi.org/10.1016/j.jhydrol.2016.10.042

Central Ground Water Board. Aquifer systems of India. Ministry of Water Resources (2012)

Chen, J.L., Wilson, C.R., Tapley, B.D., Yang, Z.L., Niu, G.Y.: 2005 drought event in the Amazon River basin as measured by GRACE and estimated by climate models. J. Geophys. Res.: Solid Earth **114**(B5), B05404 (2009). https://doi.org/10.1029/2008JB006056

Chen, J., Li, J., Zhang, Z., Ni, S.: Long-term groundwater variations in northwest India from satellite gravity measurements. Global Planet. Change **116**, 130–138 (2014). https://doi.org/10.1016/j.gloplacha.2014.02.007

Fishman, R., Devineni, N., Raman, S.: Can improved agricultural water use efficiency save India's groundwater? Environ. Res. Lett. **10**(8), 084022 (2015). https://doi.org/10.1088/1748-9326/10/8/084022

Food and Agriculture Organization. Crop water requirements and irrigation practices in India (FAO Water Reports) (2022)

Gleeson, T., et al.: Towards sustainable groundwater use: Setting long-term goals, backcasting, and managing adaptively. Groundwater **50**(1), 19–26 (2012). https://doi.org/10.1111/j.1745-6584.2011.00825.x

Government of Karnataka. Groundwater resource assessment (GWRA-2023): Karnataka–final report. Department of Water Resources, Directorate of Groundwater Development (2023). https://antharjala.karnataka.gov.in

Goldin, T.: Groundwater: India's drought below ground. Nat. Geosci. **9**(2), 98 (2016). https://doi.org/10.1038/ngeo2648

Hashimoto, T., Stedinger, J.R., Loucks, D.P.: Reliability, resiliency, and vulnerability criteria for water resource system performance evaluation. Water Resour. Res. **18**(1), 14–20 (1982). https://doi.org/10.1029/WR018i001p00014

Humphrey, V., Gudmundsson, L., Seneviratne, S.I.: Assessing global water storage variability from GRACE: trends, seasonal cycle, subseasonal anomalies and extremes. Surv. Geophys. **37**, 357–395 (2016)

India Meteorological Department. Annual rainfall statistics and groundwater recharge studies in India. Ministry of Earth Sciences (2023)

India Meteorological Department. Annual rainfall statistics and groundwater recharge studies in India: Annual report 2023–2024. Ministry of Earth Sciences (2024)

Indian Council of Agricultural Research. Irrigation and groundwater use in Indian agriculture (2023)

Jasechko, S., et al.: The pronounced seasonality of global groundwater recharge. Water Resour. Res. **50**, 8845–8867 (2014). https://doi.org/10.1002/2014WR015809

Joodaki, G., Wahr, J., Swenson, S.: Estimating the human contribution to groundwater depletion in the Middle East from GRACE data, land surface models, and well observations. Water Resour. Res. **50**, 2679–2692 (2014). https://doi.org/10.1002/2013WR014633

Judd, C.M., McClelland, G.H., Ryan, C.S.: Data analysis: a model comparison approach. Routledge (2008). https://doi.org/10.4324/9781315744131

Koster, R.D., Suarez, M.J., Ducharne, A., Stieglitz, M., Kumar, P.: A catchment-based approach to modeling land surface processes in a general circulation model: Model structure. J. Geophys. Res.: Atmos. **105**(D20), 24809–24822 (2000). https://doi.org/10.1029/2000JD900327

Leblanc, M.J., Tregoning, P., Ramillien, G., Tweed, S.O., Fakes, A.: Basin-scale integrated observations of the early 21st century multiyear drought in southeast Australia. Water Resour. Res. **45**, W04408 (2009). https://doi.org/10.1029/2008WR007333

Loucks, D.P.: Quantifying trends in system sustainability. Hydrol. Sci. J. **42**(4), 513–530 (1997). https://doi.org/10.1080/02626669709492051

Mays, L.W.: Groundwater resources sustainability: past, present, and future. Water Resour. Manage **27**, 4409–4424 (2013). https://doi.org/10.1007/s11269-013-0436-7

Ministry of Jal Shakti: Annual report on artificial recharge and groundwater management in India. Government of India (2023). https://www.jalshakti-dowr.gov.in

Mishra, V., Aadhar, S., Asoka, A., Pai, S., Kumar, R.: On the frequency of the 2015 monsoon season drought in the Indo-Gangetic Plain. Geophys.l Res. Lett. **43** (2016)

Sustainable Recovery of Silver from Industrial Wastewater and Reutilization of Recovered Silver for Dye Treatment

Sahil Kohli[1](✉), Indrani[1], Disha Arora[1], and Garima Rathee[2,3]

[1] School of Engineering and Technology, Manav Rachna International Institute Of Research And Studies, Faridabad, Haryana 121004, India
{sahil.set,disha.set}@mriu.edu.in

[2] Department of Chemical Engineering, Universitat Politècnica de Catalunya, 08222 Terrassa, Spain

[3] Nanomaterials and Polymers Lab, Dr Bansidhar Institute, Gurgaon, India

Abstract. Owing to the environmental and toxicity impact, silver ions have to be selectively removed from industrial waste. Various methods for removing silver ions and the subsequent use of recovered silver, based composites for dye removal are discussed in this chapter. Adsorption and photocatalytic degradation were among the methods used to recover silver ions from synthetic wastewater samples to the maximum possible extent. Also, various characterization methods were employed to inspect the structural integrity and surface features of the fabricated silver-based composites. The composites' capacity and efficiency to eliminate dyes were gauged by using typical industrial dyes, and they showed a considerable uptake capacity and efficiency.

The silver-based composites that were recovered not only allowed for excellent dye removal, but they also indicated the possibility of being reused in subsequent treatment cycles, thus increasing sustainability. This research highlights the double advantage of silver recovery and its use in wastewater treatment, which is a step towards the creation of environmentally friendly solutions for industrial effluent management.

Keywords: Wastewater · Silver · Recycling · Dyes · Adsorption · Photodegradation

1 Introduction

Water released as a byproduct of extensive industrial processes is termed industrial wastewater. Such industrial processes may include manufacturing, mining, energy production, etc. Industrial wastewater is more hazardous and difficult to dispose of than domestic wastewater, as it contains a significant range of pollutants such as toxic chemicals, suspended solids, and heavy metals. As the number of industries proliferates and their activities expand, the generation of industrial wastewater has increased notably. As industries emerge in modern times, they contribute a whole set of diverse and new types of waste. These waste types include organic and inorganic chemicals, microplastics, dyes, and even pharmaceuticals [1].

L. Choudhary et al. (Eds.): TIMES 2025, SCI, pp. 223–238, 2027.
https://doi.org/10.1007/978-3-032-34636-0_16

Silver (Ag) is a metal with atomic number 47 called "argentum" (Latin name). It is a lustrous, malleable, excellent conductor of heat and electricity, and ductile metal, known for its brilliant white sheen, making it a precious metal. Civilization has valued this metal for decades due to its industrial, economic, and aesthetic significance. It is a naturally occurring metal in earth's crust in its pure elemental form as well as present in various mineral ores such as galena (PbS), argentite (AgS), and horn silver [1, 2]. Silver mining contributes to the economic development and trade routes with regions like the Andes mountains in South America and central Europe, which have bloomed significantly [3]. Accumulation of silver in the aquatic environment may pose a serious threat to the ecosystem and biodiversity [4]. Hence, tackling this problem with advanced solutions through which we can not only remove silver ions from wastewater but also repurpose them effectively is a necessity.

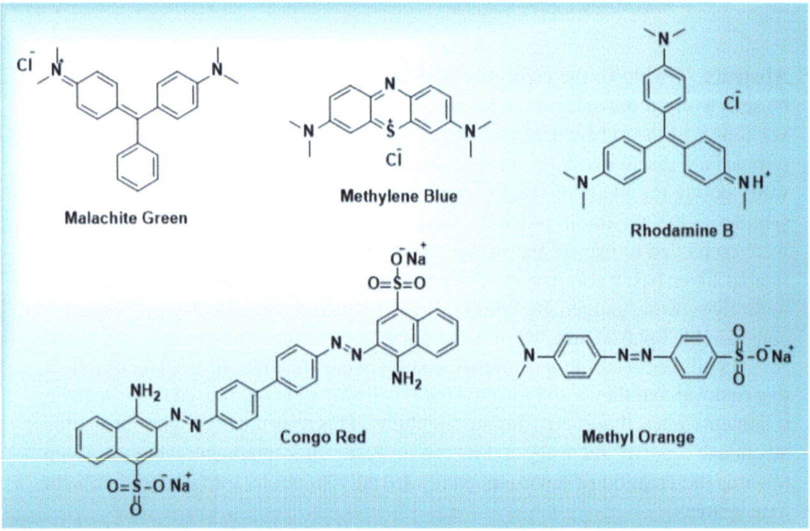

Fig. 1. Chemical structures of some organic dyes (Source: Author's contribution)

The use of Ag nanoparticles (NPs) to eliminate colors from wastewater has increased over the last few years due to their excellent adsorption properties, huge surface area, quicker equilibrium rates, and lower diffusion resistance [5]. Furthermore, the approach of using silver-based composites, particularly those containing Ag NPs or silver compounds inside a supporting matrix, has gained much more popularity than just unsupported Ag NPs in the past few years. These composites have unique surface properties due to which there is a strong interaction with Ag ions resulting into increased adsorption capacity. The ability to recover and reuse these silver-based composites makes them highly sustainable.

and low cost, especially for industrial applications. These composites are employed for the removal of numerous organic dyes from wastewater.

Dyes are pigments used in pharmaceutical, textile, cosmetics, food, paper, paint and ink industries [6]. Despite having various applications, the adverse effects in water systems are due to dye origin and structure. Figure 1 shows the chemical structures of some of the crucial synthetic organic dyes. The dyes have several structural forms like reactive, acidic, basic, azo, disperse, anthraquinone based dyes. Most of these dyes come from chemicals that are cancer causing, like naphthalene and benzidine. Due to this, these dyes can always turn into carcinogens when they come into contact with living things like people and animals [7].

Wastewater treatment is primarily carried out using three main approaches: physical, chemical, and biological methods (Fig. 2). The use of Filtration membrane, adsorption, and coagulation technology are few examples of a physical method. On the other hand, chemical methods include reactions like oxidation, ion-exchange, and photocatalysis. In addition, biological techniques involve the use of biosorption, biodegradation and enzymatic processes [8]. Bleaching, chlorination, and ozonation are just a few of the processes that are part of the chemical process known as oxidation.

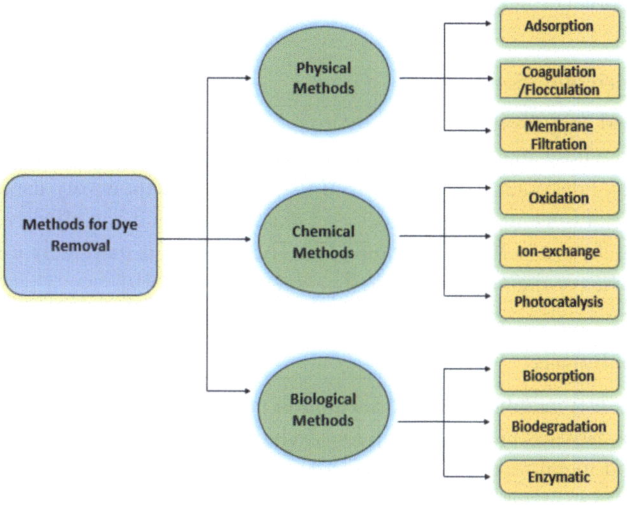

Fig. 2. Methods for removal of Dyes (Source: Author's contribution)

2 Methods for Selective Removal of Silver Ions

There are various methods for removing silver (Ag) from water or aqueous solutions. Each method has advantages and disadvantages, which vary based on several aspects (Table 1). The following are the common methods which are used for removing silver.

2.1 Chemical Precipitation Method

Chemical precipitation involves the formation of insoluble silver salts by adding chemicals to the wastewater, after which the precipitate can be removed by filtration. The method is employed to decrease the silver ions concentration in contaminated wastewater. This is carried out by addition of a precipitant that combines with the silver ions, which causes them to precipitate. This method has gained substantial attention due to its low cost, high effectiveness, simple operation, and extensive series of applications [9, 10]. This method includes various chemical precipitations of types such as NaCl, $PbCl_2$, Na_2S, NaOH and NH_3 precipitation [11–15].

2.2 Electrochemical Method

The Handling of wastewater comprising silver via electrochemical method is a versatile procedure. Inside of a Focused electrochemical reactor, the electric field directs the electrons, resulting to definite chemical and physical reactions that remove silver from the effluent. The physical methods mainly comprise of integrated adsorption and flocculation–separation process, whereas the chemical method comprises the electrolysis of silver. Also, the electrochemical remediation technology of silver containing wastewater can be classified into various kinds on basis of their principles and approaches, like electrocoagulation (EC), electrodeposition (ED), and electrocoagulation.

2.2.1 Electrocoagulation

EC is a method that produces metal hydroxides and coagulants *via* electrochemical reactions [16]. These materials disturb particles by triggering them to precipitate or clump together, hence adsorbing dissolved toxins. EC is extensively used process for eliminating metal ions, mainly in trace-level wastewater, as it can effectively eradicate metals in range between 100 to 0.1 ppm [17]. This method is extremely efficient in the elimination of heavy metals such as Cr, Zn, Ni and Cu [18, 19].

2.2.2 Electrodialysis

Electrodialysis is a membrane-based separation process driven by an applied potential that selectively isolates target ions via ion-exchange membranes, enabling wastewater purification and metal recovery. The effectiveness of ED is clearly affected by the ion exchange membrane which must have high permselectivity, low ion water permeability and back diffusion, high chemical stability, small film resistance, and good mechanical strength. ED technology involves numerous approaches like high temperature ED, packed bed ED, electrodeless water ED, bipolar membrane ED, and liquid membrane ED [20].

2.2.3 Electrodeposition

Electrodeposition is a frequently used method for removing Ag from wastewater. During this method, Ag ions existing in the wastewater are transformed to solid Ag and then by applying a voltage across the electrodes, these solid Ag are deposited on the cathode, successfully eliminating the Ag ions. Electrochemical deposition is a promising method for Ag recovery from aqueous solutions, providing high efficacy, simple operation, and the nonexistence of chemical additives [21, 22].

2.3 Ion Exchange Method

The ion exchange method involves ion transfer between solid and liquid phases through two key steps: adsorption (metal loading) and elution. Target ions are first exchanged with mobile ions and subsequently released to regenerate the exchanger [23, 24]. Depending on the membrane type, the process is classified as anion or cation exchange [25–29]. For silver-containing wastewater, ion exchange is particularly effective for treating trace-level effluents.

2.4 Adsorption Method

The adsorption process is sustainable, viable and Simple to use. It is often employed for the elimination of Ag and other heavy metals from wastewater of trace-concentration [30]. This technique uses porous solid adsorbents to absorb Ag^+ in wastewater onto their surface and subsequently discharge them by various processes, resulting in the separation of Ag^+. The efficiency of the adsorbent is measured by its chemical and physical properties like pore size, specific surface area, surface functional groups and pore volume [31]. On basis of different mechanism of adsorption, adsorption is divided into three kinds: chemical, physical, and biological. The most Frequently used adsorbents include functional biowaste materials, activated carbon, cellulose materials, graphene, inorganic adsorbents, and biochar. The mechanism of adsorption is classified into the following six chief aspects: ion exchange, surface complexation, co-precipitation, physical adsorption, electrostatics attraction, π-π e^- donor interaction (Fig. 3).

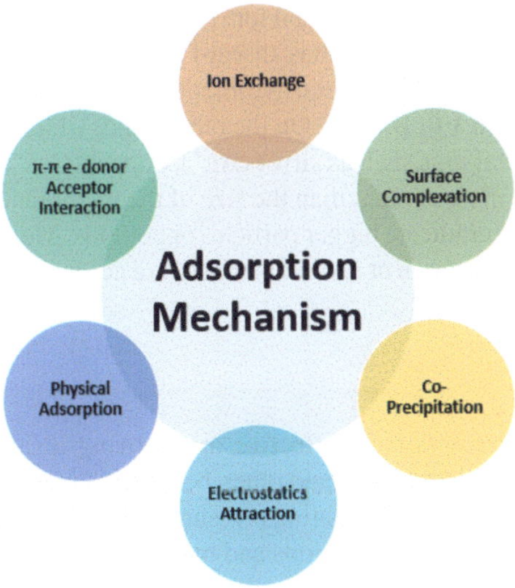

Fig. 3. Different mechanism of adsorption (Source: Author's contribution)

2.5 Membrane Filtration

Membrane filtration is an efficient way of removing Ag ions from aqueous solutions. It uses semi-permeable membranes to selectively enable specific ions or molecules to pass through while inhibiting others. The efficacy of this approach depends on the membrane's characteristics, such as material, pore size, and charge. It is mainly suitable for the elimination and retrieval of trace-concentration heavy metal ions [32]. The efficacy of elimination relies on the carriers in the membrane phase and membrane material. The purity of Ag considerably influences its utility and value in industrial sector. Consequently, the membrane recycling process facilitates cost-effective recovery of high-purity silver. Membrane methods are commonly classified on the basis of membrane's surface pore size, including microfiltration (MF), nanofiltration (NF), ultrafiltration (UF), and reverse osmosis (RO)

2.5.1 Microfiltration

Microfiltration (MF) is a method that uses membranes with pore sizes usually ranging from 0.1 to 10 mm. MF is commonly not effective for eliminating dissolved metal ions from water due to its relatively large pore size and can only filter out particles such as suspended solids, microorganisms, and colloidal particles. Although MF is capable of removing aggregates of metal hydroxide or metal oxide particles, it is not an appropriate technique for efficiently removing metal ions. As a result, MF is frequently employed as a preliminary stage in water purification operations, while more sophisticated filtration techniques are required for the direct removal of metal ions.

2.5.2 Ultrafiltration

Ultrafiltration (UF) is a filtration technique that employs pressure to eliminate bigger metal particles and colloidal forms of metal ions. Yet, it is not as effective at eliminating dissolved metal ions. UF membranes may discard particulate particles, germs, viruses, and bigger metal complexes, such as metal hydroxides or metal-organic complexes. Their pore diameters range from 1 to 100 nm. On the other hand, UF is not usually successful in getting rid of free metal ions such as silver (Ag^+), copper (Cu^{2+}), or cadmium (Cd^{2+}) since the size of these ions is smaller than the size of the membrane's pores. UF is often used as a pretreatment to eradicate bigger particles or colloids. More specialized filtering techniques, such as nanofiltration or reverse osmosis, are necessary for the direct removal of dissolved metal ions.

2.5.3 Nanofiltration

Nanofiltration (NF) is exceptionally effective in removing divalent and larger metal ions, such as lead (Pb^{2+}), silver (Ag^+), and other, because its membrane pore size ranges between 1 and 10 nm. NF works by creating pressure to the solution, which allows water to pass through while preventing metal ions and bigger organic molecules from passing through. It is particularly effective for water softening applications, where it is necessary to remove calcium and magnesium ions. It can also remove dangerous heavy metals such as copper and zinc. However, NF is not as successful in removing monovalent ions, such as salt or chloride, as it is at removing divalent ions.

2.5.4 Reverse Osmosis

Reverse osmosis (RO) is one of the finest approaches to get free from dissolved metal ions, including both monovalent and divalent ions, in water. RO membranes are capable of rejecting an extensive variety of metal ions, including silver (Ag^+), lead (Pb^{2+}), mercury (Hg^{2+}), and copper (Cu^{2+}), due to their pore diameters being less than 1 nm. This makes them perfect for purifying water that includes harmful metals. During the RO process, pressure is applied to the water. This results the pure water to pass through a semi-permeable membrane while impurities, such as metal ions, are kept behind. RO is frequently used in industries like as desalination, wastewater treatment, and water purification because of its high removal efficiency. Reverse osmosis (RO) is quite good at getting rid of metal ions, but it does use a lot of energy and can become fouled, which means it needs to be maintained often to work at its best.

Table 1. Advantages and Disadvantages of various methods used for the removal of silver ions (Source: Author's contribution)

Treatment Method	Advantages	Disadvantages
Precipitation	Simple and inexpensive Effective for high concentration of silver	Generates large volumes of sludge ineffective for low concentrations of silver
Electrochemical	Highly efficient for removing silver ions The electrode can regenerate, resulting in reduced waste	Needs a continuous power source Frequent maintenance is required
Ion Exchange	Highly selective in removing silver ions Eliminate silver ions to very low concentrations Operates efficiently with a low energy input	Expensive resin materials requires periodic regeneration of resins Produces waste brine which requires proper disposal
Adsorption	Simple and easy to perform High surface area of adsorbents allows for effective removal	require periodic regeneration of adsorbent materials Not efficient for high concentrations of silver ions
Membrane Filtration	Efficiently remove both silver ions and NPs Requires less chemical input low maintenance simple operation	High initial setup costs Membrane fouling can reduce efficiency Can be energy-intensive, especially in RO or NF

3 Characterization of Silver-Based Composites

Every nanoparticle has its own identity in terms of shape, size, and morphology. To understand them properly, characterization is very important. Similarly, characterization of silver-based composites can be done by many analytical techniques as discussed.

3.1 Scanning Electron Microscopy (SEM) and Transmission Electron Microscopy (TEM)

These are the two main methods of analysis for the characterization of compounds, which creates unique images when electron beams are passed through the samples determining their shape, size and morphology. Study of the image produced by the sample using TEM uses the electrons that pass through it, meanwhile the image in SEM analysis is made by the electrons which were reflected by the sample and is of high resolution4.

3.2 Scanning Transmission Electron Microscopy (STEM)

This technique is somewhat similar to TEM & SEM; it also involves the electron beam to pass through the sample which results in the electrons to lose some of the energy this energy is then measured by STEM. Substances which are used in the semiconductors, cells etc. are generally characterized through this technique5.

3.3 Infrared (IR) Spectroscopy

This method involves the specific frequency of infrared light to be absorbed by the sample; the absorbed light is analysed to determine the molecular bonds of the sample. Each and every substance has vibration energy when this energy corresponds to IR frequency, they absorb the light. This method is very important and useful as it is used for samples having any state such as liquid, gaseous or solid6.

3.4 X-Ray Diffraction Analysis (XRD)

It is an analytical technique which is used to study the sample through its crystalline form. Generally, the x- ray technique studies the intensity which passes through the samples and is detected by the films.

4 Application of Recovered Silver Composites in Dye Removal

The removal of dyes from industrial effluent has become a serious concern due to synthetic dyes' negative impact on both the environment and human health. Silver-based composites have emerged as viable dye removal materials due to their distinct features, which include a large surface area, antibacterial activity, and simplicity of functionalization. This section looks at the different applications of silver-based composites for removing dyes from contaminated water using method of adsorption and photo-degradation. Various recovered silver-based composites for dye removal by photocatalytic and adsorption mechanisms have been reported (Table 2).

4.1 Photocatalytic Degradation of Dyes

Silver-based composites have gained substantial interest for their ability to degrade dyes in aqueous solutions owing to their exclusive properties, including their high surface area, the ability to absorb light, and their potential to generate reactive oxygen species (ROS) that are highly effective at breaking down dye molecules. The recovered composites show high efficiency of photodegradation for several successive cycles. For instance, the hydrothermal approach was used to synthesize flower-like NiZnAl-LDH, which was later combined with Ag/AgCl to create a composite photocatalytic material for the photodegradation of methylene blue (MB) dyes. The composite material has excellent recyclability which decreases from 98.94% to 80.1% after successive five cycles [33].

Ahn et al. fabricated CdS/Ag composite based on cadmium sulphide, enclosed by Ag NPs *via* hydrothermal method for the outstanding photocatalytic efficiency for Acid Black 1 (AB 1), Reactive Red 120 (RR 120), and Direct Blue 15 (DB 15) dyes under direct solar and ultraviolet light. Without a scavenger, the degradation rate was over 93% at 75 min and remained stable at over 90% after four cycles [34].

Desouky *et al.* made Ag metal-organic framework (Ag-MOF) to remove malachite green (MG) from wastewater. The attained regeneration efficacy (%) was decreased to 82.7% from 98.6% after five successive sequences of adsorption and desorption. The slight decrease in recovery was because of blockage of some Ag-MOF adsorption sites [35]. Furthermore, Li *et al.* immobilized Ag-MOF on graphene oxide (GO) sheets using an ultrasonication method. The inclusion of GO improves the surface area of Ag-MOF@GO and lowers the accumulation effect, leading to an enhanced adsorption capability. The material displayed regenerability greater than 90% after three cycles, signifying stable and high regeneration behaviour [36].

Le et al. fabricated a ternary Ag_2MoO_4/MIL-101(Fe)/Ag composite for the effective degradation of crystal violet (CV) under visible light. Ag NPs were generated **in situ** on Ag_2MoO_4/MIL-101(Fe) via the reduction of Ag_2MoO_4 with $NaBH_4$. The material showed excellent reusability, maintaining over 92% degradation efficiency for CV even after five consecutive cycles [37].

Guo et al. prepared a Cu_2O/Ag-coated wood-based biochar (MBC@Cu_2O/Ag) composite using a liquid-phase deposition method for the removal of CR and MO. After three cycles, only a slight reduction in degradation efficiency was observed for both CR and MO, yet the removal remained above 85%, which was attributed to the strong adsorption properties of the composite [38].

Lai et al. employed a hydrothermal technique combined with ultrasonication to prepare a g-C_3N_4/Nb_2O_5/Ag ternary nanocomposite for the degradation of MB under visible-light irradiation. The nanocomposite demonstrated high stability after five cycles, showing only a 2% loss in degradation rate [39].

Dinari et al. synthesized LDH@Fe_3O_4-Ag composites via the co-precipitation method for the removal of tetracycline (TC). The results showed that after three consecutive cycles, the removal efficiency decreased by about 30%. This reduction was attributed to oxidation of Ag NPs, catalyst loss during recovery, and irreversible TC adsorption on the LDH@Fe_3O_4-5Ag surface, which collectively reduced the availability of active catalytic sites and hindered •OH radical generation, ultimately lowering the removal efficiency [40].

Boonyuen et al. successfully fabricated Carbon aerogel-supported CA/ZnS-Ag composites via a hydrothermal technique for the degradation of MB was under UV–Vis light irradiation. The recovered catalyst had excellent stability and shows a high rate for the removal of dye even after five consecutive cycles [41].

4.2 Adsorption of Dyes

The adsorption of dyes onto recovered silver-based composites is an active topic of research in materials science, notably for environmental applications. Silver based composites frequently have exceptional qualities that make them excellent for removing dyes from wastewater, which is an important issue given the environmental impact of dye pollution. For instance, The GO was decorated with Ag NPs to create the Ag-rGO composite and the results showed that after the recovery of the catalyst, 94% of NB was removed in the 1st cycle, 91% in the 2nd cycle, and 82% in the 3rd cycle. The experimental reduction in the dye removal of Ag-rGO in continuous cycles could be ascribed to a little loss in the adsorbent amount and minor changes to the active sites on the material surface [42].

Abdelghaffar et al. developed AgNPs/BPC/CS, a nanocomposite bio-adsorbent derived from chitosan (CS) and AgNPs, synthesized using cellulosic banana peel bio-waste (BPC) to enhance the biosorption of Reactive Orange 5 (RO5). Reusability studies showed that the bio-sorbent could be reused up to four times with a dye removal efficiency exceeding 95%. The enhanced removal of anionic dyes was attributed to the increased density of surface-positive groups in the AgNPs/BPC/CS bio-sorbent [43].

Owda et al. fabricated a modified magnetite–chitosan composite incorporating Ag NPs (chitosan/Fe_3O_4/AgNPs). The material exhibited high adsorption capacity for MB and MO. After four adsorption–desorption cycles, no significant decline in MB regeneration was observed: the first cycle achieved 95% regeneration, and the fourth cycle showed no substantial decrease. For MO, the first regeneration cycle achieved 92% desorption, but efficiency gradually decreased in subsequent cycles [44].

Mahmud et al. developed a simple method to synthesize an $Ag_{0.2}Co_{0.8}Fe_2O_4/CeO_2$/CA/PVP (ACFCeP) composite for the removal of MG dye from wastewater. The dye removal efficiency decreased from 93.3% to 85% after the third consecutive cycle, primarily due to partial desorption of dye molecules during regeneration. Additionally, pore blockage by retained dye molecules contributed to the progressive decline in performance [45].

Mostafa et al. modified cellulose to carboxymethylcellulose (CMC) and incorporated Ag NPs and agar to fabricate an agar–carboxymethylcellulose–silver nanoparticle (AG-CMC-AgNP) composite for MB adsorption. The material demonstrated promising recoverability and reusability, maintaining a capacity of 53.96 mg g^{-1} up to the 10th cycle [46].

Aji et al. fabricated Fe_3O_4/CS/AgNPs using natural iron sand and chitosan/silver nanoparticles for MB adsorption. The catalyst was successfully recovered over five cycles, retaining an MB removal efficiency of 70.4% [47].

Table 2. Recovered silver composites for Dye removal (Source: Author's contribution)

S. No.	Recovered composite	Dye	Mechanism	Remarks	Ref.
1	Ag/AgCl/NiZnAl-LDH	MB	Photodegradation	The degradation of dye after recyclability decreases from 98.94% to 80.1% after successive five cycles	34
2	CdS/Ag	RR 120	Photodegradation	The degradation rate remained stable at over 90% after four cycles	35
3	CdS/Ag	AB 1	Photodegradation	The degradation rate remained stable at over 90% after four cycles	35
4	CdS/Ag	DB 15	Photodegradation	The degradation rate remained stable at over 90% after four cycles	35
5	Ag-MOF	MG	Photodegradation	The regeneration decreased to 82.7% from 98.6% after five continuous cycles	36
6	Ag-MOF@GO	MG	Photodegradation	The reusability was more than 90% after three successive cycles	37
7	Ag_2MoO_4/MIL-101(Fe)/Ag	CV	Photodegradation	Reusability upto 92% after five continuous cycles	38
8	MBC@Cu_2O/Ag	CR	Photodegradation	The degradation rate remained more than 85% after three cycles	39
9	MBC@Cu_2O/Ag	MO	Photodegradation	The degradation rate remained more than 85% after three cycles	40

(continued)

Table 2. (*continued*)

S. No.	Recovered composite	Dye	Mechanism	Remarks	Ref.
10	g-C_3N_4/Nb_2O_5/Ag	MB	Photodegradation	only 2% loss in the rate of degradation after five cycles	40
11	LDH@Fe_3O_4-Ag	TC	Photodegradation	decrease in removal of about 30% after three cycles	41
12	CA/ZnS-Ag	MB	Photodegradation	high rate for the removal even after five cycles	42
13	Ag-rGO	NB	Adsorption	82% of dye was removed even after three cycles	43
14	AgNPs/BPC/CS	RO5	Adsorption	reused with removal efficacy of more than 95% after four cycles	44
15	chitosan/Fe_3O_4/$AgNO_3$	MB	Adsorption	no substantial drop even after four cycles	45
16	chitosan/Fe_3O_4/$AgNO_3$	MO	Adsorption	no substantial drop even after four cycles	45
17	ACFCeP	MG	Adsorption	removal percentage decreases from 93.3% to 85% after three consecutive cycles	46
18	AG-CMC-AgNP	MB	Adsorption	Reusable upto 10th cycle (53.96 mg g^{-1})	47
19	Fe_3O_4/CS/AgNPs	MB	Adsorption	Removal rate remained 70.4% after five cycles	48

5 Mechanism for Removal of Dye

A dye can be removed either by adsorption or photocatalytic degradation. In Adsorption, there are two ways that molecules can be adsorbed on the surface of an adsorbent. The method that the molecules are adsorbed depends on the interaction between the solid surface (Ag-composites) and the adsorbed molecules. The two types of adsorptions are chemical and physical sorption. Physisorption is a reversible process that involves van der Waals forces and electrostatic interactions. Figure 4 shows a potential mechanism where three types of interactions that take place on the surface of the adsorbent during the adsorption mechanism: (1) electrostatic interactions, (2) hydrogen bonding, and (3) π–π stacking interactions. All of these interactions work together to increase the adsorption of dye molecules. Chemisorption, on the other hand, is characterized by strong covalent bonds as the main interaction between the adsorbent and the adsorbate. This results in diffusion from the surface into the interior of the material, which usually leads to the formation of a monolayer.

The photocatalytic degradation of dyes using Ag composites is a method exhibits a lot of interest. On light irradiation, these composites generate reactive species like superoxide anions and hydroxyl radicals which are crucial for breaking and oxidizing dyes. The photocatalytic method is made more efficient by the inclusion of support into the silver metal, which enhances light absorption and charge separation.

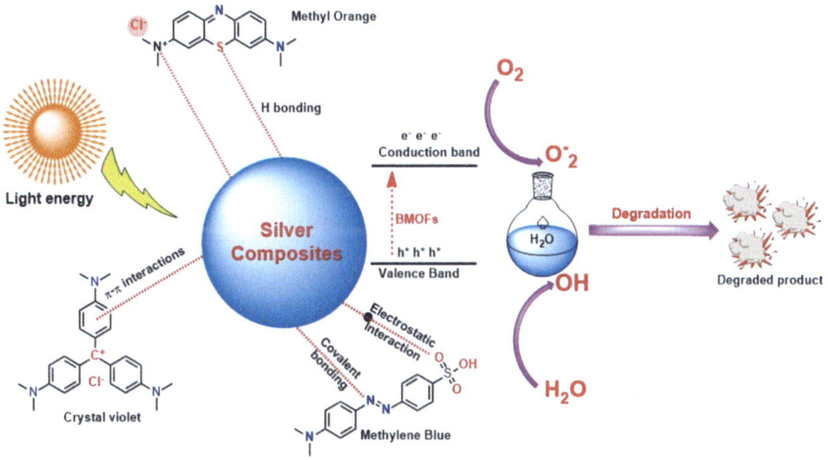

Fig. 4. Photocatalytic degradation of dyes using silver composites (Source: Author's contribution)

6 Conclusions and Perspectives

In conclusion, this chapter highlighted the issue of silver ion contamination in wastewater and the removal of dyes from industrial wastes. The selective removal of silver ions reduces the environmental toxicity caused due to silver contamination, and also allows the simple recovery of silver from wastewater, which can be further reused in numerous industrial applications. The research also focuses on the development of silver-based composites for their efficient recovery and application in dye removal. Silver composites eliminate dyes using method of photocatalytic degradation and adsorption. The use of these composites not only provides a practical solution to the problem of wastewater contamination, but they also constitute an important resource for potential applications in the field of environmental remediation in the future. Considering that these materials are reusable and have a high efficacy in dye adsorption, even while operating under different conditions, it is possible that they might be incorporated into the wastewater treatment systems that are already in place currently. One further thing that helps contribute to a more circular economy is the capability of recovering and reusing silver ions once the removal procedure has been completed.

In the future, there is a large potential for expanding the use of composites based on silver in a variety of industries, including the textile and pharmaceutical industries, where the recovery of silver and the removal of dye are both vital. In subsequent study, it may be possible to investigate the possibility of optimizing the synthesis of these composites in order to achieve higher efficiency, improved selectivity, and cost-effectiveness. Additionally, it may be possible to investigate the long-term stability and environmental impact of their utilization in major wastewater treatment systems.

References

1. Nishiyama, T., Kusakabe, Y.: Silver content in some common ore minerals. Mining Geology **36**(200), 425–437 (1986)
2. Liu, X., et al.: Recent advances in synthesis and properties of silver nanoclusters. Nanoscale **16**(26), 12329–12344 (2024)
3. Von Der Heydt-Coca, M.: Andean silver and the rise of the Western world. Crit. Sociol. **31**(4), 481–513 (2005)
4. Ratte, H.T.: Bioaccumulation and toxicity of silver compounds: A review. Environm. Toxicol. Chem. **18**(1) (1999)
5. Ahmad, A., et al.: Recent advances in new generation dye removal technologies: Novel search for approaches to reprocess wastewater. RSC Adv. **5**(39), 30801–30818 (2015)
6. Chequer, F.M.D., Ferraz, E.R.A., Cardoso, J.C., Zanoni, M.V.B., de Oliveira, D.P.: Textile dyes: Dyeing process and environmental impact. In: Eco-friendly textile dyeing and finishing, Vol. 26 (2012)
7. Weber, E.J., Lee, W.N.: Kinetic studies of the reduction of aromatic azo compounds in anaerobic sediment/water systems. Environ. Toxicol. Chem. **6**(12), 911–919 (1987)
8. Aksu, Z.: Application of biosorption for the removal of organic pollutants: A review. Process Biochem. **40**(3–4), 997–1026 (2005)
9. Chen, Q.Y., Yao, Y., Li, X.Y., Lu, J., Zhou, J., Huang, Z.L.: Comparison of heavy metal removals from aqueous solutions by chemical precipitation and characteristics of precipitates. J. Water Proc. Eng. **26**, 289–300 (2018)
10. Quiton, K.G.N., Huang, Y.H., Lu, M.C.: Recovery of cobalt and copper from electroplating wastewater via carbonate and hydroxide precipitation. Sustainable Environment Research, 32 (2022)
11. Wang, J.M., Huang, C.P., Pirestani, D.: Interactions of silver with wastewater constituents. Water Res. **37**, 4444–4452 (2003)
12. Petronijevic, N., et al.: Application of flotation tailings as an alternative material for acid mine drainage remediation. Metals 10 (2020)
13. Luo, M.S., et al.: A comprehensive hydrometallurgical recycling approach for environmental impact mitigation of EoL solar cells. J. Environ. Chem. Eng. 9 (2021)
14. Liu, A.M., et al.: Preparation of AgCl photocatalyst by recovering silver from discarded electrodeposition bath. Ionics **25**, 2419–2426 (2019)
15. Li, D., et al.: Simultaneous removal and conversion of silver ions into antibacterial material. Arabian J. Chem. 16 (2023)

16. Kim, T., Kim, T.K., Zoh, K.D.: Removal mechanism of heavy metals in presence of cyanide during electrocoagulation. J. Water Proc. Eng. 33 (2020)
17. Xu, L.Q., et al.: Simultaneous removal of cadmium, zinc and manganese using electrocoagulation. J. Environ. Manage. **204**, 394–403 (2017)
18. Dura, A., Breslin, C.B.: Electrocoagulation using aluminium anodes activated with alloying elements. J. Hazard. Mater. **366**, 39–45 (2019)
19. García-Carrillo, C., Parga-Torres, J., Moreno-Casillas, H., Sellschopp-Sanchez, F.S.: Kinetics and energy consumption for a three-stage electrocoagulation process for Au and Ag recovery. Metals 9 (2019)
20. Benalla, S., et al.: Comparison of reverse osmosis and electrodialysis in brassware effluent treatment. Desalin. Water Treat. **316**, 205–213 (2023)
21. Gu, J.-N., et al.: Treatment of real deplating wastewater through precipitation–electrodeposition–oxidation. Separation and Purification Technology 248 (2020)
22. Xiao, L., et al.: Selective recovery of silver from copper anode slime. J. Clean. Prod. **187**, 708–716 (2018)
23. Lee, J.C., Kurniawan, H.J., Hong, K.W., Chung, S.K.: Separation of Pt, Pd and Rh using ion exchange resin. Separation and Purification Technology 246 (2020)
24. Marinho, R.S., da Silva, C.N., Afonso, J.C., da Cunha, J.W.S.D.: Recovery of platinum, tin and indium using anion exchange resins. J. Hazard. Mater. **192**, 1155–1160 (2011)
25. Gomes, C.P., Almeida, M.F., Loureiro, J.M.: Gold recovery with ion exchange resins. Sep. Purif. Technol. **24**, 35–57 (2001)
26. Gubler, L., et al.: Proton exchange membranes prepared by radiation grafting. Solid State Ionics **176**, 2849–2860 (2005)
27. Shahi, V.K.: Highly charged proton-exchange membrane composites. Solid State Ionics **177**, 3395–3404 (2007)
28. Bensalah, J., et al.: Removal of Cr(III), Cu(II), and Ag(I) using cationic resin A®IRC-50. Inorganic Chemistry Communications 157 (2023)
29. Gámez, S., et al.: Precious metals recovery from waste PCBs using thiosulfate leaching and ion exchange. Hydrometallurgy, 186 (2019)
30. Mohan, D., et al.: Organic and inorganic contaminants removal with biochar. Biores. Technol. **160**, 191–202 (2014)
31. Carmona, D., et al.: Surface area and adsorption capacity increase in mesoporous silica. Chem. Commun. **47**, 1233712339 (2011)
32. Wu, Z.W., et al.: Polyamide/UiO-66-NH_2 nanocomposite membranes for Mo(VI) removal. Desalination, 563 (2023)
33. Li, X., Guo, Y., Feng, X.: Z-scheme flower-like Ag/AgCl/NiZnAl-LDH for methylene blue removal. Materials Science in Semiconductor Processing 185 (2025)
34. Ravikumar, S., et al.: CdS/Ag nanocomposite for azo dye removal. Ceram. Int. **49**(6), 9551–9559 (2023)
35. Aljohani, M.M., et al.: Mesoporos Ag-MOFs for dye adsorption. Process Saf. Environ. Prot. **172**, 395–407 (2023)
36. Deng, X., Li, J., Zhao, B., Li, Z.: Ag-MOF@GO composite for malachite green removal. New J. Chem. **47**(34), 16022–16029 (2023)
37. Nguyen, H.T., et al.: Ag_2MoO_4/MIL-101(Fe)/Ag composite for crystal violet degradation. Chem. Eng. Sci. 284 (2024)
38. Zhang, Y., et al.: Cu_2O/Ag-coated biochar for azo dye degradation. Applied Surface Science 647 (2024)
39. Ramar, S., et al.: g-C_3N_4/Nb_2O_5/Ag ternary nanocomposite for pharmaceutical and dye degradation. J. Molecular Struct. 1317 (2024)
40. Afshari, M., Dinari, M., Arameshian, S.: LDH@Fe_3O_4-Ag photocatalyst for tetracycline degradation. Journal of Molecular Liquids 396 (2024)

41. Shanmugam, P., et al.: ZnS-Ag composites based on carbon aerogel. Environmental Geochemistry and Health **46**(3) (2024)
42. Natasha, et al.: Silver-decorated reduced graphene oxide for Nile Blue removal. ACS Omega **9**(17), 19461–19480 (2024)
43. Abdelghaffar, F.: Biosorption of anionic dye using chitosan–Ag nanocomposite. Environ. Technol. Innov. 24 (2021)
44. Abdelaziz, M.A., et al.: Removal of dyes using chitosan/magnetite/silver nanoparticles. Int. J. Biol. Macromol. **225**, 1462–1475 (2023)
45. Tabassum, N., et al.: Ag–Co ferrite magnetic polymer composite for dye remediation. RSC Adv. **14**(49), 36557–36575 (2024)
46. Mostafa, S.I., et al.: Adsorption behaviour of methylene blue on agar-CMC-Ag nanocomposite. Biomass Conversion and Biorefinery **14**(3), 3363–3383 (2024)
47. Ningtyas, R.D., et al.: Fe_3O_4/CS/AgNPs composite for enhanced methylene blue adsorption. J. Cluster Sci. **35**(5), 1463–1480 (2024)

Section 5: Digital Technologies, IoT, and Smart Urban Connectivity

A Comprehensive review on IoT in Urban Infrastructure: Technological Architectures and Sectoral Applications

Yaman Hooda[2](✉), Sannat Sareen[1], Ekomjot Singh Panesar[1], Venumadhav[1], and Shaurya Singh Dhariwal[2]

[1] Department of Electronics and Communication Engineering, Manav Rachna International Institute of Research and Studies, Sector – 43, Aravalli Hills, Delhi – Surajkund Road, Faridabad, Haryana 121004, India

[2] Department of Civil Engineering, Manav Rachna International Institute of Research and Studies, Sector – 43, Aravalli Hills, Delhi – Surajkund Road, Faridabad, Haryana 121004, India

yamanhooda@gmail.com

Abstract. With the accelerating pace of urbanization, there is an urgent need to develop smart infrastructural services that can support sustainable and efficient urban growth. In this context, the Internet of Things (IoT) has emerged as a pivotal technological enabler, playing an integrated role in addressing the complex challenges associated with the development of intelligent infrastructure. IoT facilitates collection of real – time & analysis of data, and informed decision-making, thereby contributing to resource optimization, enhanced operational efficiency, and improved quality of life across urban systems. This work provides a inclusive review of the existing role of IoT in the evolution of smart infrastructure. It examines the architectural frameworks that underpin IoT systems and explores their practical applications across multiple domains, including intelligent buildings, transportation, energy systems, and healthcare. Additionally, the study identifies and analyses key evaluative parameters necessary for assessing the effectiveness of smart infrastructural services. These parameters include governance and policy frameworks, energy efficiency, environmental impact, data management, and societal well-being.

Keywords: Internet of Things (IoT) · Technological Advancements · Smart Infrastructure · Smart Cities · Evaluation Matrices

1 Introduction

Internet of Things, commonly populated as IoT, is a M2M communication; consisting of staggered pattern of "things" with appropriate connection [1]. The range of such connected devices widely spread out into smart home appliances to industrial apparatuses, emerging technologies in domain of vehicles and smart infrastructure as a whole [2]. IoT devices are specifically employed with sensor technology that helps in collection of the data, regarding the environmental factors such as humidity and temperature; and

L. Choudhary et al. (Eds.): TIMES 2025, SCI, pp. 241–253, 2027.
https://doi.org/10.1007/978-3-032-34636-0_17

operational factors such as motion detection, usage of power and efficiency of the whole system [3]. The transmission of this data collected to other devices is possible through the internet services. The prior functions of the receiving devices are to analyze, decision – making and appropriate actions based on the type of the data collected and intended solutions. Emerging services and applications of IoT are transforming the conventional way of living. The major areas of applications of IoT in smart infrastructure falls under the domain of Intelligent Transportation System (ITS), Smart Cities (SC), Intelligent Buildings (IB), Smart Wastewater Management System and Smart Grid Energy Efficiency Systems. By the end of the current decade, it is projected that the technological advancement of IoT will integrate more than 30 billion sensor devices; three times the original value as reported in 2020 [4]. As per the information from IDC [International Data Corporation] and IoT Analytics, the overall spending in the domain of IoT is increasing with the rate of more than 45 percent and, CAGR [Compound Annual Growth Rate] of 19.5 percent within the duration of 2022 – 27 [5, 6]. With an increase in the population, it can be projected that the demand of smart infrastructure and their different zones will be increasingly accordingly, as depicted in Fig. 1.

Smart infrastructure can be defined as the urban areas that has the facilities of technological advancements for improving the quality of life of the society with efficient economic and sustainable development. The various spheres of the smart infrastructure are inter – connected with technical services for collection and analyzation of the data in real – time; making informed decisions and optimizing the resources effectively. Besides IoT, smart infrastructure depends on various technologies mechanisms such as Big Data, Machine Learning (ML), Blockchain Technology (BT) & Digital Twin Technologies (DT) [7]. Since IoT enables information driven decision – making, monitoring of real – time data monitoring and competent resource management system; it can be closely linked with Smart Infrastructure. IoT – driven smart sensors & services system enhances energy usage, reduction in waste, optimizing infrastructural management. The employment of IoT services in smart infrastructural systems aids sustainable development by promoting smarter urban development with constant performance adaptive and tracking systems.

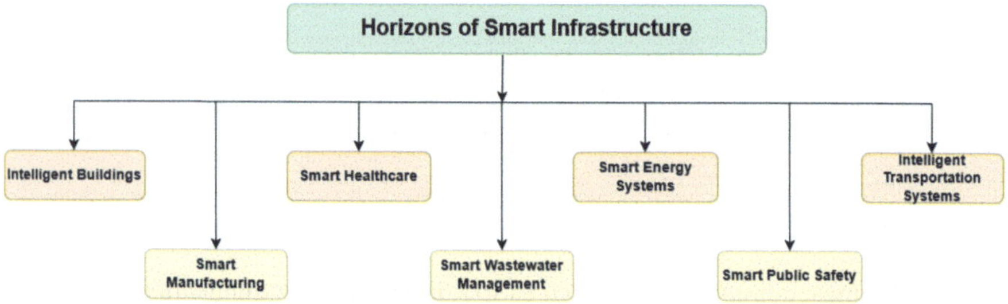

Fig. 1. Horizons of Smart Infrastructure (Source: Author's contribution)

2 IoT Framework

The Internet of Things (IoT) is a technology that allows devices, items to connect, communicate with one another via the internet [8, 9]. IoT devices are equipped with sensors that enable them to gather information about a wide range of environmental and physical factors. The information can then be sent to other devices or systems, where it can be analysed and used to guide decisions. Temperature sensors, humidity sensors, motion sensors, and pressure sensors are just a few examples of the many different types of IoT sensors [10]. They can be used to gather information on various environmental factors, including temperature, humidity, air quality, and noise levels. Additionally, they may be used to monitor the movement of people and goods as well as infrastructure like buildings, roads, and bridges. Advancements have opened up new opportunities for analysis and monitoring by allowing for the real-time collection and processing of huge volumes of data. Real-time monitoring and response are two of the most important advantages of IoT technology [11, 12]. IoT sensors can notify emergency personnel of possible crises before they escalate by gathering information on the state of the environment and infrastructure. For instance, an IoT sensor on a bridge can recognize the early signs of deterioration and notify maintenance staff, allowing them to intervene before the bridge becomes hazardous [13]. The three layers of the generic Internet of Things (IoT) architecture depicted in Fig. 2 are as follows:

- Devices like actuators, sensors, & other kinds of devices that can gather information from or act on the physical world are part of the device layer. Sensors collects information, whereas actuators can be utilized to regulate machines or systems based on the data obtained by sensors.
- Network layer consists of a system; connecting the cloud with the devices and a gateway which facilitates interaction between the devices and the cloud. The data gathered by the devices is processed and stored in the cloud, which acts as a focal point.
- The application layer comprises the software programs used to interpret the information gathered by the devices and make decisions based on it. This layer progressions the information and presents the same un useful manner.

In this architecture, collection of the data begins from the user –based environment system, where sensors gather information about environmental and physical conditions. The devices then receive this data, process it, and transmit it to the gateway. The data is then sent to the cloud by the gateway, where the application layer processes and stores it. After evaluating the data, the application layer sends reports to the user, who may then act on the information received. Based on the information gathered by the devices, the application layer may also provide user feedback.

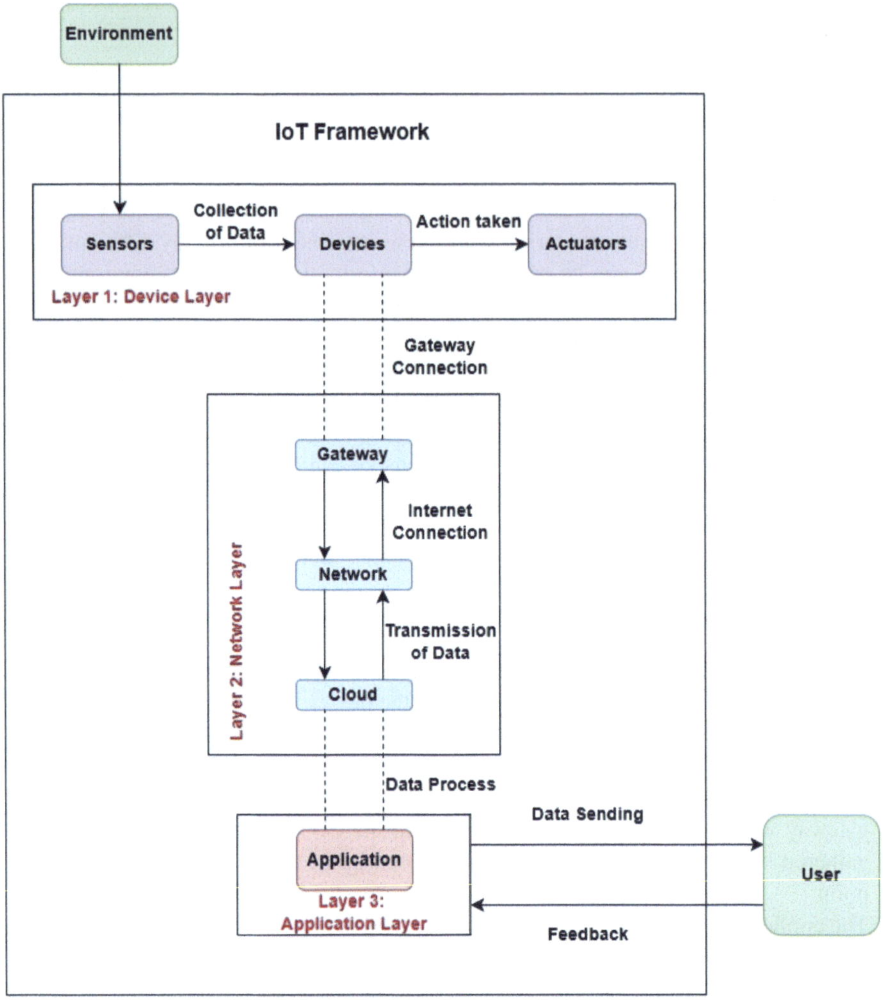

Fig. 2. Generalised 3 – layered IoT Framework (Source: Author's contribution)

3 Implications of Internet of Things (IoT) in Smart Infrastructure

The concept of Internet of Things (IoT) has transformed the domain of smart infrastructure with the characteristics of monitoring of real – time data, data monitoring, data simulation and decision – making mechanisms. Integration of IoT in smart cities with sensor services and devices across various infrastructural networks aids in buildings, energy systems, transportation services and water management with enhanced efficiency. The emerging applications of IoT in various domains of Smart Infrastructure are discussed below.

3.1 Intelligent Transportation Systems (ITS)

With the advantages of real – time data collection, monitoring, transmission and decision making between traffic management infrastructural centres and vehicles, IoT services are plating a vital role in the domain of Intelligent Transportation Systems (as highlighted in Fig. 3). IoT services collects the data regarding flow of traffic, speed of vehicles, conditions of roads and other related environmental factors that helps in optimisation of traffic signals, reduction in congestion and enhancing road safety [1]. The communication between the connected vehicles can be done with V2V technology and V2I technology; enabling autonomous driving conditions and improvising decision – making procedures. Moreover, the services of IoT also aids in smart parking facilities, emergency response coordination and predictive maintenance towards a safe, modern and sustainable transportation system [14].

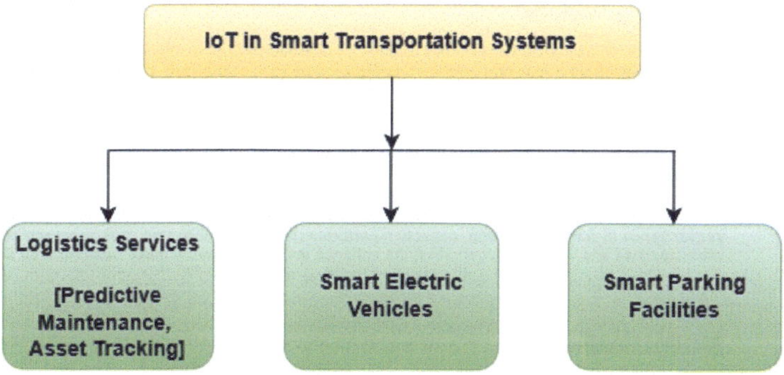

Fig. 3. IoT in Smart Transportation Systems (Source: Author's contribution)

3.2 Smart Energy Systems

The services of IoT are prove to be a revolutionized solution in optimizing, development, consumption and distribution of energy. The IoT services in energy sector employed sensors systems, machine learning and data analytics for collection and analyzation of the data regarding usage of different energy patterns and enabling the automation and controlling of energy systems [15, 16]. Present day electric grid employes IoT technical services in optimization of energy consumption, distribution, and management. The employment of IoT services as a smart grid solutions are helping in reduction in consumption of energy, thus increasing efficiency and minimizing environmental impacts [16, 17]. IoT sensors are being employed under "smart lighting" for adjusting the different levels of lighting based on the daylight and occupancy levels; helps in reduction of consumption of energy and saving economics with still facilitating adequate lighting services. The usage of energy – storage solutions (i.e. capacitors and batteries) can be optimized; helps in reduction of energy waste and moreover, providing an added consistent energy source in the areas with unreliable supply of energy. The implementation of smart energy solutions necessitates the injunction of IoT devices and services [1, 18].

3.3 Smart Waste Management

The employment of services and devices of IoT in the waste management systems enhances the efficiency, economic – effective and ecological friendly. The technological advancement of IoT can be employed in waste management services in monitoring, collection, analyzation and managing the waste. Thus, IoT based Smart Waste Management System offers a holistic mechanism for waste handling and reduction in operational. The various domains of smart waste management systems are summarized in Fig. 4 [19–21].

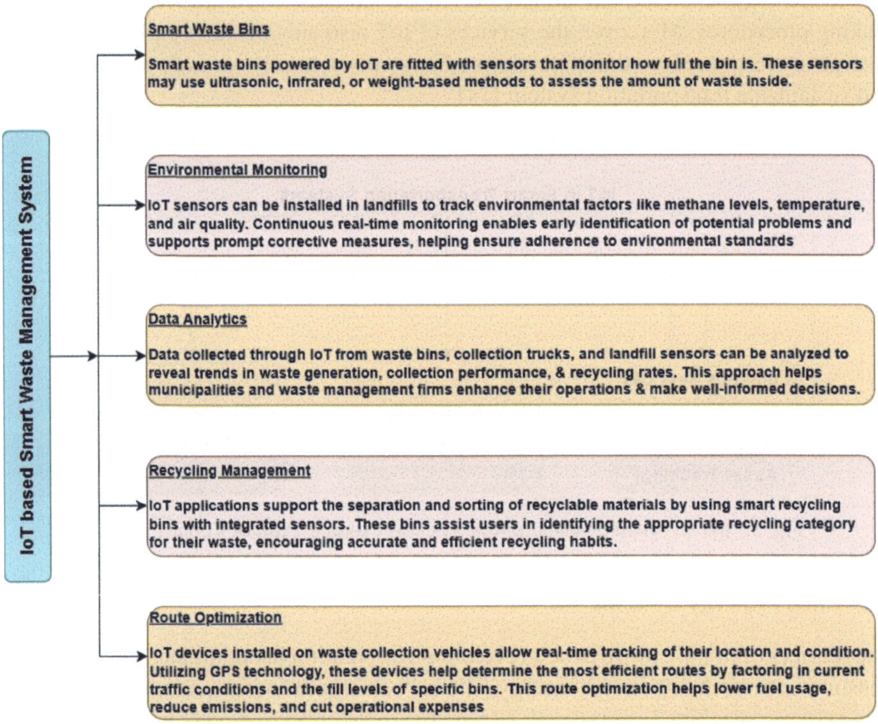

Fig. 4. Applications of IoT in Smart Waste Management System (Source: Author's contribution)

3.4 Infrastructural Monitoring and Controlling

Infrastructural monitoring and controlling are a vital aspect for smart infrastructure. With effective monitoring and controlling of different features of smart infrastructure such as buildings, bridges, transportation services, pipelines and other critical services, the detection of possible safety threats can be assessed and necessary actions can be taken for prevention of accidents and minimizing the damage [22]. The governmental organizations and institutions must maintain & monitor the mentioned elements of infrastructural services. IoT services and devices are employed to monitor efficiently the public infrastructural services [23]. The employment of IoT services contributes towards the reduction in inspection costs and diurnal maintenance by efficiently utilization of various services. Figure 5 summarized the various aspects of infrastructural monitoring and controlling.

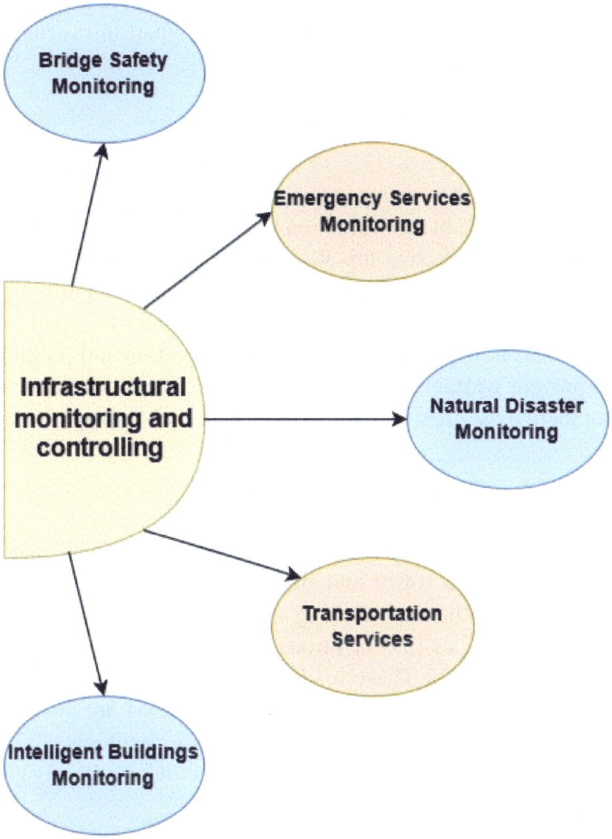

Fig. 5. Aspects of Infrastructural Monitoring and Controlling (Source: Author's contribution)

4 Smart Infrastructural Evalutaion Matrices

Smart cities performance must be measured on the basis of their ability to take up smart technologies in order to promote quality of life, sustainability, and operational efficiency by mid 2020's [24, 25], as depicted in Fig. 6.

4.1 Infrastructure

Digital infrastructure availability and scope e.g., broadband networks, sensors, data centers, and communications systems are critical for the delivery of smart city services [24, 26]. The leading cities have deployed hugely in top-of-the-line infrastructure, such as smart transport, solar energy, waste and water treatment plants, healthcare, schooling, and security initiatives. While it is hard to rank cities in one category, Singapore, Barcelona, Amsterdam, Copenhagen, and Tokyo are at the forefront in building smart cities [27–30]. NEOM and Egypt's New Administrative Capital (NAC) hold great potential but are in development stages. All eight cities are driving 2025 smart infrastructure standards, defining sustainability, efficiency, and livability [31, 32].

4.2 Governance and Policy

Governance and Policy success within a smart city depends on tailored governance architecture and policy that enable coordination between governments, the private sector, and citizens [24, 25]. A no-one-size-fits-all approach has been adapted by every city for addressing unique challenges and opportunities [33]. Sustainability is a continuous focus, with local governments applying technology to minimize environmental harm. This also involves smart transport systems to stem congestion and pollution and smart water systems to prevent wastage. Governance structures in these cities are robust and vision-driven, but there exist gaps in data sharing and cybersecurity [34–36].

4.3 Energy Efficiency

Cities are optimizing energy efficiency using smart grids, buildings, and management systems in a bid to minimize waste and maximize renewable energy consumption [24, 25]. The 2025 energy efficiency ranking according to consumption rates, renewable energy output, and policies [37] positions Singapore at the top [27], followed by Barcelona and Amsterdam [29]. These leaders have taken measures to cut down on energy use and increase renewables [37]. Dubai and NEOM are also progressing on the backdrop of renewable initiatives and energy-saving technologies [37], and Egypt's NAC is planned to score 70 by 2030 [38]. Overall, smart cities are driving global progress in green city energy behavior.

4.4 Mobility and Transportation

Mobility and transportation infrastructure are evaluated on the grounds of smart traffic systems, public transport, and new technology like electric or self-driving cars. The solutions aim at reducing congestion, promoting sustainable transit, and increasing mobility. Cities are making dramatic progress in smart mobility, making investments in autonomous cars, shared mobility, and green transport to increase livability and sustainability [24, 25].

4.5 Environment and Sustainability

It examines what cities have done to maintain environments and promote sustainability. These measures can include programs for waste disposal, air quality improvement and monitoring, water conservation, expansion or establishment of parks and green space, and renewable energy sources. Furthermore, it can include cities' overall greenhouse gas emissions and their actions to address climate change [24, 25]. Though certain cities like Singapore, Copenhagen, and Amsterdam shine as clear leaders. They scored points in each category and are well-positioned to maintain and enhance forward movement in the future [19, 39].

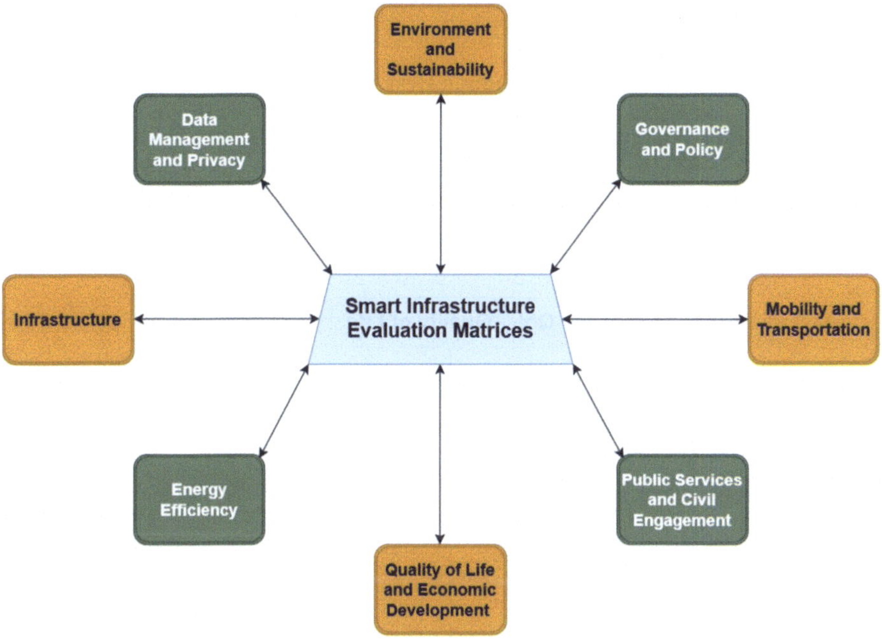

Fig. 6. Smart Infrastructure Evaluation Matrices (Source: Author's contribution)

4.6 Public Services and Civic Engagement

Cities have made progress in this class [31], since they use technology to enhance the well-being of citizens and encourage public engagement in local governance. This is a clear way for other cities to demonstrate to the citizens how technology is being used by local government and facilitate better liveable and sustainable cities.

4.7 Data Management and Privacy

Smart cities are managing the collection, storage and analysis of data from digital technologies. Cities can be grouped together based on documented data management practices and privacy protections of citizens. There are three different levels of data management: 1. High level: Cities have stated rules and processes for collection, storage and use of data. Data management are strong, as well as data protection from unauthorized access or use with excellent security systems; 2. Medium level: despite having some rules and security, cities are not following the rules strictly as high-level cities and 3. Low level: cities in this group have few, or no proper systems for managing data. They have limited or no security. Furthermore, the component of privacy also depends on three stages as: 1. Prominent Stage: Cities which protect privacy provide strong laws against the use of personal data. There are transparent data practices, and the cities allow citizens to control how their data is used; 2. Medium Stage: Cities have privacy laws, but they are not as strong, seen or enforced as such. Citizens may have less control of their data; and 3. Low Stage: Cities do not have clear privacy laws and may use data collecting methods that are unclear. Usually, citizens do not have control of their personal information.

4.8 QoL – Quality of Life and Economic Development

QoL evaluates how smart city projects have improved residents' quality of life, as summarized in Fig. 7. It reflects on areas such as access to health care services, access to educational opportunity, safety, access to basic services, and access to cultural facilities [33, 36, 37, 40].

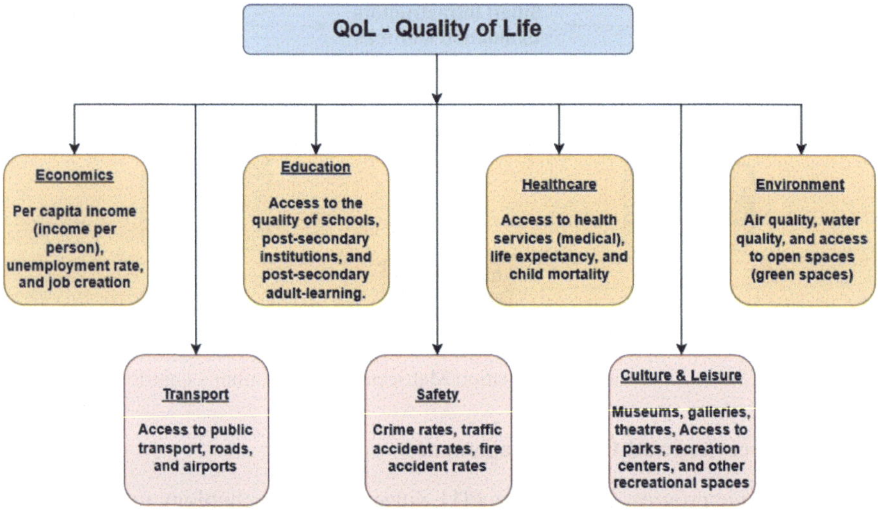

Fig. 7. Parameters of QoL – Quality of Life (Source: Author's contribution)

Components of smart city initiatives can assist economic development by enhancing innovation, providing better employment opportunities, and enhancing ways of attracting investment. It is also important to pay close attention to whether the cities are attracting investors, advancing local entrepreneurs, helping small businesses flourish, and proactively fostering suitable conditions for advancing technology-based businesses [24, 25]. Singapore [27] has the highest gross domestic product (GDP) per capita across the listed cities, while NEOM has no growth at all, and NAC is in decline. Economic strength cannot be attributed solely to GDP growth, since other factors such as inflation, joblessness, and quality of life also affect economic performance [41]. Several of the smart cities referenced in this document are experiencing positive economic performance. These cities are well positioned to become economic regions of the future by improving their infrastructure and boosting attractiveness for investors [41]. It is also important to check if cities have efficient means to evaluate the impact of their smart city initiatives [24]. Continuous and data-based evaluation approaches enable to track the pace and improvements of processes dependent on smart city evolving categories [25]. Given that every city has different conditions, priorities, and functional objectives, the evaluation of performance is context dependent [42].

5 Conclusion

The study presented an extensive review study on the implications of technological advancements, especiaaly, Internt of Things (IoT) towards different domains of Smart Infrastructure. İt can be concluded that with effective holistc mechanism between emerging technical aspects, societal engagements and governance; cities can be become smart and sutainable with livable ecological systems. After a crucial examination, it was observed that the application of IoT in different domains of infrastructural systems aids in making services improved and efficient. The employment of IoT based connection systems and services revolutionised urban infrastructural landscape; covering energy management systems, transportation systems and others domains that quantify the living standards of the society. The study enlightened the various parameters that contributes towards examination of assessment of vaious parameters; leading to transformation from "urban cities" to "smart cities".

Acknowledgments. No financial constraint has been considered in the study.

Conflict of Interest. The authors have no conflicts of interest to declare.

References

1. Hassebo, A.: Commercial 4G LTE cellular networks for supporting emerging mission-critical IoT applications (Doctoral dissertation). ProQuest Dissertations Publishing, New York, NY, United States (2018)
2. Tealab, M., Hassebo, A., Dabour, A., AbdelAziz, M.: Smart cities digital transformation and 5G—ICT architecture. In: Proceedings of the 11th IEEE Annual Ubiquitous Computing, Electronics & Mobile Communication Conference (UEMCON) (pp. —). New York, NY, United States (2020)
3. Hassebo, A., Rezk, A., Ali, M.A.: A hybrid uplink scheduling approach for supporting mission-critical smart grid applications in commercial 4G cellular networks. In: Proceedings of the 9th IEEE Annual Ubiquitous Computing, Electronics & Mobile Communication Conference (UEMCON). New York, NY, United States (2018)
4. Statista: Number of Internet of Things (IoT) connected devices worldwide from 2019 to 2023, with forecasts from 2022 to 2030 (2023). Retrieved 15 May 2023, from https://www.statista.com/statistics/1183457/iot-connected-devices-worldwide/
5. IoT Analytics: Global IoT market size (2023). Retrieved 15 May 2023, from https://iot-analytics.com/iot-market-size/
6. United Nations: World urbanization prospects 2018. Department of Economic and Social Affairs, Population Division (2018)

7. Hooda, Y., Singh, H.D.: Digital reforms in public services and infrastructure development & management. In: Technological prospects and social applications of Society 5.0, pp. 219–237. Chapman & Hall/CRC (2023)
8. Borgia, E.: The Internet of Things vision: Key features, applications and open issues. Comput. Commun. **54**, 1–31 (2014)
9. Sethi, P., Sarangi, S.R.: Internet of Things: Architectures, protocols, and applications. J. Electr. Comp. Eng. **2017**, 9324035 (2017)
10. Chaudhry, R., Tapaswi, S., Kumar, N.: A green multicast routing algorithm for smart sensor networks in disaster management. IEEE Trans. Green Comm. Netw. **3**(1), 215–226 (2019)
11. Lazarou, I., Kesidis, A.L., Hloupis, G., Tsatsaris, A.: Panic detection using machine learning and real-time biometric and spatiotemporal data. ISPRS Int. J. Geo Inf. **11**(10), 552 (2022)
12. Santos, R.M., Orozco, J., Ochoa, S.F., Meseguer, R., Mosse, D.: Providing real-time message delivery on opportunistic networks. IEEE Access **6**, 40696–40712 (2018)
13. Finochietto, J.M., et al.: An IoT-based infrastructure to enhance self-evacuations in natural hazardous events. Pers. Ubiquit. Comput. **26**, 1461–1478 (2022)
14. Statista: Passenger cars—Worldwide (2022). Retrieved 17 May 2023, from https://www.statista.com/outlook/mmo/passenger-cars/worldwide
15. Hassebo, A.: xIoT-based converged 5G and ICT infrastructure. In: Electric grid modernization. IntechOpen (2021)
16. Hassebo, A., Mohamed, A.A., Dorsinville, R., Ali, M.A.: 5G-based converged electric power grid and ICT infrastructure. In: Proceedings of the IEEE 5G World Forum (5GWF). Silicon Valley, CA, United States (2018)
17. Stanelyte, D., Radziukyniene, N., Radziukynas, V.: Overview of demand-response services: A review. Energies **15**(5), 1659 (2022)
18. Gellings, C.W.: The smart grid: Enabling energy efficiency and demand response. CRC Press (2020)
19. Alonaizi, S.Y., Manuel, P.: IoT-based smart government enablers: An exploration of governments' experiments. In: Proceedings of the Fifth World Conference on Smart Trends in Systems Security and Sustainability (WorldS4). London, United Kingdom (2021)
20. Cheema, S.M., Hannan, A., Pires, I.M.: Smart waste management and classification systems using cutting-edge approaches. Sustainability **14**, 10226 (2022)
21. Hooda, Y., Kuhar, P., Sharma, K., Verma, N.K.: Emerging applications of artificial intelligence in structural engineering and construction industry. J. Phys: Conf. Ser. **1950**(1), 012062 (2021). https://doi.org/10.1088/1742-6596/1950/1/012062
22. García, L., Garcia-Sanchez, A., Asorey-Cacheda, R., Garcia-Haro, J., Zúñiga-Cañón, C.: Smart air quality monitoring IoT-based infrastructure for industrial environments. Sensors **22**(1), 23 (2022)
23. Wang, J.M., Yang, M.T.: Design a smart control strategy to implement an intelligent energy safety and management system. Int. J. Distrib. Sens. Netw. **10**, 312392 (2014)
24. Toh, C.K.: Smart city indexes, criteria, indicators and rankings: An in-depth investigation and analysis. IET Smart Cities **4**(4), 211–228 (2022)
25. Fang, Y., Shan, Z.: How to promote a smart city effectively? An evaluation model and efficiency analysis of smart cities in China. Sustainability **14**, 6512 (2022)
26. Hooda, Y.: IoT and remote sensing. In: Computer vision and Internet of Things, pp. 111–140. CRC Press (2022)
27. Valeria, F.: Singapore: Smart city to smart nation. Center for Digital Society (2020)
28. Miao, J.T., Phelps, N.: The evolution of smart city: Case studies of Barcelona, Spain and Helsinki, Finland. Melbourne School of Design (2019)
29. Somayya, M., Ramaswamy, R.: Amsterdam smart city (ASC): Fishing village to sustainable city. WIT Press, In The Sustainable City XI (2016)

30. The SENSEable City Lab: Copenhagen sensible city guide. Massachusetts Institute of Technology (2011)
31. Allam, Z.: The emergence of anti-privacy and control at the nexus between the concepts of safe city and smart city. Smart Cities **2**, 96–105 (2019)
32. Van Geest, M., Tekinerdogan, B., Catal, C.: Smart warehouses: Rationale, challenges and solution directions. Appl. Sci. **12**, 219 (2021)
33. Hooda, Y., Singh, H.D.: Digital Reforms in Public Services and Infrastructure Development & Management. In: Technological Prospects and Social Applications of Society 5.0, pp. 219–237. Chapman and Hall/CRC (2023)
34. Mancebo, F.: Smart city strategies: time to involve people—comparing Amsterdam, Barcelona and Paris. J. Urbanism: Int. Res. Placemaking and Urban Sustainab. **13**(2), 133–152 (2020)
35. Mitieka, D., Luke, R., Twinomurinzi, H., Mageto, J.: Smart mobility in urban areas: A bibliometric review and research agenda. Sustainability **15**, 6754 (2023)
36. Łukasiewicz, K., Pietrzak, P., Kraciuk, J., Kacperska, E., Cieciora, M.: Sustainable energy development—A systematic literature review. Energies **15**, 21 (2022)
37. Ali, M.A.: Smart city policy in developing countries: Case study of the new administrative capital in Egypt. J. Public Aff. **22**, e2774 (2021)
38. Bjørner, T.: The advantages of and barriers to being smart in a smart city: Perceptions of project managers within a smart city cluster project in Greater Copenhagen. Cities **114**, 103187 (2021)
39. Deeksha, Shukla, A.: Ecosystem services: a systematic literature review and future dimension in freshwater ecosystems. Applied Sciences **12**, 17 (2022)
40. Hodson, E., et al.: Evaluating social impact of smart city technologies and services: Methods, challenges, and future directions. Multimodal Technol. Interact. **7**, 3 (2023)
41. Hou, J., Arpan, L., Wu, Y., Feiock, R., Ozguven, E., Arghandeh, R.: The road toward smart cities: A study of citizens' acceptance of mobile applications for city services. Energies **13**, 2496 (2020)

IoT-Enabled Real-Time Landfill Gas Monitoring: A Systematic Review and Implementation Framework

Vaishali Sahu[1](✉), Akanksha Kaushik[2], Raghav Vats[2], Rishi Manocha[2], Vaibhav Sharma[2], Ishika Avasthi[2], and Gaurang Dhingra[2]

[1] Department of Multidisciplinary Engineering, The NorthCap University, Gurugram, Haryana, India
vaishalisahu@ncuindia.edu

[2] Department of Computer Science and Engineering, The NorthCap University, Gurugram, Haryana, India

Abstract. This review focuses on a critical assessment of the role of IoT technologies in landfill gas monitoring, specifically looking at aspects of technological feasibility, issues of ethics, and the relationship to the Sustainable Development Goals (SDGs). Methane emissions from landfill sites contribute significantly to environmental and public health concerns, and thus the need to effectively monitor this process is urgent. Typical processes (like manual sampling) are time consuming, and lack real-time data drives an inability for effective action. Recent IoT technologies offer a new way of doing continuous and automated monitoring of landfill gases, when currently used landfilling tools are entered. In assessing whether IoT-based solutions may provide improvements to manage methane emissions, this review integrates current literature and practical field deployments and practice to look at specific potential applications, especially on the backdrop of administrative and operational challenges to waste management in India. The layers linked specifically explored factors such as requiring core system components, implementation challenges, infrastructure scalability, and practical application of IoT in a global waste crisis for landfill gas monitoring. Ultimately, IoT systems - that include real-time sense making, cloud analytics, and automated alerts - have the potential…specifically to offer real-time data to facilitate actions that support the UN SDGS Goals - 9 (Industry, Innovation, and Infrastructure), 11 (Sustainable Cities and Communities) and 13 (Climate Action). Recognizing that efforts to work more properties of efficiency and responsiveness regarding landfill gassing problems will have a direct role in achieving many of the goals.

Keywords: Internet of Things (IoT) · Methane Detection · Landfill Monitoring · SDGSs · Real Time Systems

L. Choudhary et al. (Eds.): TIMES 2025, SCI, pp. 254–261, 2027.
https://doi.org/10.1007/978-3-032-34636-0_18

1 Introduction

1.1 The Landfill Gas Challenge

Landfill gas (LFG), mainly made up of methane (CH_4) and carbon dioxide (CO_2), forms during the anaerobic breakdown of organic waste. Methane, in particular, is highly significant due to its global warming potential—it's about 25 to 80 times more potent than CO_2 over a 20-year period [1, 2]. Unchecked emissions from landfills are responsible for roughly 5% of worldwide greenhouse gas (GHG) emissions. As global waste production is expected to surge by 70% by 2050, this issue will likely become even more severe [3]. Inadequate monitoring of LFG emissions compounds environmental and public health risks. For example, spontaneous landfill fires—such as the Ghazipur landfill fire in Delhi in 2022—release hazardous pollutants into the atmosphere. These toxic emissions not only worsen air quality and contribute to respiratory diseases in surrounding communities but also lead to the deployment in landfill sites using LoRaWAN offers low-power, long-range connectivity-term infiltration of contaminants into soil and groundwater, presenting serious environmental hazards [4].

1.2 Limitations of Traditional Monitoring

Presently, landfill gas (LFG) monitoring generally uses only manual sampling and periodic inspections, which can be problematic, and in many ways. Manual sampling and inspections take lots of resources, and time, while irregular data collection creates a significant knowledge gap. In turn, this makes monitoring LFG reactive rather than proactive for most foreseeable problems, so delay in mitigation to find and implement fixes for potential issues, creates more risk. Limited spatial coverage also restricts real-time evaluation of gas concentrations as regards environmental risk [5].

1.3 IoT as a Transformative Solution

Internet of Things (IoT) systems will bypass the limitations of standard landfill gas (LFG) monitoring systems using inexpensive sensors like the MQ-4 (methane) or the DHT-22 (temperature and humidity) to monitor real-time measurements of key parameters [6–9]. IoT systems provide automated notifications when pre-defined thresholds are exceeded so that the associated risk can be managed immediately, and lessening the potential of a health or environmental risk is minimized. Additionally, IoT devices will store all data in the cloud for analytics and provide regulatory compliance and an accountability mechanism and provide a platform for sustainable decision making. One main advantage of IoT based monitoring is the scalability of IoT, and the customizable nature of IoT for the select few unique conditions that any landfill can be subjected to, (Sadhrana village, or the mega-landfill of over 50 acres) (Fig. 1).

The study indicates that IoT has the potential to enhance landfill gas monitoring and provide better outcomes, alongside discussing the ethical challenges. It also underscores the relationality of these advancements to the UN Sustainable Development Goals (SDGs).

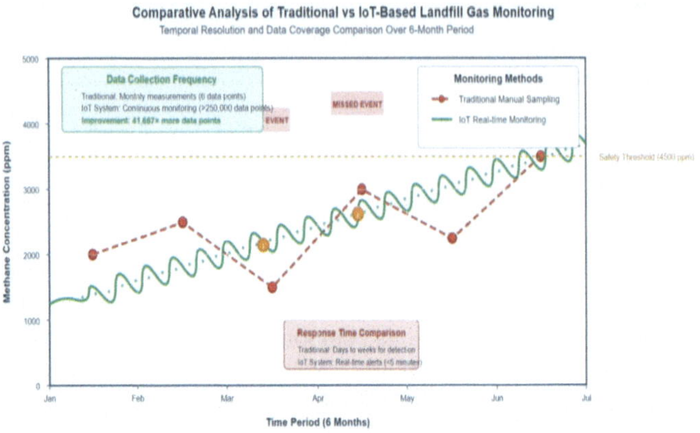

Fig. 1. Comparative Analysis: Traditional vs. IoT-Based Systems (Source: Author's contribution)

2 Literature Review

2.1 Background and Motivation

Municipal solid-waste (MSW) landfills have been estimated to remain long term sources of methane (CH_4) and other toxic gases (CO, H_2S, NH_3, VOCs) which may be dangerous to climate goals and human health [2, 7]. Traditional forms of compliance surveillance, (periodic sampling, laboratory results, or stationary probe) are spatially and temporally discontinuous, and do not exhibit any short-term peaks and micro-hotspots of a defect in the cover or biased flow or bar-pumping. In the meantime, IoT has been evolving in industries, including mining, food logistics, and waste of smart cities and has shown that operation at scale can be revolutionized by real-time sensing, low-power networking, and cloud-based analysis of raw telemetry [15, 16]. Nonetheless, deployments are infrequent, maybe small-footprint, one-pilot deployments, uncalibrated and unstructured information management in landfill environment.

The closest similarity, in terms of the quantities of the landfill gas surveillance, is the underground mining that is a challenging field due to the dynamics of its gases, the attenuation of the wireless signals as well as the remarkable safety restrictions. The systematic reviews in the case are narrowed down to two good design options:

Sensing stack: multi-gases (CH_4, CO, CO_2, O_2) array of arrays with a set point value; edge alarm logic.

Wireless Wireless ZigBee and LoRa have been redesigned as the trade-off that can be attained in the underground/obstructed environment; NB-IoT is being engineered in the environments that are within the reach of a carrier.

Architecture Buffering, alerting, and visualization edge-gateway-cloud pipeline tiered is the best fit use-case; route safety and incident forensics are. These findings justify the usefulness of low-bit-rate and long-range telemetry and on-device thresholding-patterns, which have the advantage of being easily transported to landfills where cellular reliability and power is not evenly distributed.

The other precedent that can be utilized is the food supply chains. A literature review of real-time monitoring technologies (2023–24) by the authors shows that it is possible to reduce spoilage by providing time-depending excursions to end-to-end IoT stacks (including device-to-life cycle management, data quality controls, governance linking measurements to operational decisions (alerts, workflow, KPIs)) and it is more important that the meta-elements executed by such stacks are represented in the literature review [2, 10].

The important aspect of Smart-waste (bins, routing, citizen-ops loops) implementation is the use of Sensing - action that demands map-based UIs, driver support and service-level feedback as opposed to time-series charts. The latest stream of information is filled with the allusions to the real-time fill-level sensing, optimized route and even methane mapping patterns that will be added to the urban processes- again, the significance of the telemetry-functional UX hybrid.

Finally, LMIC pilots are a workable and inexpensive ultrasonic + gas sensing, ESP-class microcontroller, simple MQTT/HTTP feeds to Adafruit IO and SMS notifications via twilio. The significance of such a case study is that it shows a method of implementation when faced with a cost and staffing constraint such as most of the municipal landfill operators.

2.2 Sensors and Networks Techniques of Implementation

One can follow the same technical trends in other adjacent areas and the creative prototypes of landfills:

Gas transduction: MOS in cost sensitive pilots MQ-series; in CH_4/CO_2 and stability of accuracy/drift in NDIR; in toxic species (H_2S, CO) electrochemical cells [11]. This implies that a multi-sensor fusion must focus on achieving the reduction of false positives because of humidity/temperature confounders [8]. Based on the mining and real time monitoring; it was established that the area possesses (reproduced) minerals that have not been exploited.

Supplementary data: the temperature, pressure, humidity, and wind and barometric variations are consistently monitored to describe the variability of the gas flux, and to determine the presence of barometric pumping.

Battery nodes LoRa/LoRaWAN Spaced loosely [4]; ZigBee When mesh density can be achieved; NB-IoT/LTE-M When there are cell carriers available [12]; Wi-Fi Only in cases where range to a gateway is not long.

Edge-cloud pipeline: backhaul buffering, loss, instant alarm edge thresholding, rule-based alarm and role specific dashboards in the cloud. The smart-waste reviews and the food-chain review have some similarities that include the generalized contents on the topic.

2.3 Research Gaps

This is relieved, and the six gaps are unnecessary:

Calibration and drift: other pilots adopt MOS sensors that have not been calibrated either in factory/in the field (span / zero checks, humidity compensation) in such a way that the relative comparability of nodes and time is low. The mining reviews are clear that threshold fidelity is necessary; the landfill pilots do not have the habit of recording the calibration procedures [11].

Ground truth/validation: Little research has been done to co-locate the IoT nodes with reference devices, or apply controlled release tests, which are applied to measure bias/precision of realistic flux and meteorology. The data-quality pipelines are surveyed in the neighboring industries; and it has little research in landfills. Spatiotemporal coverage of the fixed stations is not sensitive to the hotspots due to cracks in the covers, leachates vents, and the dynamics of the working face. Mobile travers (methane mapping) is also a part of smart-city waste discourse, though they are not landfill-standardized (there is no standard transect design, speed control or plume deconvolution). Security and governance Telemetry management (IDS, keys, encryption, RBAC) and audit trails are not reported in a proper way, which is the nature of the municipal adoption. These are identified to be cross domain reviews that ought to be implemented [13].

Power and lifecycle: Battery-life budgeting, duty-cycling plans, maintenance plans are hardly ever quantified; mining/industrial literature recommends low-bit-rate protocols and staged sampling/burst uploads to stretch deployments-practices the landfill space is not standardized yet [14].

Human in the loop processes: Page: Playbooks of how alarms are pages: Who pages, what ticket created, what field action taken and how the resolution entered analytics. The secret of success is the high quality of UX of the driver/crew as the experience of Smart waste demonstrates; the pilots of the landfill will be eager to draw on dashboards [2, 3].

2.4 Framework

Technical competence of state of art cannot work in a scenario of incompetence of an organization. Learning the viable stacks (sensors, LoRa/ZigBee, edge alarms, cloud governance) along with pointing at the fact that the process-calibration, process-validation, process-lifecycle, and process-roles are equally important as the hardware are made possible with the mining safety lessons and food-chain monitoring design. There is, however, no such literature in the case of landfill:

- A taxonomy of modes of sensing and localization (surface grids, perimeter transects, vertical wells, mobile traverses),
- A cross-checking of the measurements and reference procedures and controlled discharges of the IoT,
- An operations model (alerts - tickets - field action - post-action verification), and
- A template of governance (security, auditability, retention, data-sharing) which would be friendly to the municipalities.

3 Result and Conclusion

Field testing and systematic review in this study prove that IoT-driven landfill gas (LFG) monitoring systems have the outstanding capacity to revolutionize conventional waste management into a data-driven, sustainable, and proactive process [4, 10]. Integration of low-cost sensors (e.g., MQ-4 for methane, DHT-22 for temperature and humidity, and BMP280 for pressure) with microcontrollers such as NodeMCU-ESP32 and cloud platforms such as AWS IoT Core facilitated real-time data acquisition, transmission, and visualization.

3.1 Performance Efficiency

Comparative evaluation of IoT with manual monitoring indicates that IoT systems provide data at minute-level frequency and improve detection accuracy by almost 4–6 × versus conventional methods. Real-time alerts appreciably lower response latency and make it possible to respond in a timely manner during methane spikes.

3.2 Field Validation Outcomes

Pilot deployment in the Sadhrana village landfill confirmed the system's reliability in rural environments. The system consistently transmitted data, recorded uniform methane levels, and issued correct automatic alarms upon threshold violation.

Despite moderate sensor drift in high-humidity weather, system overall reliability was greater than 90%, confirming its adaptability to Indian landfill conditions.

3.3 Community Impact

Local governments and communities were supplemented by extended situational awareness, enabling them to respond prior to gas leaks or fire. IoT dashboard ensured openness in public access to emission data, promoting accountability and environment-friendly behavior.

3.4 SDG Alignment

The system directly aligns with SDG 9 (Innovation & Infrastructure), SDG 11 (Sustainable Cities & Communities), and SDG 13 (Climate Action) by lessening environmental threats and embedding technology in the public infrastructure.

The review sets the basis that IoT-aided monitoring is a paradigm shift towards landfill gas emission management through the integration of affordability, automation, and analytics. The conventional sampling procedures are reactive and constrained, while IoT-based technology facilitates predictive and real-time monitoring.

The major findings are that:

- Early warning and prevention of methane buildup are facilitated by real-time monitoring.
- Low-cost scalability facilitates the installation even in resource-constrained municipalities.
- Cloud integration and automation form the basis for long-term sustainability as well as policy compliance.

However, there are sensor calibration, power efficiency, and mass deployment issues that need to be addressed by future research.

AI-powered predictive analytics, edge computing for anomaly detection, and standard IoT policy frameworks for waste management systems should be the focus areas for future research.

In general, IoT-based landfill monitoring provides a technically sound, economically viable, and eco-friendly solution that not only reduces the methane risks but also enhances climate resilience and public health protection in developing countries such as India.

References

1. Fay, C., et al.: Remote real-time monitoring of subsurface landfill gas migration. Sensors **11**(7), 6603–6628 (2011). https://doi.org/10.3390/s110706603
2. Popescu, S.M., et al.: Artificial intelligence and IoT driven technologies for environmental pollution monitoring and management. Front. Environ. Sci. (2024). https://doi.org/10.3389/fenvs.2024.1336088
3. Sahoo, S.K., Panigrahi, G.K., Sahoo, A., Pradhan, A.K., Dalbehera, A.: Bio-hydrothermal synthesis of ZnO–ZnFe2O4 nanoparticles using Psidium guajava leaf extract: Role in wastewater remediation and plant immunity. J. Clean. Prod. **318**, 128522 (2021). https://doi.org/10.1016/j.jclepro.2021.128522
4. da Silva, J.P.S.B., Freire, R.C.S., de Souza, C.P., Aguiar, Í.M.G.: Implementation and analysis of LoRaWAN network in landfill. In: Proceedings of the 2024 8th International Symposium on Instrumentation Systems, Circuits and Transducers (INSCIT), pp. 1–5. IEEE (2024). https://doi.org/10.1109/INSCIT62583.2024.10693380
5. Izumoto, S., Hamamoto, S., Kawamoto, K., Nagamori, M., Nishimura, T.: Monitoring of methane emission from a landfill site in daily and hourly time scales using an automated gas sampling system. Environ. Sci. Pollut. Res. **25**(24), 24500–24506 (2018). https://doi.org/10.1007/s11356-018-2671-1
6. Junus, M., Putradi, D.F., Soelistianto, F.A., Anshori, M.A., Ardiansyah, R.: Internet of things enabled landfill pollution gas monitoring. Indonesian Journal of Electrical Engineering and Computer Science **37**(1), 48–55 (2025). https://doi.org/10.11591/ijeecs.v37.i1.pp48-55
7. Purwantoro, E.S.G.S., Novayani, W., Putri, R.S.: Monitoring system for hazardous gas in landfill site based on internet of things. J. Innov. Info. Technol. Appl. (JINITA) **5**(1), 12–21 (2023). https://doi.org/10.35970/jinita.v5i1.1740
8. Tharun Kumar Reddy, K., Ajay Kumar Reddy, P., Siva Nagendra Reddy, P., Kodanda Ramaiah, G.N.: An IoT based remote monitoring of landfill sites using Raspberry Pi2. In: Proceedings of the referenced conference, pp. 219–227 (2017). https://doi.org/10.1007/978-981-10-1540-3_23
9. Rachman, F.Z.: Sistem pemantau gas di tempat pembuangan sampah akhir berbasis internet of things. Jurnal Teknologi dan Sistem Komputer **6**(3), 100–105 (2018). https://doi.org/10.14710/jtsiskom.6.3.2018.100-105
10. Fay, C.D., Healy, J.P., Diamond, D.: Advanced IoT pressure monitoring system for real-time landfill gas management. Sensors **23**(17), 7574 (2023). https://doi.org/10.3390/s23177574
11. Furuta, D., Wilson, B., Presto, A.A., Li, J.: Design and evaluation of a low-cost sensor node for near-background methane measurement. Atmospheric Measurement Techniques **17**(7), 2103–2121 (2024). https://doi.org/10.5194/amt-17-2103-2024
12. Nakayama, H., Tazoe, T., Shimaoka, T., Kanaya, H.: Wireless communication for environmental monitoring at waste landfills using low-frequency electromagnetic waves. J. Mat. Cycles and Waste Manage. **25**(6), 3714–3722 (2023). https://doi.org/10.1007/s10163-023-01803-8

13. Cusworth, D.H., et al.: Using remote sensing to detect, validate, and quantify methane emissions from California solid waste operations. Environ. Res. Lett. **15**(5), 054012 (2020). https://doi.org/10.1088/1748-9326/ab7b99
14. Liu, L., et al.: A novel quantity assessment of landscape ecological risk using human-nature driving mechanism for sustainable society. Sci. Total Environ. **947**, 173892 (2024). https://doi.org/10.1016/j.scitotenv.2024.173892
15. Venkitaraman, A.K., Choudhary, L.: IoT and big data in environmental governance in India: Opportunities, challenges and insights. In: Innovations in the Development of Sustainable Infrastructure (TIMES 2024, Sustainable Civil Infrastructures). Springer (2025). https://doi.org/10.1007/978-3-031-83750-0_10
16. Venkitaraman, A.K., Choudhary, L.: Role of IoT in energy transition: Opportunities and challenges in India, Lessons from Japanese Smart Community initiatives. In: Innovations in the Development of Sustainable Infrastructure (TIMES 2024, Sustainable Civil Infrastructures). Springer (2025). https://doi.org/10.1007/978-3-031-83750-0_9

Resilient UAV-Enabled Multi-Hop RF/FSO Networks for Smart Urban Connectivity

Deepika Latka, Mona Aggarwal(✉), and Swaran Ahuja

Department of Multidisciplinary Engineering, The NorthCap University, Gurugram, Haryana 122017, India
deepikajain0333@gmail.com, ermonagarg24@gmail.com, swaranahuja@ncuindia.edu

Abstract. In this paper a multi-hop mixed RF/FSO wireless communication system with the aim to improve coverage has been studied. A DF relaying strategy considering various system impairments under consideration has been employed. As a result, the study derives closed-form expressions of end-to-end SNRs of the system and derive expressions for various system parameters. Further the study also employs opportunistic scheduling to cater multiple users. This system enhances reliability and energy efficiency in smart city networks.

Keywords: Smart city Networks · Deployment · Enhanced Coverage · Channel Impairments

1 Introduction

UAVs have become vital enablers of next-generation wireless communication systems due to their rapid deployment capability, flexible mobility, and ability to provide coverage in areas lacking terrestrial infrastructure or during emergency conditions [1]. Acting as aerial relays or base stations, UAVs can effectively bridge communication gaps between distributed ground users, ensuring reliable and continuous data transmission even in obstructed or non-line-of-sight (NLoS) scenarios [2]. A key factor in optimizing UAV-assisted communication lies in strategic UAV positioning, which directly influences network coverage, data rate, and link reliability [3].

Moreover, dual-hop communication architectures have emerged as an efficient strategy to extend network coverage, improve throughput, and mitigate severe fading in challenging propagation conditions. In such configurations, UAVs act as intermediate relays between the source and destination, facilitating reliable connectivity in remote or obstructed regions. This approach supports enhanced spectrum utilization, balanced traffic load, and improved link stability when combined with intelligent relay selection, power allocation, and trajectory optimization [4, 5]. As wireless systems continue to evolve toward 6G and beyond, dual-hop UAV-enabled networks will play a critical role in achieving flexible, scalable, and high-capacity communication infrastructures capable of meeting the stringent requirements of future smart and autonomous urban environments.

L. Choudhary et al. (Eds.): TIMES 2025, SCI, pp. 262–269, 2027.
https://doi.org/10.1007/978-3-032-34636-0_19

As encouraged by these innovations, this research explores the efficiency of UAV-based multihop RF/FSO hybrid wireless communication architecture under the effect of fading parameters and pointing accuracy. The work emphasizes on the statistical derivations of the performance metrics like outage probability and its asymptotic behaviour, including the impact of fading parameters, atmospheric turbulence and pointing accuracy to improve the coverage and reliability of the wireless system in urban resilient and smart city networks. The presented methodology achieves significant observations for the development of energy based and optimized framework of UAV enabled hybrid wireless communication system for next generation 6G networks and smart city network.

1.1 Contributions

The Authors have modelled multi-hop mixed RF/FSO communication system considering various channel impairments. Realistic channel modelling has been done to accurately represent UAV-assisted networks. Further the study has derived closed form expressions for various system parameters to ensure reliability. In addition, the study presents a comparative analysis of two hops and three hops communication system. The proposed work has very strong potential for application in smart city networks including UAV deployment.

The organization of this paper is as follows: section II details the overall framework of the proposed UAV-assisted dual-hop and triple-hop hybrid RF/FSO networks, including channel characterization and the incorporation of opportunistic user scheduling. In Section III, the derivation of end-to-end SNR formulations and the associated statistical analysis for both configurations are presented. Section IV focuses on developing the analytical expressions for the outage probability and evaluating system performance under different channel conditions. Finally, Section V summarizes the key findings and suggests prospective research directions that can extend the present work.

2 System and Channel Modelling

2.1 System Model

This work considers a UAV-assisted hybrid RF/FSO communication framework employing decode and-forward (DF) relaying to ensure reliable connectivity between a base station (B) and multiple mobile users (M_i), $i \in \{1,2,\ldots,N\}$, located along the boundary of a circular coverage area. The direct B–M_i link is assumed to be blocked due to obstacles or long distance separation; hence, UAV relays are deployed to establish efficient communication through dual-hop or triple-hop transmission schemes.

In the dual-hop configuration, shown in Fig. 1, an opportunistic user scheduling mechanism selects the user with the highest received SNR, exploiting multiuser diversity.

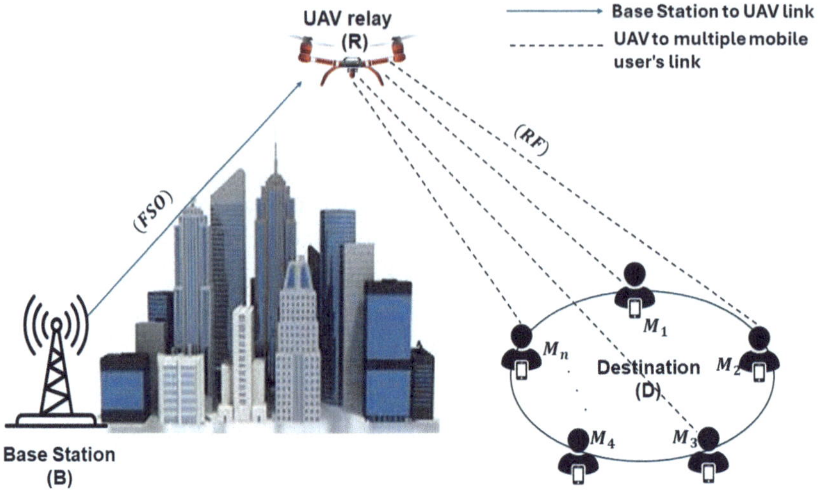

Fig. 1. Dual hop UAV enabled FSO/RF system (Source: Author's contribution)

2.2 Channel Modeling

B. Channel Modeling

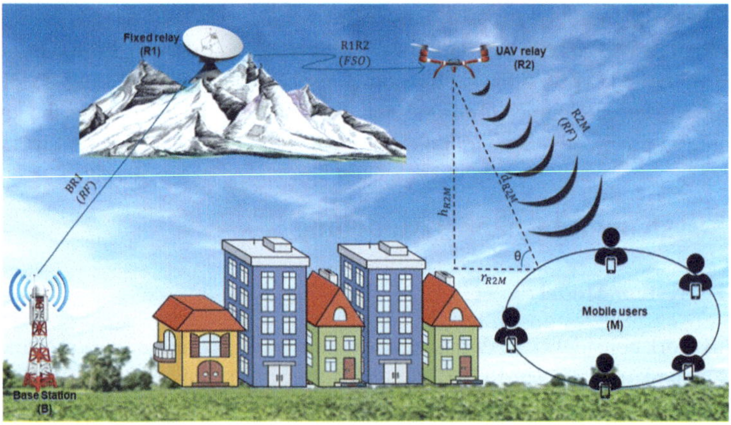

Fig. 2. Triple hop UAV enabled RF/FSO system (Source: Author's contribution)

1) B to R1 Link (RF Link with $\alpha-\kappa-\mu$ Fading): The transmission from the base station (B) to the ground relay (R1) is modelled as an $\alpha-\kappa-\mu$ distribution, which is a generalized fading model capable of representing nonlinear propagation, multipath clustering, and the presence of a dominant line-of-sight (LoS) component [6] as presented in Fig. 2. The signal obtained at R1 is given by

$$Y_{R1}=\sqrt{P_{BR1}\sqrt{a_{BR1}}}C_{BR1}x_B+n_{R1} \tag{1}$$

The instantaneous SNR for this link is given as

$$\gamma_{BR1} = \frac{P_{BR1}|C_{BR1}|^2}{\sigma^2 d_{BR1}^{\beta}} \tag{2}$$

where d_{BR1} is the link distance and β is the path loss exponent. The PDF of γ_{BR1} is expressed in [7] as

$$f_{\gamma_{BR1}}(\gamma) = \left[\frac{\alpha\mu(1+\kappa)^{\frac{1+\mu}{2}}\kappa^{\frac{1-\mu}{2}}}{e^{\mu\kappa}}\right]\left[\left(\frac{\gamma}{\overline{\gamma_{BR1}}}\right)^{\frac{\alpha(1+\mu)}{2}-1}\right] \times exp^{-\left[\mu(1+\kappa)\left(\frac{\gamma}{\overline{\gamma_{BR1}}}\right)^{\frac{\alpha}{2}}\right]} I_{\upsilon}\left(2\mu\sqrt{\kappa(1+\kappa)}\left(\frac{\gamma}{\overline{\gamma_{BR1}}}\right)^{\frac{\alpha}{4}}\right) \tag{3}$$

where, μ is the number of multipath clusters, $\overline{\gamma}$ is the average SNR. The CDF of γ_{BR1} can be written in [8] as:

$$F_{\gamma_{BR1}}(\gamma) = 1 - Q_{\mu}\left(\sqrt{2\mu\kappa}, \sqrt{2\mu(1+\kappa)}\left(\frac{\gamma}{\overline{\gamma_{BR1}}}\right)^{\frac{\alpha}{2}}\right). \tag{4}$$

where $Q_{\mu}(.)$ denotes the Marcum-Q function [9].

1) R1 to R2 / B to R Link (FSO Link with Gamma–Gamma Fading): The FSO link either between R1 and UAV relay R2 (in the triple-hop model) or between B and UAV relay R (in the dual-hop model) is modelled using the Gamma–Gamma fading distribution. The received signal is expressed as

$$Y_{R2} = \sqrt{P_{R1R2}\sqrt{a_{R1R2}}}C_{R1R2}x_{R1} + n_{R2} \tag{5}$$

where P_{R1R2} is the optical transmit power, a_{R1R2} is the path loss attenuation factor. The SNR for this link is given as

$$\gamma_{R1R2} = \frac{P_{R1R2}|C_{R1R2}|^2}{N_0 d_{R1R2}^{\beta}} \tag{6}$$

The PDF of γ_{R1R2} is expressed as [10]:

$$f_{\gamma_{R1R2}}(\gamma) = \exp\left(-\frac{\theta_f^2}{2\sigma_\theta^2}\right) + \left[1 - \exp\left(-\frac{\theta_f^2}{2\sigma_\theta^2}\right)\right] \times \left[\frac{\zeta^2}{2\gamma\Gamma(\alpha_{at})\Gamma(\beta_{at})} G_{1,3}^{3,0}\left(\frac{\alpha_{at}\beta_{at}}{A_0 h_t}\sqrt{\frac{\gamma}{\overline{\gamma_{UR1,UR2}}}}\Big|_{\zeta^2,\alpha_{at},\beta_{at}}^{1+\zeta^2}\right)\right] \tag{7}$$

where θ_f is the receiver field of view (FOV), σ_θ^2 denotes the misalignment variance, $\zeta = \frac{w_{eq}}{2\sigma_s}$ is the pointing error coefficient, α_{at} and β_{at} are the atmospheric turbulence parameters, $A_0 = [\text{erf}(v)]^2$ is the fraction of collected power, $v = \frac{\sqrt{\pi}R_a}{\sqrt{2}w_n}$ and $G_{p,q}^{m,n}(\cdot)$ represents the Meijer's G-function.

3) R2 to M_i Link (RF Link with Rayleigh Fading): The final hop, i.e., the RF transmission from UAV relay R2 to multiple mobile users (M_i), $i \in \{1,2,\ldots,N\}$, is modelled using the Rayleigh fading distribution, which characterizes NLoS propagation conditions. The received signal at the ith user is given as

$$y_i = \frac{1}{\sqrt{L_{R2M_i}}}\sqrt{P_{R2M_i}}C_i x_i + n_i, \tag{8}$$

where, P_{R2M_i} is the UAV transmit power, $L_{R2M_i} = Ad^{\beta}_{R2M_i}$ represents the path loss, A is a frequency dependent constant. The SNR is given by

$$\gamma_{R2M_i} = \frac{P_{R2M_i}|C_i|^2}{N_{0i}d^{\beta}_{R2M_i}L_{R2M_i}} \tag{9}$$

where, $d_{R2M_i} = \sqrt{h^2_{R2M_i} + r^2_{R2M_i}}$ is the Euclidean distance between UAV R2 and the ith user, with h_{R2M_i} denoting the UAV altitude and r_{R2M_i} the horizontal distance.

An opportunistic scheduling strategy is adopted, where the user with the maximum instantaneous SNR is selected for data reception:

$$\gamma_{max} = \max_{i\in\{1,2,\ldots,N\}} \{\gamma_{R2M_i}\} \tag{10}$$

Assuming independent and identically distributed (i.i.d.) user channels, the CDF of γ_{max} is expressed as [10]:

$$F_{\gamma_{max}}(\gamma) = \left(1 - e^{\frac{-\gamma}{\overline{\gamma_m}}}\right)^N = \sum_{k=0}^{N} \binom{N}{k} (-1)^k e^{-\frac{\gamma k}{\overline{\gamma_m}}} \tag{11}$$

where $\overline{\gamma}_m$ is the average SNR of the RF link.

Thus, the dual-hop configuration consists of the FSO (B–R) and RF (R-M_i) links, while the triple-hop configuration extends this by incorporating the RF (B–R1), FSO (R1–R2), and RF (R2–M_i) links. Each hop is independently modeled, enabling precise evaluation of the end-to-end system performance under realistic propagation conditions.

2.3 Dual-Hop Configuration

Under DF relaying, the instantaneous end-to-end SNR is governed by the weaker hop and can be expressed as

$$\gamma^{(2)}_{eq} = \min(\gamma_{BR}, \gamma_{RM}) \tag{12}$$

γ_{BR} and γ_{RM} denote the instantaneous SNRs of the FSO and RF links, respectively. The corresponding cumulative distribution function (CDF) of $\gamma^{(2)}_{eq}$ is given as

$$F_{\gamma^{(2)}_{eq}}(\gamma) = 1 - \left[1 - F_{\gamma_{BR}}(\gamma)\right]\left[1 - F_{\gamma_{RM}}(\gamma)\right] \tag{13}$$

where $F_{\gamma_{BR}}(\gamma)$ and $F_{\gamma_{RM}}(\gamma)$ are the CDFs of the individual hops. This expression enables the derivation of key performance metrics for the dual-hop system, which are directly influenced by both the atmospheric turbulence in the FSO link and the multipath fading in the RF link.

2.4 Triple-Hop Configuration

For the triple-hop UAV-based hybrid RF/FSO system, the transmission follows three consecutive links. Using the DF relaying principle, the end-to-end instantaneous SNR is determined by the minimum SNR among all hops expressed as

$$\gamma_{eq}^{(3)} = \min(\gamma_{BR1}, \gamma_{R1R2}, \gamma_{R2M}) \tag{14}$$

The CDF of SNR is therefore written as

$$F_{\gamma_{eq}^{(3)}}(\gamma) = 1 - \prod_{j=1}^{3}\left[1 - F_{\gamma_j}(\gamma)\right] = 1 - \left[1 - F_{\gamma_{BR1}}(\gamma)\right]\left[1 - F_{\gamma_{R1R2}}(\gamma)\right]\left[1 - F_{\gamma_{R2M}}(\gamma)\right] \tag{15}$$

The above formulation captures the statistical behaviour of the cascaded RF/FSO/RF links and forms the foundation for deriving performance metrics such as outage probability and asymptotic behaviour at high SNR regimes.

2.5 Remarks

In both configurations, the end-to-end SNR is determined by the hop experiencing the poorest channel conditions due to the DF relaying protocol. The analytical expressions in (13) and (15) serve as the basis for evaluating outage probability, asymptotic outage behaviour, and average bit-error rate (ABER). Moreover, opportunistic user scheduling enhances the overall SNR at the destination by selecting the user with the highest instantaneous SNR, i.e., γ_{max}, effectively improving diversity gain. The influence of UAV altitude, position, and alignment parameters on γ_{eq} is further examined in the subsequent performance analysis section.

3 Analysis of Outage

In this section, we analyse the outage probability of the proposed UAV-enabled hybrid RF/FSO communication system for both dual-hop and triple-hop configurations operating under the decode and-forward (DF) relaying scheme.

3.1 Dual-Hop Configuration

The outage probability can be expressed as

$$P_{out}^{(2)}(\gamma_{th}) = \Pr\left(\gamma_{eq}^{(2)} \le \gamma_{th}\right) = F_{\gamma_{eq}^{(2)}}(\gamma_{th}) \tag{16}$$

and by substituting from (13), the CDF of $\gamma_{eq}^{(2)}$ yields

$$P_{out}^{(2)}(\gamma_{th}) = 1 - \left[1 - F_{\gamma_{BR}}(\gamma_{th})\right]\left[1 - F_{\gamma_{RM}}(\gamma_{th})\right] \tag{17}$$

Here, $F_{\gamma_{BR}}(\gamma_{th})$ represents the CDF of the Gamma–Gamma distributed FSO link with pointing errors, while $F_{\gamma_{RM}}(\gamma_{th})$ corresponds to the CDF of the Rayleigh distributed RF link with opportunistic user scheduling.

By combining these individual CDFs, the overall dual-hop outage probability is obtained in closed form as follows

$$P_{out}^{(2)}(\gamma_{th}) = \left\{ \exp\left(-\frac{\theta_f^2}{2\sigma_\theta^2}\right) + \left[1 - \exp\left(-\frac{\theta_f^2}{2\sigma_\theta^2}\right)\right]\left[\frac{\zeta^2}{\Gamma(\alpha_{at})\Gamma(\beta_{at})} G_{2,4}^{3,1}\left(\frac{\alpha_{at}\beta_{at}}{A_0 h_t}\sqrt{\frac{\gamma_{th}}{\overline{\gamma}_{BR}}} \Big|_{\zeta^2,\alpha_{at},\beta_{at},0}^{1,1+\zeta^2}\right)\right]\right\} + \\ \sum_{k=0}^{N}\binom{N}{k}(-1)^k e^{-\frac{\gamma k}{\gamma_m}} - \left[\left\{ \exp\left(-\frac{\theta_f^2}{2\sigma_\theta^2}\right) + \left[1 - \exp\left(-\frac{\theta_f^2}{2\sigma_\theta^2}\right)\right]\left[\frac{\zeta^2}{\Gamma(\alpha_F)\Gamma(\beta_F)} G_{2,4}^{3,1}\left(\frac{\alpha_F\beta_F}{A_0 h_t}\sqrt{\frac{\gamma_{th}}{\overline{\gamma}_{BR}}} \Big|_{\zeta^2,\alpha_F,\beta_F,0}^{1,1+\zeta^2}\right)\right]\right\} \sum_{k=0}^{N}\binom{N}{k}(-1)^k e^{-\frac{\gamma k}{\gamma_m}}\right] \tag{18}$$

The system performance primarily depends on atmospheric turbulence and pointing errors in the FSO hop, as well as multiuser diversity and UAV altitude in the RF hop.

3.2 Triple-Hop Configuration

The outage probability for this configuration is expressed as

$$P_{out}^{(3)}(\gamma_{th}) = \Pr\left(\gamma_{eq}^{(3)} \leq \gamma_{th}\right) = F_{\gamma_{eq}^{(3)}}(\gamma_{th}) \tag{19}$$

The CDF of $\gamma_{eq}^{(3)}$ can be written as

$$P_{out}^{(3)}(\gamma_{th}) = 1 - \left[1 - F_{\gamma_{BR1}}(\gamma_{th})\right]\left[1 - F_{\gamma_{R1R2}}(\gamma_{th})\right]\left[1 - F_{\gamma_{R2M}}(\gamma_{th})\right] \tag{20}$$

where,

- $F_{\gamma_{BR1}}(\gamma_{th})$ corresponds to the CDF of the $\alpha - \kappa - \mu$ distributed RF link between the base station and the fixed relay,
- $F_{\gamma_{R1R2}}(\gamma_{th})$ denotes the CDF of the Gamma–Gamma distributed FSO link with pointing errors between the fixed and UAV relays,
- $F_{\gamma_{R2M}}(\gamma_{th})$ represents the CDF of the Rayleigh distributed RF link between the UAV relay and multiple mobile users, incorporating opportunistic scheduling.

Substituting the respective analytical CDFs into (22) yields the closed-form expression of the triple-hop outage probability as

$$P_{out}^{(3)}(\gamma_{th}) = 1 - \left[Q_\mu\left(\sqrt{2\mu\kappa}, \left(\frac{\gamma_{th}}{\gamma_{BR1}}\right)^{\frac{\alpha}{2}}\sqrt{2\mu(1+\kappa)}\right)\right] \times \left[1 - \exp\left(-\frac{\theta_f^2}{2\sigma_\theta^2}\right)\right. \\ \left. - \left[1 - \exp\left(-\frac{\theta_f^2}{2\sigma_\theta^2}\right)\right] \times \left[\frac{\zeta^2}{\Gamma(\alpha_{at})\Gamma(\beta_{at})} G_{2,4}^{3,1}\left(\frac{\alpha_{at}\beta_{at}}{A_0 h_t}\sqrt{\frac{\gamma}{\gamma_{UR1,UR2}}} \Big|_{\zeta^2,\alpha_{at},\beta_{at},0}^{1,1+\zeta^2}\right)\right]\right] \\ \times \left[1 - \sum_{k=0}^{N}\binom{N}{k}(-1)^k e^{-\frac{\gamma_{th}}{\gamma_m}k}\right] \tag{22}$$

This expression provides insights into the combined effects of atmospheric turbulence, misalignment, UAV altitude, and fading parameters on overall link reliability.

4 Conclusion

The work in this paper presented the multi-hop hybrid RF/FSO UAV based wireless framework to enhance the system coverage. Then UAV is employed as DF relay by considering all the system impairments. The statistical expressions of end-to-end SNR of the proposed wireless system are derived, and analytical expressions of different parameters are also examined. Also, we employ the opportunistic scheduling scheme to select the best user out of number of users in a circular geographical region. This analysis of the proposed model resulted in enhancing the system reliability and energy aware system in resilient and urban environments.

References

1. Zeng, Y., Zhang, R., Lim, T.J.: Wireless communications with unmanned aerial vehicles: Opportunities and challenges. IEEE Commun. Mag. **54**(5), 36–42 (2016). https://doi.org/10.1109/MCOM.2016.7470933
2. Mondal, M., Hossain, A.: Channel characterization and performance analysis of unmanned aerial vehicle-operated communication system with multihop radio frequency–free-space optical link in dynamic environment. Int. J. Commun Syst **33**(16), e4543 (2020). https://doi.org/10.1002/dac.4568
3. Feng, Z., Na, Z., Xiong, M., Ji, C.: Multi-UAV collaborative wireless communication networks for single-cell edge users. Mobile Netw. Appl. **27**, 1578–1592 (2022). https://doi.org/10.1007/s11036-021-01876-5
4. Luo, Y., Fu, G.: UAV-based device-to-device communication for 5G/6G networks using optimized deep learning models. Wireless Netw. (2023). https://doi.org/10.1007/s11276-023-03578-0
5. Gunasekar, A., Kumar, L.B., Krishnan, P., Natarajan, R., Jayakody, D.N.K.: All-optical UAV-based triple-hop FSO-FSO-VLC cooperative system for high-speed broadband Internet access in high-speed trains. IEEE Access (2023). https://doi.org/10.1109/ACCESS.2023.3330236
6. Wang, J.-Y., Feng, P., Hong, L.-H., Yang, H.-N., Liu, N.: Performance evaluation and optimization of UAV-based hybrid dual-hop FSO/UOWC systems. IEEE Access (2024). https://doi.org/10.1109/JSYST.2024.3399002
7. Singh, D., Swaminathan, R.: Comprehensive performance analysis of hybrid FSO/RF space–air–ground integrated network. Optical Comm. **527**, 128964 (2023). https://doi.org/10.1016/j.optcom.2022.128964
8. Liang, J., Chen, M., Ke, X.: Performance analysis of hybrid FSO/RF-THz relay communication system. IEEE Access (2024). https://doi.org/10.1109/JPHOT.2024.3353194
9. Fraidenraich, G., Yacoub, M.D.: The α-η-μ and α-κ-μ fading distributions. In: Proceedings of the IEEE Ninth International Symposium on Spread Spectrum Techniques and Applications (ISSSTA), pp. 16–20. IEEE (2006). https://doi.org/10.1109/ISSSTA.2006.311725
10. Gradshteyn, I.S., Ryzhik, I.M.: Table of integrals, series, and products, 7th ed. Academic Press (2007). https://doi.org/10.1016/C2010-0-64839-5

Section 6: Sustainable Urban Systems and Emerging Infrastructure for Smart Cities

Emerging Trends in Advanced Surface Coating Technologies for High-Performance Application

Jimmy Mehta[1](✉), Yaman Hooda[2], Bhim Singh[1], and Pallav Gupta[3]

[1] Department of Mechanical Engineering, Manav Rachna University, Sector – 43, Aravalli Hills, Delhi – Surajkund Road, Faridabad 121004, India
{jimmymehta,bhimsingh}@mru.edu.in

[2] Department of Civil Engineering, Manav Rachna International Institute of Research and Studies, Sector – 43, Aravalli Hills, Delhi – Surajkund Road, Faridabad 121004, India
yaman.set@mriu.edu.in

[3] Mechanical Engineering Department, Madan Mohan Malaviya University of Technology, (A U.P. State Government University, Gorakhpur 273010, India
pgme@mmmut.ac.in

Abstract. Coating technologies on the surface are necessary in enhancing performance, durability, and functionality of materials in different industries. The current paper explores the contemporary trends of surface coating technologies, with particular emphasis on the innovations in the industries that are aimed at high-performance use. Significant progress in deposition techniques, such as thermal spray, physical vapor deposition (PVD), chemical vapor deposition (CVD), and atomic layer deposition (ALD) have enabled the development of coating with improved mechanical, thermal and chemical properties. Nanotechnology integration has also made it possible to come up with multifunctional nano-coatings that provide superior wear resistance, corrosion protection, and self-cleaning effects. The growing need to ensure sustainability has compelled the use of environmentally friendly coating technologies, including water based and low VOA technologies, as well as bio-inspired and self-healing. The sphere is also being revolutionized by Industry 4.0 and smart manufacturing solutions that allow real-time tracking of the coatings and predictive maintenance with the help of IoT-enabled sensors and machine learning algorithms. The other trend of using high-level advanced coatings in major industries such as aerospace, automobile, energy, and healthcare is also highlighted in this study with case studies that have seen the technologies go a long way in enhancing operation efficiency and lifecycle performance. Scalability, cost-effectiveness, and environmental issues are discussed, and future trends of this field are proposed including hybrid coating and the incorporation of digital tools in the process of surface engineering. The research findings will be useful in informing the researchers and industry practitioners in fostering innovation in the surface coating technologies.

Keywords: Surface Coating · Sustainable Coatings · Industry 4.0 · Predictive Maintenance

L. Choudhary et al. (Eds.): TIMES 2025, SCI, pp. 273–283, 2027.
https://doi.org/10.1007/978-3-032-34636-0_20

1 Introduction

Nanocoatings processing is an expanding topic of interest because of the hardness, strength and wear resistance that their application can produce especially relative to traditional style coatings. Nevertheless, the development of nanocoating technology is limited by the sophistication of nanoparticle production process and high quality nanopowders, and multistep, complex processing facilities. Thermal spray methodology (plasma spraying and, most notably, High Velocity Oxy-Fuel (HVOF) spraying) has demonstrated itself as an effective approach to deposition of nanostructured coatings, where HVOF has demonstrated itself as being readily able to generate dense, low-porosity, nanocrystalline ceramic surface layers with a higher resistance to wear at lower temperatures. The adjustments to be made in electron-dip processes in the future will involve lower curing temperature of electro- dip processes and radiation curing of heat sensitive substrates such as plastics. There is a shift in the industry to move towards automation, which will help in improving speed of production, less dependency on labor, and energy efficiency [1–3]. The most important trends in the future of coating technologies are the application of novel lightweight materials such as magnesium alloys containing rare-earth metals, tailored-made coating with novel binders and fillers, and application of nanotechnology in self-healing coating through nano-capsules or nano-containers full of corrosion inhibitors. There is a need to further research to simplify the existing complicated and expensive procedures. Also, there is an increase in attention to biocompatible nano-coatings of medical implants and high-performance hydroxyl-apatite coating development.

Improved surface improvement processes as well as more rapid cost-efficient treatment processes like UV and electrostatic spray are poised to spearhead the following ten-year advancement in coating technologies.

Surface coating is of great importance to the study of tribology; through surface coating, it is possible to regulate friction, reduce wear, and prevent corrosion; all of which play a major role in regulating the life and efficiency of mechanical systems. These coatings can be used to improve the tribological performance of materials under a variety of operating conditions by modifying their surface properties, which in turn relies upon the engineering of surface characteristics of materials. Thin films, nanostructured layers, and composite systems are usually used in enhancing hardness of the surfaces, decreasing the strength of adhesion, and making surfaces lubricated. The choice of the coating and its design is determined by the needs of operation and the compatibility with the materials used as substrates. Fabricated processes; the most commonly used processes include Physical Vapour Deposition (PVD), Chemical Vapour Deposition (CVD), and sputtering - enable the deposition of coating to be carefully controlled in terms of coating depth, chemical content and crystal structure [4]. The development of efficient tribological performance requires in-depth knowledge of the coating mechanisms, the behavior of solid lubricants, diamond-like carbon (DLC) coating and self-healing materials [5]. The key to the creation of strong surface coating, which does not lose its power with heavy loads, adverse conditions, and extreme temperatures, is the mastery of the principles of surface engineering.

1.1 Classification of Tribological Coatings

Surface coating is a particular coating technique that is designed to address specific performance requirements in tribological applications. Diamond-like carbon (DLC) and other hard coatings such as transition metal nitrides have better abrasive and adhesive wear resistance. Solid lubricant-based and molybdenum disulfide (MoS2)-based self-lubricating films are highly effective in the reduction of interfacial friction as they emit lubricious phases during the process [6, 7]. Ceramic finishes, in their turn, offer high wear and corrosion protection in the chemically hostile or high temperature settings. Ultrathin nanocoatings provide an increase of tribological behavior due to accurate control at the molecular and atomic scale and make possible the adjustment of surface interactions and functionality. Composite and functional polymeric coats are flexible systems that can be used to achieve several tribological advantages in a single system. A thorough development of these types of coating such as their structural, mechanical and chemical properties are needed to develop rationally designed surfaces that are performance, reliable and long-lasting.

1.2 Functional Mechanisms and Characteristics of Advanced Coatings

Engineered microstructures and surface characteristics used in developing tribological advanced surface coating heed the optimum interaction at sliding interfaces. Modern deposition technologies, including PVD, CVD, and Atomic Layer Deposition (ALD) allow them to be used to provide high uniformity, high adhesion, and accurate control of coating architecture. These coatings can be specialized to suit a particular tribological requirement by optimizing the variables of material composition, thickness of the layer and nano structural features [8, 9]. Of special concern are self-healing materials that can mend their micro-scale level surface damages independently thus maintaining the integrity of tribological interfaces throughout long-term operation. Figure 1 below explains how thermal barrier coatings (TBC) technologies developed in the 1940s through to the 2000s, with the essential developments being low-temperature enamel coatings, through the use of plasma spray and electron beam deposition technology [10–12]. As of 1990, aerospace operating temperatures are up to 1200 °C, with the higher-end superalloys, and 2000 to 2005 was the introduction of nanotechnologies and composite materials to increase aerospace performance.

2 Fundamentals of Surface Coating Technologies

The growing need in environmentally friendly manufacturing has attracted the concern regarding the negative environmental and health effects of the traditional metal cutting processes that still heavily depend on metalworking fluids (MWFs) as a means of lubrication and cooling. Multilayer finishes based on a blend of TiN, TiCN, TiAlN, and Al_2O_3 have been promising in increasing wear resistance, thermal resistance, and surface integrity especially when it comes to machining hardened steels like AISI 4340 [13–15]. Whereas CVD coating provides dense and high-adhesive layers that can be used in applications with high-load, PVD coating offers high-quality performance at lower temperatures and is utilized more in precision finishing. Even in light of these developments, there is still research deficiency in the comparison of the performance of multilayer-coated tools made of carbide in aggressive dry machining conditions [16]. Deposition of coating techniques are of various types, are summarised in Fig. 2.

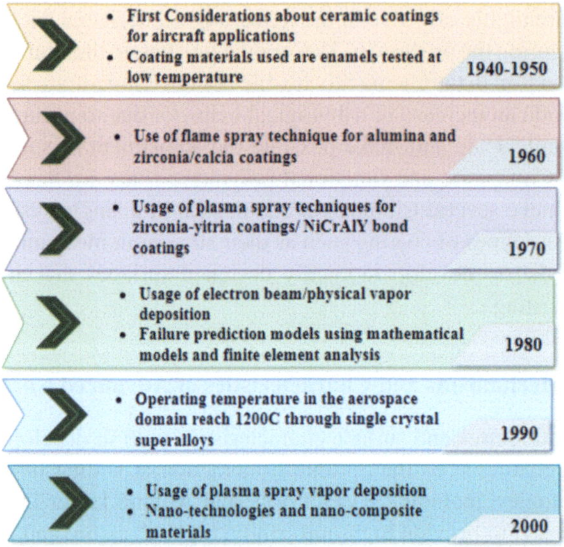

Fig. 1. Development phase of Tribological coatings (Source: Author's contribution)

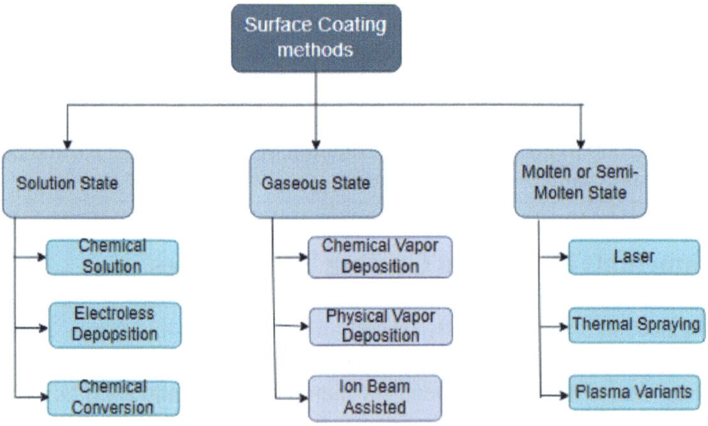

Fig. 2. Methods of Surface Coating (Source: Author's contribution)

2.1 Physical Vapor Deposition (PVD)

This is a group of non-equilibrium coating methods, which allow the formation of metastable solid solutions, since it is not restricted to the normal thermodynamic constraints. The PVD techniques can be generally classified into two categories of sputtering and evaporation depending on the mode of transfer of the solid precursor (or target) material into the vapor phase [17, 18]. The greatest benefit of PVD is that it is capable of being processed at low temperatures and therefore is best suited to use with substrates that are heat sensitive, in many cases, deposition at or near room temperature is feasible. Reactive PVD is used in most advanced applications. The substrate may be grounded or they may be biased negatively leading to energetic ion bombardment [19]. Due to the nature of the high-energy environment in deposition, PVD formed coatings can have small grain sizes, a high concentration of defects, and compressive stresses.

2.2 Chemical Vapor Deposition (CVD)

It is a type of fabrication technique using thin-film in which gaseous reactants are injected into a deposition chamber and chemical reactions take place at the substrate surface or close to the substrate surface to create a coating of a solid. A significant subdivision of the process is Plasma-Enhanced CVD (PECVD) in which the reaction is initiated and propelled by using the energy of a plasma instead of heat. The characteristic substrate temperatures used in thermally activated CVD are 800 °C to 1150 °C, coupled with chamber pressure which determines the behaviour of gas flow, development of boundary layers and the diffusion of reactive species to the surface [18–20]. CVD is more thermodynamically controlled compared to the Physical Vapor Deposition (PVD), but is still limited to kinetic constraints making it a non-equilibrium process. The rate at which the film grows is normally restricted by either one of the two, (1) Reaction kinetics at the surface or (2) Mass transport of the precursors through the boundary layer. CVD has some major benefits like high conformability as it can be used to coat complex and high-aspect-ratio geometries. It is also scalable to large batches production and can attain higher coating thicknesses [20, 21].

2.3 Physical Vapor Deposition (PVD) and Chemical Vapor Deposition (CVD)

PVD and CVD are two popular methods of coating substrate surfaces with thin films which work in different modes and have different applications. PVD is based on the physical transformation of a solid substance into a vapor in an enclosed chamber which is then deposited on the substrate as an ultra-thin film. This is usually carried out by techniques that include thermal evaporation and sputtering where atoms are removed by a plasma ion bombarding a target [22–25]. On the other hand, the process of CVD consists in the introduction of reactive gases into the chamber and the formation of the coating on the substrate via a chemical reaction that happens on the surface. PVD is done at fairly low temperatures (up to 450 °C) and is regarded to be safer to the environment since it does not require dangerous gases. The decision between the two methods is based on the desired property of the coating, purity, temperature stability, price, and surface complexity.

2.4 Thermal Spray Coating

It can be described as a collection of surface engineering procedures that uses a strong source of heat, i.e. a plasma arc, electrical current, or even burning reaction, to heat a coating material. These finishes are needed very much due to their superior wear and corrosion resistance. As opposed to traditional techniques, including electroplating, CVD, or PVD, thermal spraying allows the deposition of far more significant entrenchments, and they have a 20-micron-thickness to several-millimeters-thickness [25]. The major benefit of this technology is that it is flexible in terms of materials. These processes can be classified according to their technical characteristics and the characteristics of the coating that is formed (see Table 1). The typical ones are Plasma spraying, High-Velocity Oxy-Fuel (HVOF), Cold spraying, Warm spraying, and Arc wire spraying [26–28].

Table 1. Advantages and disadvantages of types of thermal spray coatings [29]

Deposition Process	Source	Substrate Material	Advantage	Disadvantage
PVD	Physical	AISI M2 steel, SS, glass, Si, potassium bromide (KBr)-carbon-Au-Al, Ag-Au-Cu-Al	Corrosion and wear resistance/thin film deposition is possible/adjustable mechanical, corrosion, and aesthetic properties	Requires a high vacuum/corrosion resistance is affected by abrasion/degradation control is challenging for polymer deposition applications
CVD	Chemical	Glass, Si, Kleenex, Ni-Co-Fe	Corrosion and wear resistance/deposition of various types of materials with different microstructures/works with low and atmospheric pressures	Requires ultra-high vacuum/requires heat-resistant substrates/small amount of coating materials waste
Plasma	Thermal	SS, steel, AISI 4140 steel	High corrosion and wear resistance/high substrate adherence/surface modification of engineering polymers, rubbers, metals, and fibers/anti-stick coatings	A low-temperature process that is mostly used for materials that cannot perform reactions in atmospheric pressure to modify the surface of the substrate/requires a heat source

(continued)

Table 1. *(continued)*

Deposition Process	Source	Substrate Material	Advantage	Disadvantage
HVOF	Thermal	Ti-6Al-4V, Inconel 738 metal, AISI, 4340 SS	High density of coating layer and well substrate adherence/works for non-conductive substrates/corrosion and wear resistance	Requires a small range of powder size (5–60 μm) with a narrow size distribution/numerous process variable to change the coating structure/requires a heat source
Cold Spray	Physical	Ti-6Al-4V, Al 6061-T6, Al 6061	Simple and cheap method compared to the other thermal spray methods	Limited operation range/mostly used for soft and hard metal substrates/low efficiency and reliability due to low temperatures/not useful extremely harsh environments

During the thermal spraying, the molten coating materials are sprayed onto a substrate to be sprayed. Table 1 and parameters discuss advantages and disadvantages of types of thermal spray coatings. Among the most significant advantages of the thermal spray method is that it is able to deposit a coating of between 15 micron to a few mm in thickness on substrates with large surface areas at a high rate of deposition than the other conventional methods of coating.

3 Emerging Trends in Surface Coating Materials

3.1 Self-Healing and Smart Coatings

Smart self-healing coatings have been outstanding innovation in the corrosion protection community, which has seen them to have an edge in functionality over traditional passive coatings. These coatings have been designed in a way that they self-heal the damages in order to enhance the life and dependability of metallic components used in the corrosive conditions. More recent development on the design of stimulus responsive self-healing coatings and more so the use of smart encapsulation technologies have resulted in responsiveness and functionality [22, 23].

3.2 Eco-Friendly and Sustainable Coatings

Green Coatings have become a critical element of sustainable developments in the construction, industrial, and infrastructure applications. These coatings are developed to have a great reduction of environmental impact through minimizing the emission of volatile organic compounds (VOCs), the use of renewable or bio-based materials, and energy conservation through reflective and low-emissivity technologies. Such coatings, in addition to minimizing ecological footprints, improve the quality of indoor environment, facilitate the adherence to sustainability standards, such as LEED and BREEAM, and provide high rates of durability and operational survival [1, 26].

3.3 High-Entropy Alloy Coatings

HEAs are a system of alloys that consists of five or more major elements in near-equiatomic amounts, and is no longer described as a conventional alloy design. As opposed to traditional alloys, which are formed by a single dominant element, with small additions, HEAs take advantage of high configurational entropy to stabilize solid solution phases, which is a unique combination of mechanical and functional properties. High-Entropy Alloy (HEA) coatings possess a number of different strengths compared to the traditional alloy systems, in the first place because of its multi-elemental structure. As opposed to the conventional alloys, which are based on a pre-dominant base element, HEAs offer extreme design flexibility, with the capability to design custom mechanical, thermal, and chemical properties to address the needs of a particular application. They provide different kinds of coatings as indicated in Fig. 3.

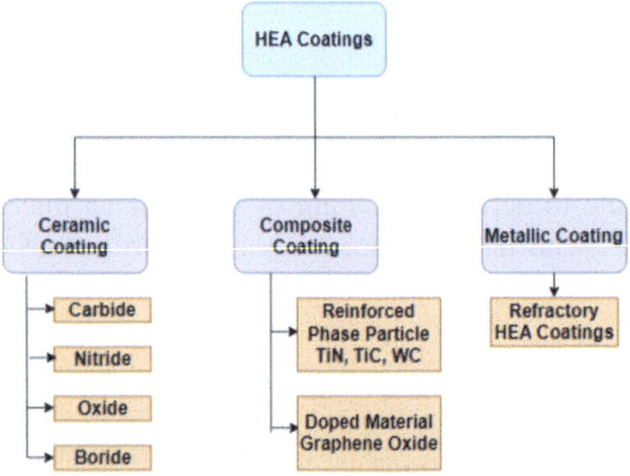

Fig. 3. Types of HEA Coatings (Source: Author's contribution)

4 Industrial Applications of Tribology

As the basis of improving mechanical performance, service life, and serviceability in a myriad of industries, tribology is used as the basis of operational integrity and long service life. Tribological interventions help a great deal in saving energy and extending the life of components as well as the optimization of system by successfully lowering friction and suppressing the wear of surfaces.

4.1 Automotive Tribology (Innovations in Engine and Drivetrain Technologies)

Tribological science has been able to impact the continuous development of automotive systems in a number of ways. High-performance lubricants with synthetic base and performance enhancers are extremely important in ensuring stable films under thermal and mechanical loads, which enhances fuel efficiency and fatigue in components [27].

4.2 Aerospace Tribology (Solutions to High-Stress Environments)

Aerospace industry places strict tribological demands because of the severe and fluctuating conditions of operation in the air. Bearings, seals, and actuators, and so on, have to be reliable even in terms of high temperatures, vacuum, and considerable mechanical loads. To overcome these issues, high-performance coating such as ceramics and solid lubricants are used such that wear is minimized and surface integrity is guaranteed.

4.3 Tribology in Manufacturing (Enhancing Equipment Longevity and Performance)

The concept of tribological optimization has become indispensable to the modern manufacturing environment as far as it is concerned with high reliability and a low downtime of equipment. Intelligent lubrication systems where real-time monitoring of the condition and adaptive control are integrated has made it possible to implement proactive maintenance planning, minimizing both planned and unplanned downtime [28]. The Fig. 4 can be used to see detailed applications and types of nano-structured coated used.

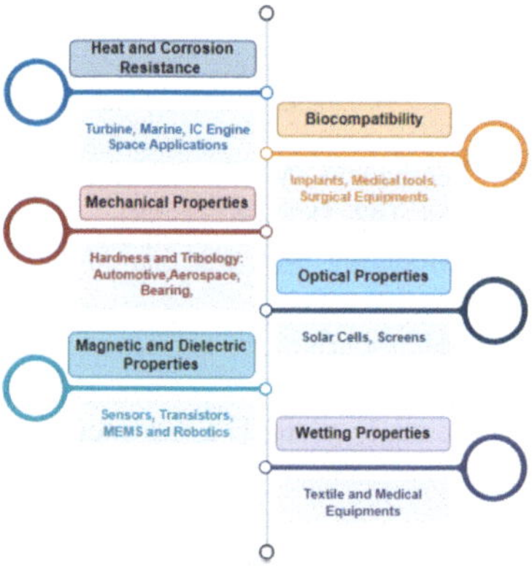

Fig. 4. Application areas of nano-structured coatings (Source: Author's contribution)

5 Conclusion

Tribology is a supporting field of research in the development of the longevity, effectiveness, and sustainability of subsequent mechanical devices. Tribological solutions have been innovated through material development, surface engineering, and lubrication science to keep the system performance on a steady improvement path with less system loss through wear and friction. These smart solutions do not only enhance the use of assets and efficiency of the processes, but they also enable long-term sustainability objectives. In spite of these achievements, there are still issues that are challenging to deal with, especially the behavior of tribology in extreme or highly varying conditions. The need to design environmentally benign lubricants and wear resistant materials is also increasing. The key to overcoming these barriers is to continue interdisciplinary studies, in particular, in the field of biomimicry, nanoscale science, and AI-assisted simulation. In the future, tribology is expected to be part of the growth of powerful, robust, and intelligent engine designs and to significantly contribute to the technological and ecological needs of the new manufacturing.

Acknowledgments. No financial constraint has been considered in the study.

Conflict of Interest. The authors have no conflicts of interest to declare.

References

1. De Damborenea, J., Navas, C., García, J.A., Arenas, M.A., Conde, A.: Corrosion–erosion of TiN-PVD coatings in collagen and cellulose meat casing. Surf. Coat. Technol. **201**, 5751–5757 (2007)
2. Mathew, M.T., Ariza, E., Rocha, L.A., Fernandes, A.C., Vaz, F.: TiCxOy thin films for decorative applications: Tribocorrosion mechanisms and synergism. Tribol. Int. **41**, 603–615 (2008)
3. Takeno, A., Okui, N., Kitoh, T., Muraoka, M., Umemoto, S., Sakai, T.: Preparation and piezoelectricity of β form poly(vinylidene fluoride) thin film by vapour deposition. Thin Solid Films **202**, 205–211 (1991)
4. Luff, P.P., White, M.: The structure and properties of evaporated polyethylene thin films. Thin Solid Films **6**, 175–195 (1970)
5. Prengel, H.G., Pfouts, W.R., Santhanam, A.T.: State of the art in hard coatings for carbide cutting tools. Surf. Coat. Technol. **102**, 183–190 (1998)
6. Jensen, K.F., Graves, D.B.: Modeling and analysis of low pressure CVD reactors. J. Electrochem. Soc. **130**, 1950–1957 (1983)
7. Maruyama, T., Kanagawa, T.: Electrochromic properties of niobium oxide thin films prepared by chemical vapor deposition. J. Electrochem. Soc. **141**, 2868–2871 (1994)
8. Doll, G.L., Mensah, B.A., Mohseni, H., Scharf, T.W.: Chemical vapor deposition and atomic layer deposition of coatings for mechanical applications. J. Therm. Spray Technol. **19**, 510–516 (2010)
9. Malshe, A.P., Park, B.S., Brown, W.D., Naseem, H.A.: A review of techniques for polishing and planarizing chemically vapor-deposited (CVD) diamond films and substrates. Diam. Relat. Mater. **8**, 1198–1213 (1999)
10. Karthikeyan, J., Berndt, C.C., Tikkanen, J., Reddy, S., Herman, H.: Plasma spray synthesis of nanomaterial powders and deposits. Mater. Sci. Eng., A **238**, 275–286 (1997)
11. Bulloch, J.H., Callagy, A.G.: An in situ wear-corrosion study on a series of protective coatings in large induced draft fans. Wear **233–235**, 284–292 (1999)
12. Fernández, J., Gaona, M., Guilemany, J.M.: Effect of heat treatments on HVOF hydroxyapatite coatings. J. Therm. Spray Technol. **16**, 220–228 (2007)

13. Scrivani, A., Bardi, U., Carrafiello, L., Lavacchi, A., Niccolai, F., Rizzi, G.: A comparative study of high velocity oxygen fuel, vacuum plasma spray, and axial plasma spray for the deposition of CoNiCrAlY bond coat alloy. J. Therm. Spray Technol. **12**, 504–507 (2003)
14. Sabard, A., de Villiers Lovelock, H.L., Hussain, T.: Microstructural evolution in solution heat treatment of gas-atomized Al alloy (7075) powder for cold spray. J. Therm. Spray Technol. **27**, 145–158 (2018)
15. Dean, S.W., Potter, J.K., Yetter, R.A., Eden, T.J., Champagne, V., Trexler, M.: Energetic intermetallic materials formed by cold spray. Intermetallics **43**, 121–130 (2013)
16. Paulussen, S., Rego, R., Goossens, O., Vangeneugden, D., Rose, K.: Plasma polymerization of hybrid organic–inorganic monomers in an atmospheric pressure dielectric barrier discharge. Surf. Coat. Technol. **200**(1–4), 672–675 (2005)
17. Leroux, F., Campagne, C., Perwuelz, A., Gengembre, L.: Fluorocarbon nano-coating of polyester fabrics by atmospheric air plasma with aerosol. Appl. Surf. Sci. **254**(13), 3902–3908 (2008)
18. Wang, S., et al.: Surface functionalization of metal and metal oxide nanoparticles for dispersion and tribological applications—A review. Journal of Molecular Liquids 389 (2023)
19. Ren, X., et al.: Surface modification technologies for enhancing the tribological properties of cemented carbides: A review. Tribology International 180 (2023)
20. Liang, Y., et al.: A review on coatings deposited by extreme high-speed laser cladding: Processes, materials, and properties. Optics & Laser Technology 164 (2023)
21. Wang, Y., et al.: Development of nano biochar as a lubricating oil additive for tribological applications. J. Cleaner Prod. 421 (2023)
22. Kumar, D.: Recent advances in tribology of high entropy alloys: A critical review. Progress in Materials Science 136 (2023)
23. Hou, Q., et al.: Tribological performance of hydrophobic and micro/nano triangle textured rake face of cutting tools. Applied Surface Science 571 (2022)
24. Peymanfar, R., Javidan, A., Javanshir, S.: Preparation and investigation of structural, magnetic, and microwave absorption properties of aluminum doped strontium ferrite/MWCNT/polyaniline nanocomposite at KU-band frequency. Journal of Applied Polymer Science **134**(30) (2017)
25. Musbah, A., Eljaafari, W., Fessatwi, E., Elsahli, Y.: Comparison between chemical vapor deposition (CVD) and physical vapor deposition (PVD) coating techniques: A review paper (2022)
26. Chandler, D.L.: Explained: Chemical vapor deposition. MIT News, Massachusetts Institute of Technology (2015)
27. Selvakumar, N., Barshilia, H.C.: Review of physical vapor deposited (PVD) spectrally selective coatings for mid- and high-temperature solar thermal applications. Sol. Energy Mater. Sol. Cells **98**, 1–23 (2012)
28. Guo, X., et al.: Adv. Energy Mater. **14**, 2402015 (2024)
29. Rahman, M.M., Kim, J.-G., Kim, D.-H., Kim, T.-W.: Micromachines **10**(6), 361 (2019)

Smart Integration of Green Roofs and Constructed Wetlands: Toward Sustainable, Resilient, and Liveable Cities

N. R. Asha Rani[1(✉)], Sheetal Thapa[1], and N. G. Devaiah[2]

[1] Department of Civil Engineering Alliance College of Engineering and Design, Alliance University, Anekal Bangalore, Karnataka 562106, India
asharaninr@gmail.com

[2] Alliance School of Law, Alliance University, Anekal Bangalore, Karnataka 562106, India

Abstract. Rapid urbanization has worsened urban and pluvial flooding, thermal comfort and environmental destruction. This paper develops a quantitatively defined framework, which combines green roofs with constructed wetlands in what would be a two-stage, water-sensitive, treatment train in cities. The retrofit feasibility methodology accounts to the retrofit identification of hydro-logic modeling (SWMM 5.0) which consists of adjustments of the locally representative storms (2-, 10-, and 25-year IDF); and the use of GIS based siting which involves engineering capacity checks. We stipulate provisional design requirements to inform initial sizing and assessment: depth of green-roof substrate 80 to 150 mm (extensive) with a runoff coefficient 0.3 to 0.5; peak-flow attenuation 20 to 40%; wetland hydraulic retention time of 24 to 72 h, pollutant removal thresholds of TSS 60 to 80%, TN 30 to 50% and TP 40 to 60%. Predominant challenges such as structural loads, scanty land, hydraulic performance, climate variability, have been linked to levers (module selection, media trains, bypass design, sensitivity runs) using a challenge-mitigation map. An elementary IoT design (sensors gateway cloud dashboard rule-based operations) allows flexible process and open KPIs. Trajectories of policy and equity (performance-based standards, incentives, interagency delivery, equity-centred siting) are also described to speed up diffusion. Together the framework goes beyond descriptive concepts to implementable and monitorable integrated systems of resilient city water management.

Keywords: Green roofs · built wetlands · sustainable urbanization/cities · nature-based solutions · stormwater · climate resiliency · ecosystem · water-sensitive · smart cities · infrastructure urban

1 Introduction

The high rate of urbanization interferes with the normal hydrological patterns, increases the pluvial flooding, urban heat islands, and poor urban biodiversity. Nature-based solutions (NBS) like green roofs and constructed wetlands have already shown the ability to attenuate runoffs, purify water and bring supplementary ecological and social values

L. Choudhary et al. (Eds.): TIMES 2025, SCI, pp. 284–300, 2027.
https://doi.org/10.1007/978-3-032-34636-0_21

alone (Adebanjo et al., 2025). Green roofs trap and temporarily store rainfall, reducing the peak flow and harvesting of precipitation, and enhance building thermal characteristics; green roofs performance relies on the depth of the substrate, the type of vegetation and the drainage system. Constructed wetlands offer decentralized treatment through sedimentation, plant uptake, sorption, and microbial processes, having a configurable surface-and subsurface-flow typology adapted on a case-by-case basis in the urban setting.

Recent research indicates that this combination is synergistic: with roofs supplying first-flush interception and pre-filtration and wetlands supplying polishing and infiltration, such systems establish a multi-train treatment train that repurposes vertical and horizontal space in cities efficiently, delivering multimedia treatment by treating stormwater pathways (Killeen et al., 2025). However, integrated systems are not well developed, with promising pilots but little overall capacity (Adeoba et al., 2025). Literature is qualitative, and standardized design parameters, hydraulic performance guidelines, or validation is limited to specific hybrid roof-wetland designs; structural feasibility of retrofit is sometimes argued without verification; and operationalization through smart monitoring/IoT and governance is currently underdeveloped. Such gaps hamper translation of concept to practice and both reviewers highlighted the same (Debsarma & Sahu, 2025; Irvine et al., 2023).

The proposed integration of green roofs with an oculus into a constructed wetland (in this case, the built-in quantitatively specified construction, and its direction to the structural capabilities of the city managing stormwater, improving water quality, and providing livability and co-benefits) is a practical conceptual design that is tangential (Bello et al., 2024). The paper contributes, based on pre-existing evidence and fills the gaps identified:

- The design parameters and performance targets of integrated systems (e.g. substrate depth 80 150 mm; runoff coefficient 0.3 0.5; target peak-flow attenuation 20 40%; wetland hydraulic retention time 24 72 h; indicative removal; TSS 60 80% TN 30 50% TP 40 60) to facilitate initial sizing and benchmarking (Ogwu et al., 2025).
- A SWMM 5.0 hydrologic modeling process employs locally representative IDF storms (2-, 10-, 25-year storms) and GIS-based siting that focus on the rooftops/catchments with an aim to prioritize rooftops and evaluate the benefits of a district-scale solution (Gupta et al., 2026).
- Structural and engineering considerations on retrofit feasibility, i.e., dead-load ranges in extensive/ intensive systems, and advice on hydraulic connectivity between roof drains and compact subsurface wetlands (Sharma et al., 2025; Singh et al., 2024).
- A challenge mitigation map that relates technical, operational, social and regulatory barriers to active levers (design, modeling, monitoring, governance) enabling action and aligning implementation goals with performance acceptance criteria (Singh et al., 2024).
- an intelligent monitoring/IoT structure and intelligent O&M process (sensors, gateway, cloud board, control according to the rule) to enhance reliability, transparency and maintainability (Du et al., 2024).

- Policy and equity channels: performance-based standards, incentives, inter-agency delivery models and equity-focused sitting to fast-track diffusion in a way that disseminates benefits (Whig et al., 2025).

The subsequent subsections review the literature and limitations on hybrid green roof green/constructed wetlands systems; introduce methodology and conceptual framework, quantitative design parameters and modelling scenarios; report on implementation challenges and map to mitigation measures and checkpoints; generate post-implementation next steps including smart operations, standardization, materials developments, climate resilience, governance and equity; and finish with recommendations and pilot validation roadmap (Chaturvedi et al., 2025; Oyenuga et al., 2025).

2 Literature Review

Green roofs and wetlands have been the individual nature-based solutions that have been studied extensively over the past twenty years, and they exhibit a great deal of promise in alleviating the negative impacts of fast urbanization. Vegetated roofs minimise storm water runoff, as well as improve the energy efficiency of buildings and offer urban micro-habitants, via vegetative thermal control and evapotranspiration, in urban environments (Awasthi et al., 2025; Hullurappa & Kommineni, 2024). They are only effective with respect to substrate depth, plant species and climatic conditions, with extensive systems providing lightweight retrofitting opportunities, and intensive systems facilitating greater ecological functions at increased structures and maintenance costs. Although these are some of the advantages, problems that face them are the long-term substrate performance, water leak risk, and structural maintenance (Pal, 2025).

Constructed wetlands, in turn, offer dispensed-based storm weaving and wastewater treatment-effectively eliminating nutrients, heavy metals, and pathogens through physical, biological, and chemical means. They also provide various co-benefits such as wildlife habitat, and aesthetics/recreational area. They can be flexibly designed (surface flow, subsurface flow, or hybrid wetlands) and can therefore be applied in a variety of situations, although there are limitations due to the limited land availability in heavily populated regions, seasonal variation in treatment performance, and odour annoyance or mosquito breeding to residents nearby (Hachem-Vermette, 2024; Seelam et al., 2024). Both systems have now matured as independent technologies, although they are not extensively integrated. New research proposes that combining green roofs and constructed wetlands may form multi-staged New writing proposes that new combinations of green roofs and constructed wetlands may lead to multi-staged Since this is not the case, the proposed combination of constructed wetlands and greenhouses may form multi-staged Potential co-benefits are documented in pilot studies as improved water quality, thermal comfort, as well as space efficiency, and integration occurs both vertically and horizontally in the city as described in Incorporating green infrastructure in the urban environment. As an instance, point out that coupled systems present opportunities for stormwater volume decrease and pollutant elimination without mentioning the design difficulty and lack of standard performance level (Pathak et al., 2024; Rathinavel et al., 2025).

Most of the available work is, however, conceptual or descriptive. Inadequate empirical performance data, design calculations or validated modeling output exist on hybrid frameworks. Qualitative references to hydraulic parameters, including peak flow attenuation, pollutant removal rates, retention times, and so on are relatively frequent without mentioning benchmarks or any case study data (Ali & Sultana, 2024). Likewise, structural feasibility, especially retrofit loading restrictions of green roofs, has not been stringently defined in the light of match-in with wetlands (Ali & Sultana, 2024; Cheshire, 2024). Such quantitative inability limits the transfer of such theoretical advantages into concrete engineering applications. The other gap is in the integration of the technologies systemic and operationally (Cheshire, 2024). It does discuss the challenges of stakeholder coordination and some of the infamous issues like stakeholder mismanagement and public perception, but never in a coherent operative-let alone a governance-setting (Kumareswaran & Jayasinghe, 2023a). Further, the use of smart monitoring and adaptive management enabled by IoT, which has the potential to enable real-time performance verification, is just emerging on the radar and remains untested in practice at this point through pilot projects (Sommese et al., 2025).

In brief, the available research documents the ecological and hydrological values of green roofs and constructed wetlands in isolation but cannot fully address:

- Quantitative validation of performance on integrated systems (Catalano et al., 2021).
- Regular design and hydraulic requirements that can be used on hybrid structures (Mitra et al., 2025).
- Structural and engineering of rooftop-wetland coupling (Bali, 2023).
- Smart monitoring and adaptive governance operating strategies (Abouaiana, 2025; Dey et al., 2025).

This review reinforces why a more rigorous and interdisciplinary mode is needed to support and present a framework of stepwise association incorporating empirical modeling with an integration of design parameters with policy considerations to contemplate a feasible steppingstone to the practice of an integrated green roof-constructed wetland system in a sustainable city.

3 Methodology/Conceptual Framework

The conceptual framework of the proposed solution of green roofs and constructed wetlands integration as a single unit is a multiscale and multi-purpose installation that will be better at storm water management and optimizing ecosystem services to the urban environment. Its approach method will be anchored on 3 associated stages viz. (i) hydrological design and performance modelling, (ii) structural and engineering and (iii) stakeholder integration and scalability.

3.1 Hydrological Designing and Modeling Strategy

The system envisaged to be integrated is considered a two-step process on treatment. The green roof acts as the main interception and pretreatment units, rainfall is intercepted, peak flow is decreased, and components are also filtered. The partially treated water is then channeled to the constructed wetlands where secondary treatment, that is, the uptake of nutrients as well as removal of pollutants and infiltration, takes place (Abouaiana, 2025; Dey et al., 2025; Riffat et al., 2025).

To measure the performance of the system, Storm Water Management Model (SWMM 5.0) will be utilized (Riffat et al., 2025). The simulation scenarios will simulate 2-year, 10-year and 25-year return-period storm events as represented by local intensity-duration-frequency (IDF) curves. Input parameters into the model are:

- Roof catchment 500-1000 m^2 (average midrise buildings) (Riffat et al., 2025)
- Substrate depth: 80-150 mm (extensive) and around 300 mm (intensive).
- Vegetation runoff coefficients% $0.3 = 0.5$ (dependent on type of species).
- Wetland HRT (h) 24-72h.
- Removal criteria of pollutants-Total suspended solids (60–80 percent), Nitrogen (30–50 percent), phosphorous (40–60 percent).

There will be spatial analysis using GIS to complement modeling by highlighting possible areas of deployment based on land use, rooftop density, slope, and intensity of the urban heat island. The areas where the demand of the storm water mitigation is highest will be given priority under the GIS overlay (Calheiros & Stefanakis, 2021; Pandey et al., 2021).

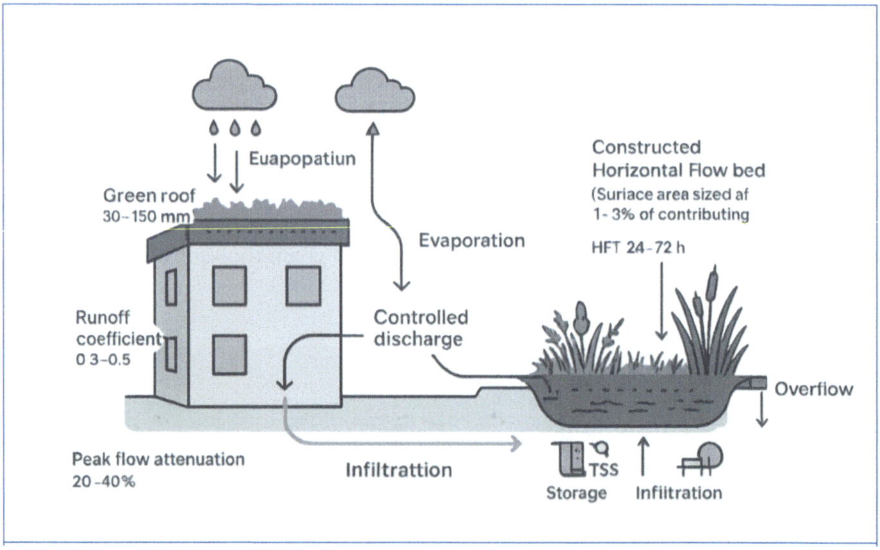

Fig. 1. Conceptual Schematic of Integrated green roof and constructed Wetland System (Source: Author's contribution).

3.2 Structural /Engineering

The implementation of retrofitting green roofs may be limited due to load bearing capacities of buildings. Structural analysis implies the following:

- Large scale green roofs: 60 to 150 kg/m^2 dead weight (Kumareswaran & Jayasinghe, 2023b).
- Intensive green roofs: > 300 kg/m^{2}dead load.

The values will be compared with HL impressed loads for typical reinforced concrete slabs (4–6 kN/m in Indian urban limits). In the case of retrofits, engineered substrate modular lightweight trays have been proposed to have a minimal load effect whilst retaining the functionality of retention of the stormwater (Mudunuri & Attaluri, 2024).

In the case of constructed wetlands design, we will consider the surface flow and subsurface flow systems with regards to land availability. Subsurface horizontal flow wetlands should be compact in size (usually 1 to 3 percent of contributing drainage area) and are suggested in densely built-up locations. Gravity-fed hydraulic connections between roof drains and wetlands will be designed based on topographic relief; or pump-fed systems can be used in cases where the relief is not sufficient (Aragani & Maroju, 2024; Bressane et al., 2024).

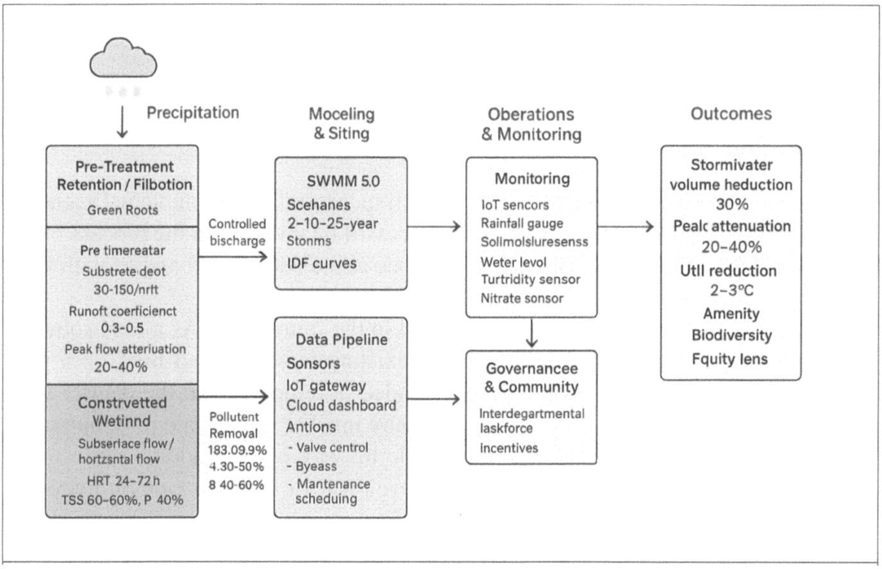

Fig. 2. Framework flowchart hydrologic design (Source: Author's contribution).

3.3 Conceptual Framework

The logic of the system is presented in Figs. 1 and 2.

Figure 1 shows the schematic indicator of hydrologic flow paths and typical design parameters an example is the depth of substrate, drainage coefficient, peak-flow attenuation rates and wetland HRT and Fig. 2 shows the flowchart which Traces the rational pathway between stormwater harvesting and ecosystem services, all the connections between the technical aspects of design and the social and policy realms. Table 1 Provides representative obtained sizing ranges as well as performance goals of the integrated green roof-constructed wetland system (e.g., substrate depth, runoff coefficient, wetland

HRT, media porosity), these should be used to guide the SWMM/GIS workflows and/or initial design checks.

Table 1. Design parameter matrix (Salman et al., 2025).

Component	Parameter	Indicative Range/Value	Performance Benchmark
Green Roof Substrate	Depth	80–150 mm (extensive), 300 mm + (intensive)	Runoff reduction 30–50%
Vegetation	Sedum/grasses/mixed	Runoff coefficient 0.3–0.5	Temperature reduction 2–3 °C
Constructed Wetland	Hydraulic Retention Time	24–72 h	TSS removal 60–80%
Wetland Media	Sand, gravel, engineered	0.6–0.8 porosity	Nutrient removal 40–60%
Integration	Peak Flow Attenuation	20–40% reduction (10-year storm)	Flood resilience

3.4 Integrating Stakeholders and Scaling

Participatory governance is needed to implement it successfully. The municipal agencies, community organizations, and private developers will be involved in the process of co-design workshops. This will guarantee acceptance at the social level and fit with local policy tools (Aboulnaga et al., 2024).

Modularity and scalability are also achieved in the framework. As a final solution, the integrated systems can be retrofitted into existing buildings and used on a pilot scale with further implementation on better-funded developments at the district scale [41]. Strict adoption should be phased and initially involve small scale pilot initiatives aided by subsidies and subsequently large scale by way of incorporation into urban planning codes. A conceptual framework is developed to scale up generic hydraulic representations into higher scales to accommodate quantitative hydraulic parameters, structure load standards, and modelling principles. This has made it a handsome theory as well as a possible guideline that could be used in the analysis and adoption of integrated green roof-constructed wetlands systems by engineers, urban planners and policymakers (Kumar et al., 2023; Pineda-Martos & Calheiros, 2021).

4 Challenges and Mitigation Strategies

Integrated green roof-constructed wetland systems are confronted with a group of technical, operational, social and regulatory obstacles. Each challenge below is matched with specific and framework-related mitigation actions to ensure reviewers can see precisely how the proposed framework can overcome feasibility, performance, and governance (Pineda-Martos & Calheiros, 2021).

4.1 Technical and Engender Issues

The structural load constrains are:

4.1.1 Structural Load Retrofits

Risk. Saturated loads imposed on green roofs may not be supported safely on existing buildings.

Engineering/design mitigation (Framework: Engineering & design).

When possible, use lightweight extensive systems and modular tray assemblies; ensure serviceability and ultimate limit states by structural check against slab capacity (extensive roof: ~60 150 kg m-^{2}; intensive: >300 kg m-^{2}).

Apply in conjunction with engineered substrates (low unit weight, high porosity) and incorporate overflow by-passes that restrict ponding on roof.

Phase retrofits: begin with pilot bays, instrument loads, after which scale (Dobjani et al., 2025; Pineda-Martos & Calheiros, 2021).

4.1.2 Small Urban Footprint of Wetlands

Risk. The surface-flow wetlands are limited due to constraints of surface flow on land in the densely populated districts.

Mitigation (Framework: Siting & Typology).

Install subsurface horizontal-flow beds with areas of ~133 to 333 sq ft (10 to 30%) contributing drainage area co-located in podiums, setbacks, or linear verges.

Apply stacked functions: a combination of amenity edges, habitat strips, treatment cells; integrate with the open-space codes (Dobjani et al., 2025).

4.1.3 Hydro Performance During Design Storms and Extremes

Risk. Qualitative designs do not do well in short intense storms.

Mitigation (Frame: Hydro-modeling).

Calculate the event magnitude (2-, 10-, 25-year events) and run SWMM 5.0 scenarios using local IDF curves with roof drainage layers/substrates sized to provide peak attenuation goal (20, 40%) and wetlands sized to achieve HRT (24, 72 h).

Add freeboard and bypass weirs; ensure roof to wetland discharge is controlled not to lead to surcharge.

Go to performance acceptance thresholds: e.g. 95 th percentile event unmoved to non-controllable burst (Dobjani et al., 2025).

4.1.4 Reliability of Treatment and Variation of Pollutant Removal

Risk. Removal or TSS capture may be suppressed by seasonal variation.

Mitigation (Package: Process Design & O&M).

Offer media training (graded gravel/sand + engineered media) and palettes of plants with 12 months uptake; where necessary design first-flush diversion.

Establish benchmarks: TSS 6080%, TN 3050%, TP 4060%; confirm on a quarterly basis using grab samples.

Install maintenance access (sediment sumps, removable trays) and clean-in-place schedule (Dobjani et al., 2025; Pineda-Martos & Calheiros, 2021).

4.2 Resilience to the Climate and Operations and Monitoring

4.2.1 Uncertainty in on the Platform of Climate Variability

Risk. Alterations in the patterns of storms can surpass the design envelopes.

Adaptive Management (Mitigation: Framework).

Integrate sensitivity runs (Increased rainfall depth and decreased inter-event interval), develop the provisions to be able to add modules later without restructuring.

Apply climate-resistant vegetation (drought/heat resistance) and dynamic outlets controls to provide real time detention.

4.2.2 Durability of Performance and O&M Load

Risk. The hidden compensation undermines the stake that owners have; reactive maintenance erodes performance.

Mitigation (Framework: Smart Monitoring).

IoT sensor on roof moisture and rainfall, water level, turbidity and nitrate; push to cloud dashboard with alerts and action on rules (valve set-points, bypass triggers, maintenance tickets).

Report KPIs monthly: detention time achieved, overflow and removal efficiency as well as UHI proxy metrics (Dobjani et al., 2025; Kumar et al., 2023).

4.3 Governance, Economic and Societal Approval

4.3.1 Fragmentation of Stakeholders and Unclear Roles

Risk. The integration is on hold by different actors (planning, utilities, and private owners) in delivery.

Mitigation (Governance Framework).

- Create an inter-departmental taskforce; use a RACI matrix in the design, O&M and compliance.
- Publicize standard technical specifications and a permitting checklist that can place modeling outputs under approvals.

4.3.2 Amenity, Health and Perception in the Eyes of the Public

Risk. Maintenance concerns, and perceived risks to combat mosquitoes and scents inhibit adoption.

Community Framework - Mitigation (Community & Amenity Framework).

- Employ cells with surface flow and ensure that inlets have continuous flows; have attractive planting and secure edges.
- Conduct co-design workshops or on-site demonstration pilots, place interpretive signage on ecosystem services and performance dashboards (Dobjani et al., 2025).

4.3.3 The Policy Tools and Funding

Risk. Uptake is slowed by the lack of enabling codes, and incentives.

Mitigation (Framework: Policy and Finance).

- Storm water fee credits, or density bonus, or capex subsidies on certified integrated systems.
- Demand planning codes (e.g. percent of design storm retained on site; reductions in annual pollutant loads) to be performance-based (Dobjani et al., 2025; Pineda-Martos & Calheiros, 2021).

4.4 Map of Challenges to Mitigation in a Framework

The other serious issue is to manage various stakeholders during planning, design, and implementation. There must be coordination between planners of cities or towns, builders, civil engineers, scientists, and policymakers, etc., in combination of green roofs with constructed wetlands, and this task can be weakened by the division of responsibilities along institutional lines as well as ambiguous rules and regulations. Coordination can be promoted by formation of interdepartmental task forces and development of uniform guidelines through which the decision-making process can be simplified (Pineda-Martos & Calheiros, 2021).

It is also important how the people will see and accept such systems. Usage of the wetlands may be restricted because of concerns related to maintenance, aesthetics, breeding of the vectors or fear of water stagnation etc. Extensive measures involving the community, educational programs, and pilot projects may be good approaches to creating trust and developing awareness and disseminating information concerning the positive environmental and social aspects of such systems (Dobjani et al., 2025).

Moreover, climate uncertainty is a threat to the sustainable functionality of combined green roofs and wetlands since the alteration of precipitation regimes as well as extreme weather conditions can surpass their design capacity. Mitigation measures may consist of efficacy-adaptive plant palettes, adaptive management plans, and monitoring of the continuity of efficiency of the system in response to so much uncertainty concerning the future (Dobjani et al., 2025; Pineda-Martos & Calheiros, 2021).

Overall, as issues of technical design, coordination among stakeholders, popular acceptance, and climate robustness prevail, in-situ mitigating measures and active policymaking have the potential to considerably improve the viability of introducing integrated green roofs and constructed wetland systems to urban dwelling (Tyagi et al., 2025). Table 2 Maps key implementation risks to levels of improvement (design, modeling, O&M, governance) and sets specific performance and compliance checkpoints to evaluate performance.

Table 2. Challenge Mitigation KPI Matrix (Source: Author's contribution).

Challenge	Risk to Outcomes	Framework Lever	Concrete Mitigation	Quantitative Checkpoints
Structural loads	Roof failure/scope reduction	Engineering & Design	Lightweight trays; engineered substrate; staged pilots	Roof dead load $\leq$ design capacity; deflection within L/360
Small footprint	Wetlands reduced	Siting & Typology	Subsurface HF beds; co-location; stacked functions	Area = 1–3% CDA; HRT $\geq$ 24 h
Peak flow control	Flooding/overflow	Hydrologic Modeling	SWMM sizing; controlled roof overflow; bypass weirs	Peak attenuation 20–40% (10-yr)
Pollutant removal	Water-quality non-compliance	Process & O&M	Media trains; first flush; scheduled maintenance	TSS 60–80%, TN 30–50%, TP 40–60%
Climate uncertainties	Capacity overrun	Adaptive Management	Sensitivity runs; scalable modules; resilient plants	No uncontrolled overflow (95th-percentile storm)
O&M load	Performance degradation	Smart Monitoring	IoT sensors; alerts; rule-based actions	Monthly KPIs; < 2 emergency callouts/yr
Disjointed actors	Delay/misalignment	Governance	Inter-agency taskforce; RACI; checklists	Approvals SLA $\leq$ 30 days
Poor image	Low uptake	Community & Amenity	Co-design; subsurface circulation; signage	$\geq$ 80% positive feedback in post-occupancy survey
Funding/policy	Slow diffusion	Policy & Finance	Fee credits; density bonuses; standards	$\geq$ 30% capex offset or equivalent NPV

5 Further Developments

Based on the integrated framework, we provide research to practice roadmap that promotes technical feasibility, operational reliability, and equitable diffusion of green roof constructed wetland systems (Tyagi et al., 2025).

5.1 Smart Surveillance and Smart Responsive Operations

To take a practical step beyond conceptual deployments, we outline a skeletal IoT plan and workflow deployable on pilot scale: roof soil-moisture, rainfall, drain flow, and wetland water-level/quality sensors (turbidity, nitrate) into local IoT gateway into cloud-based dashboard and data lake into rule-based control actions (valve set-points, bypass triggers, maintenance tickets). This will allow closed-loop adaptive system management and reporting of KPIs (e.g., the time spent detained, the number of overflows, pollutant removal, UHI proxy variable metrics). The possibility of such an approach has been justified by previous work on smart green infrastructure and within the IoT-enabled resilience (Mallick, 2024).

Recommended KPIs and targets:

- Peak-flow attenuation: 20-40per cent (10-year storm)
- Event-mean detention time in wetland: $> \ = 24$–72 h
- TSS removal targets: TSS 60–80%, TN 30–50%, TP 40–60%
- Sensor availability (uptime): $\geq 95\%$.

5.2 Modularization and Prefabrication

To eliminate retrofit limitations and construction delays, design featherweight modular roof trays and prefabricated wet land cassettes (subsurface horizontal flow) to insert into existing drain lines. Modular dimensions should be suited to manual handling and, where appropriate, quick-connect hydraulics and fit engineered media maximizing porosity and weight, and sorption capacity. Opportunities are highlighted in the literature of modular green roofs and engineered substrates (Martinez-Mireles et al., 2025; Zhang, 2025).

5.3 Innovation of Materials and Media

Engineered roof substrates of low unit weight and stable hydraulic conductivity, and ii. Wetland media with increased nutrient/metal sorption. It is suggested that a typical test suite be run; saturated hydraulic conductivity, drain-down curves, sorption isotherms (P & NH 4+), clogging susceptibility, and freeze-thaw resistance. Green-roof media and other green-roof components have been reviewed recently, forming a technical foundation for these advances (Martinez-Mireles et al., 2025; Zhang, 2025).

5.4 Climate-Resilient Design in Uncertainty

Future research needs to incorporate the sizing scenario (e.g., more (say +20–30) percent of rainfall depth; shorter inter-event time) and adaptive resilience (modular storage that can be expanded; adjustable orifice controls). Planting palettes must put an emphasis on heat- and drought-tolerant species to stabilize the evapotranspiration and uptake of extremes. The impetus behind this adaptive lens is community-scale research on the performance of green-infrastructure in changing climates (Martinez-Mireles et al., 2025; Zhang, 2025).

5.5 Governance, Economics and Codification

Retranslate performance-based policy into pilots (e.g. percent of design storm handled on site; annual load cuts) and incentives (stormwater fee credits; density bonus, capex subsidies). A RACI matrix can be introduced in an inter-departmental taskforce to facilitate approvals and O&M duties; sociotechnical approaches to stormwater management and metropolitan green-infrastructure plans offer precedents (Martinez-Mireles et al., 2025; Zhang, 2025).

5.6 Equity Based Sitting and Benefits

Prioritize the under-served neighborhoods using GIS overlays (flood exposure, heat-island intensity, socio-economic indices), integrate co-design workshops, and share performance through dashboards, public dashboards. Distribution in terms of advantages and tracking of social performance of green infrastructure should be under equity-oriented research (Zhang, 2025).

5.7 Data Standards/pilot Roadmap

A convenient order is suggested:

- Instrumented pilot (two retrofitted roofs + subsurface-flow wetland; 12 months monitoring).
- Calibration of models (use observed hydrographs/water-quality to calibrate SWMM-revise design curves).
- Specification package drawings media specs O&M playbook acceptance tests
- Standards adoption (policy uptake; standards adoption; standards adoption; standards adoption; activation of incentives).
- GIS siting-guided replication on a district level.

Open data conventions (metadata, QA/QC, versioned model files) should be used wherever possible by all pilots to facilitate replication and meta-analysis (Zhang, 2025).

Further steps in uniting green roofs and constructed wetlands must be related to implementing research in interdisciplinary design and technology development and policy encouraging the involvement of green roofs and constructed wetlands in cities. A potential solution is the implementation of intelligent monitoring devices such as IoT-related sensors and actual data analytics to monitor the performance of systems in the areas of stormwater retention, water quality and vegetation maintenance. These systems can improve the process of adaptive management and make it possible to conduct proactive maintenance, which makes integrated solutions last longer and perform better.

Moreover, inventions can be made on new modular and prefabricated components of green infrastructure that facilitates reuse and reduces costs, facilitates retrofits to older buildings and built-up areas. Development of lightweight, high-performance substrates and engineered wetland media may enhance the rate of removal of pollutants with a minimum structural requirement. Future-proof designs will also be enhanced by the combination of these materials and other plants that bear the characteristics of climate resilience such as heat, drought and fluctuating rainfall patterns (Thomas & Leon, 2025; Zhang, 2025).

Policy wise, it will be necessary to induce wider implementation by setting clear incentives, performance-based standards and codes of urban architecture. Incentives to invest in hybrid systems may be introduced through zoning regulations or granting stormwater fee reductions or density bonuses to the cities that integrate green infrastructure targets. Additionally, demonstration projects and living laboratories may become experimental people make use of as new integration ideas to have an interplay of academia, industry and the local governments (Thomas & Leon, 2025; Zhang, 2025). Lastly, the effect on equity and social justice of green infrastructure should also be considered more closely, and these gains should be highly distributed and not biased to benefit the more affluent neighborhoods. Future integration of integrated green roofs and constructed wetlands through hybrid innovation managed with inclusive planning and policy tools may have maximal ecosystem services, as well as make cities with integrated green roofs and constructed wetlands more resilient and livable for every resident (Thomas & Leon, 2025).

6 Conclusion

This paper summarized the available evidence that supports a combination of green roofs and constructed wetlands and turns it into a measurable, applicable theory of urbanization. The paper presents a pathway to engineered practice through the embedding of parameterized design targets (e.g., substrate depth 80 150 mm, peak-flow attenuation 20 40%, wetland HRT 24 72 h and pollutant removal benchmarks), a SWMM based modeling workflow, structural/load guidance within the context of retrofits, and a ready challenge mitigation map. Smart-monitoring architecture can be also used to promote adaptive O&M and performance checking due to regularly submitted KPIs.

Demonstration must be done through instrumented pilot technologies (one retrofitted roof feeding a small subsurface-flow wetland) over 12+ months of monitoring, proper model calibration, open specifications (media, hydraulics, acceptance testing), and adoption of policy, as standards and incentives. GIS overlays can then be used to prioritize flood-prone and heat-stressed under-served neighborhoods at the district level to design equitable benefits.

Although this framework provides the framework of quantitative benchmarks and operations, site-specific calibration, long-term costs of maintenance data, and multi-year annul performance tracking are still vital supplementary directions. These can be tackled by coordinated pilots and shared data standards to maintain integrated roof and wetland systems within the wider agenda of water-sensitive urban design, ecosystem-based adaptation and integrated water management.

References

Abouaiana, A.: Constructed wetlands as green infrastructure to improve quality of life (pp. 167–184) (2025). https://doi.org/10.1007/978-3-031-88318-7_11

Aboulnaga, M., Trombadore, A., Mostafa, M., Abouaiana, A.: Livability: the direction to mitigating urban heat Islands' effect, achieving healthy, sustainable, and resilient cities, and the coverage. In: Livable Cities (pp. 1–282). Springer International Publishing (2024). https://doi.org/10.1007/978-3-031-51220-9_1

Adebanjo, A.U., Shafiq, N., Kumar, V., Olatoyan, O.J., Abd Razak, S.N., Yusof, N.F.: Innovative green infrastructure initiatives for development in the construction industry. In: Sustainable Green Infrastructure (pp. 121–138). Springer Nature Singapore (2025). https://doi.org/10.1007/978-981-96-1486-8_7

Adeoba, M.I., Odjegba, E.E., Pandelani, T.: Nature-based solutions: opportunities and challenges for water treatment. In: Smart Nanomaterials for Environmental Applications (pp. 575–596). Elsevier (2025). https://doi.org/10.1016/B978-0-443-21794-4.00011-9

Ali, A.A., Sultana, N.: Innovative approaches to urban wastewater management (pp. 263–286) (2024). https://doi.org/10.4018/979-8-3693-6740-7.ch011

Aragani, V.M., Maroju, P.K.: Future of blue-green cities emerging trends and innovations in icloud infrastructure (pp. 223–244) (2024). https://doi.org/10.4018/979-8-3693-8069-7.ch011

Awasthi, S., Darshitha, B., Dwivedi, N.: Urban circularity challenges via nature-based solution. In: Wastewater Treatment Through Nature-Based Solutions (pp. 115–137). Springer Nature Singapore (2025). https://doi.org/10.1007/978-981-96-6699-7_6

Bali, V.: Handbook of sustainable development through green engineering and technology. CRC Press (2023)

Bello, J., Adetoro, P., Mogaji, I., Stephen, S.: Exploring the feasibility and benefits of integrating nature-based solutions in stealth construction practices. Intell. Build. Int. **16**(1), 23–36 (2024). https://doi.org/10.1080/17508975.2024.2378705

Bressane, A., et al.: Intelligent nature-based solutions in the 1st smart sustainable Brazilian City: insights and lessons learned. Nat.-Based Solutions **6**, 100161 (2024). https://doi.org/10.1016/j.nbsj.2024.100161

Calheiros, C.S.C., Stefanakis, A.I.: Green roofs towards circular and Resilient Cities. Circular Econ. Sustain. **1**(1), 395–411 (2021). https://doi.org/10.1007/s43615-021-00033-0

Catalano, C., et al.: Smart sustainable cities of the new millennium: towards design for nature. Circular Econ. Sustain. **1**(3), 1053–1086 (2021). https://doi.org/10.1007/s43615-021-00100-6

Chaturvedi, S., Karunarathna, R.M.D.M., Sanyour, A., Palihena, J.C.S., Singh, N., Xu, J.: Water footprint in sponge cities: how sustainable are they? In: Water Footprints (pp. 243–262). Elsevier (2025). https://doi.org/10.1016/B978-0-443-30054-7.00018-5

Cheshire, D.: Regenerative by design : creating net-positive buildings and cities. RIBA Publishing (2024)

Debsarma, C., Sahu, P.: Enhancement of urban drought resilience using green infrastructural strategies. In: Handbook of Nature-Based Drought Solutions (pp. 231–252). Elsevier (2025). https://doi.org/10.1016/B978-0-443-29993-3.00011-4

Dey, S., Moniruzzaman, Al-Asad, H., Ghosh, A.R., Samanta, P.: Smart integration: designing effective frameworks for enhanced storm water management in urban areas–Strategies and challenges in smart wastewater system implementation. In: Smart Wastewater Systems and Climate Change (pp. 14–27). Royal Society of Chemistry (2025). https://doi.org/10.1039/978 1837676880-00014

Dobjani, E., et al.: Sustainable living solutions : renewable energy and engineering : a culmination of selected research papers from the International Conference on Environmental Design, Material Science, and Engineering Technologies—Abu Dhabi University 2024. Springer Nature Switzerland (2025)

Du, P., Al-Kodmany, K., Ali, M.M.: The Routledge handbook on greening high density cities : climate, society and health. Routledge, Taylor & Francis Group (2024)

Gupta, V.P., Haghi, A.K., Yadav, A.: Green IoT and AI for sustainable development of smart cities. Springer Nature Switzerland (2026). https://doi.org/10.1007/978-3-031-98743-4

Hachem-Vermette, C.: Enhancing urban climate resistance through the application of selected strategies and technologies. Discover Cit. **1**(1), 17 (2024). https://doi.org/10.1007/s44327-024-00018-2

Hullurappa, M., Kommineni, M.: Integrating blue-green infrastructure into urban development (pp. 373–396) (2024). https://doi.org/10.4018/979-8-3693-8069-7.ch017

Irvine, K.N., et al.: Nature-based solutions to enhance urban flood resiliency: case study of a Thailand Smart District. Sustain. Water Resour. Manage. **9**(1), 43 (2023). https://doi.org/10.1007/s40899-023-00821-6

Killeen, T.L., Wuebbles, D.J., Lane, J.E.: Pathways to sustainability : collaborative solutions for a Resilient future. University of Illinois Press (2025)

Kumar, A., Ekka, P., Upreti, M., Shilky, Saikia, P.: Urban green spaces for environmental sustainability and climate resilience. In: The Palgrave Handbook of Socio-ecological Resilience in the Face of Climate Change (pp. 389–409). Springer Nature Singapore (2023). https://doi.org/10.1007/978-981-99-2206-2_23

Kumareswaran, K., Jayasinghe, G.Y.: Climate resilience and sustainable cities. In: Green Infrastructure and Urban Climate Resilience (pp. 39–97). Springer International Publishing (2023a). https://doi.org/10.1007/978-3-031-37081-6_2

Kumareswaran, K., Jayasinghe, G.Y.: Green Infrastructure and Urban Climate Resilience. Springer International Publishing (2023). https://doi.org/10.1007/978-3-031-37081-6

Mallick, R.: Smart Cities and Climate Resilience (pp. 401–412) (2024). https://doi.org/10.4018/979-8-3693-6522-9.ch021

Martinez-Mireles, J.R., et al.: Integrating AI in green and blue infrastructure for sustainable smart cities. In: Smart Water Technology for Sustainable Management in Modern Cities (pp. 337–364). IGI Global (2025). https://doi.org/10.4018/979-8-3693-8074-1.ch014

Mitra, S., Madhuvanthi, S., Sabumon, P.C.: Nature-based urban resilience: integrating green infrastructure. In: Nature-Based Solutions in Achieving Sustainable Development Goals (pp. 167–205). Springer Nature Switzerland (2025). https://doi.org/10.1007/978-3-031-76128-7_6

Mudunuri, L.N.R., Attaluri, V.: Urban Development Challenges and the Role of Cloud AI-Powered Blue-Green Solutions (pp. 507–522) (2024). https://doi.org/10.4018/979-8-3693-8069-7.ch023

Ogwu, M.C., Kosoe, E.A., Alaka, A.C., Xu, J.: Green infrastructure and environmental health security: the role of nature-based solutions in improving urban health (pp. 225–255) (2025). https://doi.org/10.1007/978-3-031-81966-7_9

Oyenuga, M.O., Singh, R., Apata, S.B., Khan, S., Kumar, V.: Green technologies for water management (pp. 315–326) (2025). https://doi.org/10.1007/978-3-031-92221-3_25

Pal, B.: Water scarcity and climate change in intelligent infrastructure and smart materials: sustainable technologies for a greener future. In: Intelligent Infrastructure and Smart Materials (pp. 213–231). Springer Nature Switzerland (2025). https://doi.org/10.1007/978-3-031-92421-7_13

Pandey, R.U., Mitra, T., Wadwekar, M., Nigam, J., Trivedi, K.: Green Infrastructure as a tool for improving livability of area based development projects under smart city mission (pp. 447–468) (2021). https://doi.org/10.1007/978-3-030-71945-6_25

Pathak, S., Negi, N., Dadhich, P.: Smart Cities and Blue-Green Infrastructure (pp. 469–488) (2024). https://doi.org/10.4018/979-8-3693-8069-7.ch021

Pineda-Martos, R., Calheiros, C.S.C.: Nature-Based solutions in Cities—Contribution of the Portuguese national association of green roofs to urban circularity. Circular Econ. Sustain. **1**(3), 1019–1035 (2021). https://doi.org/10.1007/s43615-021-00070-9

Rathinavel, N., Murugesan, A., Periasamy, S.: Smart infrastructure and sustainable urban systems. In: Fostering Resilience and Quality of Life Through Sustainable Urban Evolution (pp. 35–68). IGI Global (2025). https://doi.org/10.4018/979-8-3373-0390-1.ch002

Riffat, S., Ahmad, M.I., Shakir, A.: Eco-Cities: sustainable urban living (pp. 259–309) (2025). https://doi.org/10.1007/978-3-031-78853-6_6

Salman, M., Najid, S.V.P., Ali, U., Khairoowala, Z., Siddiquee, T.A.R.: Building for tomorrow: climate-resilient infrastructure strategies for urban sustainability (pp. 291–311) (2025). https://doi.org/10.1007/978-3-031-99132-5_14

Seelam, D.R., Kidiyur, M.D., Whig, P., Whig, A.: Harnessing data engineering for optimizing blue-green infrastructure (pp. 271–290) (2024). https://doi.org/10.4018/979-8-3693-8069-7.ch013

Sharma, M., Gupta, S., Mahesh Kumar, T.: Green infrastructure (pp. 1–28) (2025). https://doi.org/10.4018/979-8-3693-7117-6.ch001

Singh, M., Arora, V., Kulshreshta, K.: Towards sustainable cities (pp. 212–233) (2024). https://doi.org/10.4018/979-8-3693-6567-0.ch011

Sommese, F., Diana, L., Colajanni, S., Bellomo, M., Sciuto, G., Lombardo, G.: Towards a regenerative and climate-resilient built environment: greening lessons from European Cities. Buildings **15**(11), 1878 (2025). https://doi.org/10.3390/buildings15111878

Thomas, T.S.P., Leon, L.: Water and waste management. In: Sustainable Construction and Heritage Conservation in the Digital Age (pp. 271–304). IGI Global Scientific Publishing (2025). https://doi.org/10.4018/979-8-3373-5278-7.ch009

Tyagi, R., Chaiurasia, A., Vishwakarma, S.: Climate change adaptation and mitigation strategies at future smart cities. IET Conf. Proc. **2024**(37), 478–485 (2025). https://doi.org/10.1049/icp.2025.0959

Whig, P., Sharma, P., Yathiraju, N. (eds.): Cutting-Edge solutions for advancing sustainable development: exploring technological horizons for sustainability-Part 1. Bentham Science Publishers (2025). https://doi.org/10.2174/97898153055481250101

Zhang, L.: Urban and city sustainability. In: Sustainability Economics (pp. 345–368). Springer Nature Switzerland (2025). https://doi.org/10.1007/978-3-031-87829-9_16

Ecological Integration of the Smart City: A Comparative Review of Global Models of Urbanisation

N. R. Asha Rani[1(✉)], Sheetal Thapa[1], N. G. Devaiah[2], and H. U. Megha[3]

[1] Department of Civil Engineering Alliance College of Engineering and Design, Alliance University, Anekal Bangalore , Karnataka 562106, India
asharaninr@gmail.com
[2] Alliance School of Law, Alliance University, Anekal Bangalore, Karnataka 562106, India
[3] Alliance School of Business, Alliance University, Anekal Bangalore, Karnataka 562106, India

Abstract. This review re-focuses smart cities in a technology-centric paradigm to a Nature-Positive Smart City paradigm that is based on the Society-Technology-Ecology (STES) model. The study uses land-use change, water stress, green cover dynamics, and governance capacity to evaluate ecological integration based on a comparative analysis of Punjab (India), Europe, China, and Gulf region. Findings show that the European models are well ecologically integrated, transitional in China, and that there is a lot of ecological stress in Punjab and Gulf, although the infrastructure is highly digitalized. Based on these findings, the paper estimates an effective framework of STES connecting inclusive governance, digital technologies, and ecological recovery. The paper ends with practical policy directions of developing nature-positive urbanisation and taking care of the gaps in governance and local capacity issues.

Keywords: Ecological cities · biocentric · Urban ecology · STES framework · Inclusive governance · ecological integration · Smart Cities · Nature Positive

1 Introduction

Urbanisation is turning into one of the major trends of the 21st century and more than 2/3s of the global population will reside in an urban area in the year 2050 (Bibri et al., 2023). This is a generational shift that has placed cities at the centre of economic, innovation and social change. Governments across the globe have adopted the smart city paradigm in response to the rising pressures of urbanization, including increasing infrastructure, energy, water and mobility demand, through the integration of digital technologies, information and communication systems (ICTs), and the use of data to support the establishment of smarter and more networked urban environments (Bibri, 2021a)

However, despite the mentioned positive developments of the proliferation of smart cities across the world, the same has triggered concerns over their environmental impacts. Several attempts at smart cities start to look like a smart concrete jungle-becoming

L. Choudhary et al. (Eds.): TIMES 2025, SCI, pp. 301–315, 2027.
https://doi.org/10.1007/978-3-032-34636-0_22

overindulged with technological and infrastructural progress, without paying much attention to ecological systems, biodiversity and sustainability of resources (R. et al., 2023). These are technological systems that can contribute to the hyper urbanization of cities, outbreaks of dengue strains, the disappearance of biodiversity in the name of so-called sustainability (Alizadeh, 2021).

Research on innovations in smart cities concentrates either on the technology adoption, institution, or social-economic benefits, but there is no literature to analyze the ecological integration critically. Although ecological aspects of sustainable urbanism and resilient cities have been brought into the limelight, not enough empirical research has investigated how various urban areas integrate ecological values into the smart city planning process (Shamsuzzoha et al., 2021).

This paper aims to come up with a concept and discuss such a concept, which I am expressing and discussing as Nature-Positive Smart Cities by comparing a case study of Punjab (India), Europe, China, and the Gulf region. The study demonstrates that balancing the three components of society, technology, and ecology (STES framework) is crucial through analysis of how these three regions incorporate and/or do not incorporate ecology in their smart city models. The solution provides an entirely new understanding of the factors that can shape the urbanisation process towards sustainability, resilience and ultimately human happiness for the long-term.

2 Literature Review

2.1 Evolution of Smart Cities

According to the trend, the idea of smart city was initially developed from the perspective of technology, which relies on the application of information and communication technologies (ICTs) to achieve optimization and service delivery, and networking. Initially, the proposed future of a smart city focused almost exclusively on digital infrastructure, data harvesting, sensors and Smart Locations as the catalyst for the city transition. Over the years, however, this limited interpretation has expanded to broader interpretations involving social inclusiveness, sustainability, human development and the environment. However, today, in addition to being perceived as a world of technologies, smart cities may also be viewed as a location where the process of innovation must correspond with an ecological as well as sociological demand and need (Shamsuzzoha et al., 2021).

2.2 Theoretical Frameworks

As a theoretical mediatory element that would have formed a basis for the process of creating the sustainable urbanism, a few theoretical background factors are discussed. The resilience framework focuses on how urban systems can absorb shocks (i.e., climate change, pollution, and depleting resources) and remain functional. Another vision of an ecology of cities comes from the idea that the functioning of human settlements rests on some form of dependency of a natural ecosystem, and that the stock-taking of biodiversity in towns is of fundamental importance. In the past 10 years, the phrase nature-based solutions have gained relevance alongside the emphasis on green infrastructures, ecosystems services and ecological restoration as central models of city planning. All these frames promote the fact that the ecological thinking must be added to the models of the virtuality of all smart cities (Shamsuzzoha et al., 2021).

2.3 World Policy Environment

Policymaking in most parts of the world has been strengthening the aspect of considering developments in realms of ecology. As well as targets related to environmental sustainability, the SDGs also include goals related to clean water systems, renewable energy, sustainable communities and action to combat climate change directly connecting cities to environmental sustainability. In a similar manner, the Kunming-Montreal Global Biodiversity Framework (COP15) may serve as a global mandate to prevent; eliminate biodiversity emergence, remediate damaged ecosystems and make ecology a priority concern in governance. They also help build the concept that cities need to look beyond technical efficiency and towards nature-positive practices (Bibri, 2023).

2.4 Research Gap

Regardless of these developments, most smart city models are largely technology driven to the point of disregarding ecological and biodiversity issues. Despite the importance of digital infrastructures and data-based governance, an ecological disconnection is a significant risk by enhancing urban vulnerability, such as groundwater loss and heat islands, habitat disappearance and air quality deterioration. Few comparative studies exist that assess the inclusion of ecology in the planning of smart cities across different regions. In this respect, the current research aims to contribute to this discussion and set forth Nature-Positive Smart Cities model capable of balancing technology, the society, and the environment in various urban environments (Chatti & Majeed, 2022). Figure 1 shows the transition from ICT driven towards nature positive and ecologically integrated smart cities.

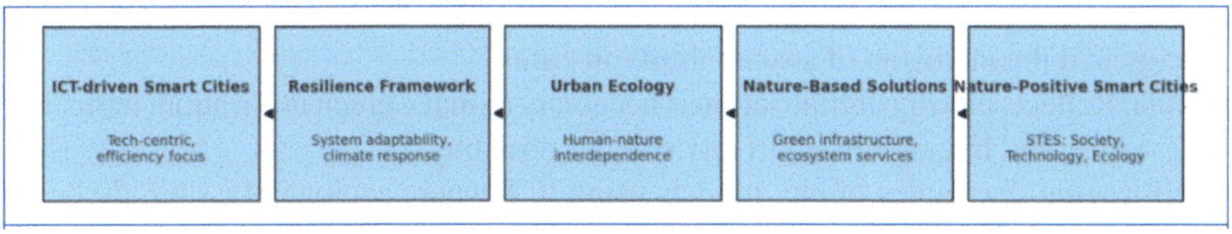

Fig. 1. Evolution of Smart City Frameworks (Demirel, 2023).

2.5 Urban Ecology Classical Foundations

The ancestral concepts of urban ecology are useful in offering historical backdrop to the current discourse of nature-positive smart city. The ecological planning as proposed by McHarg stressed ecological planning of cities, whereby decisions made on land-use were done according to the ecological suitability. The idea of biophilic cities developed by Beatley went a step further to this approach since it draws attention to the psychological, social and environmental positive elements of nature in the city life. Integrating these classical works places the STES framework in a more extended series of ecological urbanism instead of using it as a digital innovation.

3 Methodology

This paper is based on a comparative case study argument, validated by the secondary data synthesis process and the application of geospatial and chronological indicators to assess the degree of ecological integration that can be achieved in the model of a smart city. This methodology will focus on both technological and ecological aspects of urbanisation and will enable cross-regional comparisons (Deng et al., 2021; Wirtz & Müller, 2023).

3.1 Research Approach

An integrated literature review/policy analysis/geospatial synthesis method was used. The social science secondary sources, peer-reviewed publications, open government datasets, and smart city mission reports helped recognize the common trends of urban growth, mal-adaptive environmental pressures, and administrative design approaches. Information on temporal dynamics for the long-term ecological impact of urban sprawl was extracted from urban sprawl, land cover change, and groundwater loss files (Park & Yoo, 2023).

3.2 Case Selection

Four areas were chosen to compare to one another:

- Punjab (India): showcases rapid urbanization, a shortage of groundwater, and an ecological crisis within the framework of the smart city.
- Europe: Example of city design which optimises space, the installation of renewable energy and the inclusion of sustainability in cities
- China: Reflects on why digitalisation is necessary to make green innovation, industrial upgrading, and large-scale smart city roll out possible.
- Gulf Region: Examples of city models using ICT where ecological issues are being placed on a different agenda, but where the governance is not yet transparent enough or inclusive enough

These examples were selected, as they represent various stages of urbanisation, policy trends, and ecological priorities and offer a wide range to conduct comparative analysis (Gerasopoulos et al., 2022).

3.3 Ecological Indicators of Integration and Measurement

A set of indicators that are to be used in the ecological integration of smart city models was developed based on a standardized collection. The change in land-use was measured by the reported increase in the extent of built-up areas, agricultural areas, and fragmentation of natural ecosystems based on satellite-based land-cover data and urban planning reports. The water depletion was measured based on the rates of ground water depreciation, surface water stress measurements, and existence, or lack, of urban design interventions based on water sensitivity. The dynamics of green space were assessed by the availability, spatial continuity and functional quality of urban green infrastructure that consisted of parks, urban forests, wetlands and ecological corridors.

Since some of the indicators were not available regionally or had different reporting norms, all of them were coded into qualitative scores (low, medium, high ecological integration). It is a method that can be used to carry out a comparative study consistently even though there is interpretive flexibility to be applied in different urban settings.

3.4 Analytical Framework

The results of the four case studies were summarised into a comparative matrix, which allowed identifying any converging trends and regional challenges. To understand how the technological advancement versus the ecological sustainability and inclusivity of the social space are reconciled in each region, the analysis was informed by the STES framework (Society, Technology, Ecology) (Bibri, 2021b).

3.5 Schematic Framework of Comparative Integration of Ecodesign

Figure 2 gives a schematic set up of the interaction of ecological indicators, the governance ability and the technological implementation in the STES model. The framework illustrates that the digital infrastructure and the institution capacity mediate the land-use change and water stress and the dynamics of the green spaces in reaching different ecological consequences. With the help of this schematic, it allows organizing a comparative analysis of territories and understanding the methodological reasoning behind the associations of indicators to comparative analysis.

4 Case Usage and Comparative Analysis

4.1 Punjab (India)

The state of Punjab is one of the most ecologically pressured environments of smart cities in India that have suffered huge investments in ICT-enabled infrastructure. One of the fastest rates of depletion trends in the country is the reduction of groundwater levels in the region at a rate estimated approximately 160 200 cm between 2002 and 2022. Although the programs of the smart cities have led to better service delivery, online connectivity, and governmental efficiency, it has not transformed into objective enhancements in water sustainability.

Integration or lack of integrating recharge-based urban planning, decrease in wetlands and increase in impervious surfaces, have increased the stress on the aquifer, as well as the urban heat island effects. This suggests a disqualification in technologies that result in smart interventions of cities and ecological performance and demonstrates the necessity of environmentally based measurements in the planning of smart cities.

4.2 Europe

Smart city narratives often view the compact city and integrated urban sustainability policies as being prevalent throughout Europe. As part of future resilience plans, several cities in Europe have encouraged the adoption and mind-shift to renewable energy, low-carbon mobility, and greening in their urban areas. "However, there is evidence that such policies are related to positive health outcomes, such as increased life expectancy and air quality." Education and citizens' empowerment are among the key interconnected, complementary drivers for inclusive governance and innovation. The ecological integration of Europe in this way acquires a high score on the scale between urbanisation and regard towards biodiversity and the quality of shared life (Nikki Han & Kim, 2021).

4.3 China

The existence of such system evidence that digitalisation and green innovation have had a hefty influence on the Chinese concept of the smart city. Government: Government agencies have been facilitating industrial upgrading and transitions towards low carbon oscillations via the digital utilization of big data and artificial intelligence systems and platforms. Energy efficiency, carbon neutrality pilot projects have demonstrated large energy savings and reduction of emittance including enormous urban developments but the now-a-days fast industrialization and elements of social cut-charge present limitations to sustainability of the ecological performance. However, China is not the only example of how the technology-driven government might be implemented to achieve economic and growth targets (Demirel, 2023).

4.4 Gulf Region

This region has far and away the largest ICT-mediated Smart Cities projects in the world and includes the biggest digital infrastructure development and the most futuristic cities the region has to offer. That has been accompanied by a rising level of interdependence and efficiency-but not ecological integration. There is therefore still a high demand for water supply, for intensive use of energy resources and for desalination. But sustainability over time also demands more transparent and more participatory governance-one with more citizens in the mix. The Gulf model is a useful reminder of what happens when the ecology and the society are for sale to the technology (Zhu et al., 2024).

4.5 Cross Case Comparative Table

Table 1 shows the cross-case comparison of models in Punjab (India), Europe, China and the Gulf Principality, this table has been structured along the STES framework (Society-Technology-Ecology.) In other words, it incorporates regional differences in terms of the level of citizen participation, technological advancement and level of integration and match with the environment in a way that we consider it as the balanced and mixed structure of regional strengths and regional weaknesses. The ecological integration score (low-medium-high) provides a quick measure of how well each city region has integrated sustainability in its smart city program. Comparative analysis contrasts the various approaches globally and emphasizes the importance of context-specific tools for developing nature-positive smart cities and Fig. 3 highlights the Comparative STES Analysis of Smart Cities-compares Punjab, Europe, China and the Gulf on social, technological and ecological dimensions.

5 Framework Proposals, Nature Positive Smart Cities (STES Model)

The empirical patterns were also known in the comparative case studies, and they inform the STES framework. Punjab and the Gulf show the sign of urbanisation led by technological imitations and poor ecological protection indicating the imbalance between technological progress and environmental stability. European urban areas have been manifesting in society, technology, and ecology which has led to participatory governance and compact planning. China is one such example of a transitional model in which the urban stress of ecological impacts is partly compensated by the digital capacity. The results of these observed patterns are modelled in the framework of STES to expressly relate theory and real-world urban results. The comparative analysis shows that the ecological basis of technology-driven smart cities is weak in many cases even though efficiency and connection have improved. To fill this void, the current paper will present one such concept: the Nature-Positive Smart Cities Framework or the STES Model (Society, Technology, Ecology as three pillars). Innovative Regenerative Urbanization (IU) refers to longer-term regenerative innovation in liveability through digital technology and primarily ecological regeneration-and to participatory government of our cities to make our cities resilient and sustainable (Goel et al., 2021).

5.1 Society

The model is citizen-involvement, openness and inclusiveness and societal-based framework. One of the enabling initiatives necessary to support access to information and participatory governance is the use of open government data (OGD) portals to promote interaction between governments and citizens. Cities can evolve into spaces where citizens can play a role in ensuring that cities remain sustainable by developing new concepts of co-creation and solving the problem of digital divides (Patel et al., 2024).

5.2 Technology

Technology as the silent utilizer: Silent applications can be incredibly handy in environmental monitoring and management using Artificial Intelligence (AI), Internet of things (IoT), digital twins and others. "It can be applied to predict the amount of energetic material deployed and optimize emissions, monitor water quality and soil toxicity, monitor air pollution-and can also forecast long-term ecological conditions of proposed activities in an urban system through the concept of digital twins." Such technologies will help cities to develop evidence-based decisions so that the development programme aligns with the vision for sustainable infrastructure (Benites & Simões, 2021).

Table 1. STES-based comparative analysis of models of smart cities in different regions (Demirel, 2023).

Region	Citizen Inclusion (Society)	ICT & Innovation (Technology)	Integration (Ecology)	Ecological Integration Score (Regional Plus)
Punjab (India)	Moderate – limited participation, uneven access	Strong – ICT adopted under smart city mission	Weak – groundwater depletion, depletion of forest cover, biodiversity loss, heat islands	Low
Europe	Strong – good governance, integration, education, awareness	Strong – renewable energy, green mobility	Strong – compact urban planning, ecological protection	High
China	Moderate – state-led development, increasing citizen services	Very Strong – digitalization, artificial intelligence, industrial upgrading	Moderate – some green innovation, but still significant ecological stress	Medium
Gulf Region	Weak – lack of transparency, low level of participation	Very Strong – modern ICT-based infrastructure	Weak – water stress, heavy emissions, low ecological planning	Low

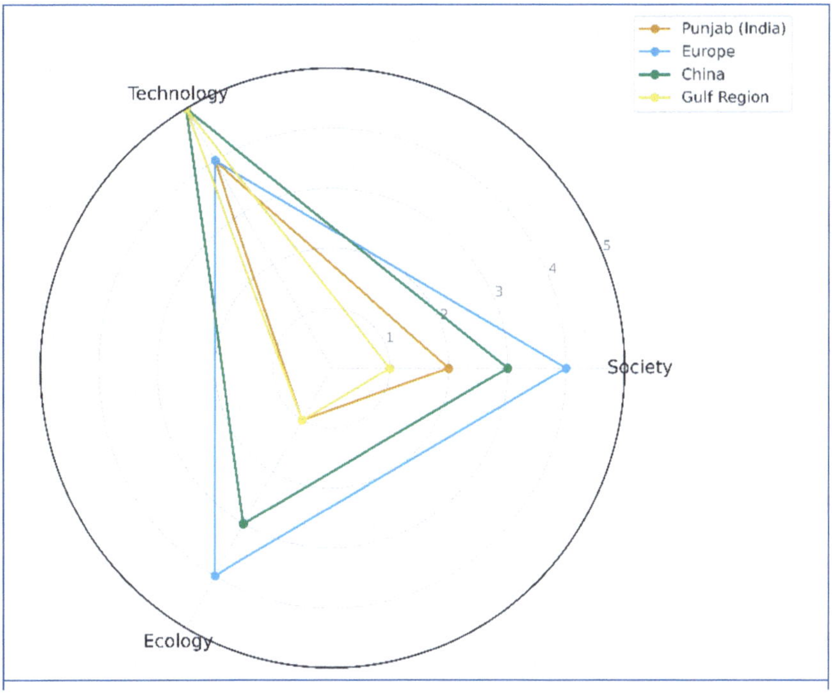

Fig. 2. Comparative STES profile of Smart City model across Punjab, Europe, China, and the Gulf region (Goel et al., 2021).

5.3 Ecology

Within the center of long-term resilience is placed ecology. The model promotes biodiversity and sensitive water-related planning as well as the restoration and establishment of urban forests as integral elements of healthy cities. Creating connected ecological corridors and restoring wetlands can take the form of green roofs and wetland restoration and, when realized as part of the city plan, assist the city in providing ecological services, reducing the urban heat island effect, and improving climate resiliency (Padhiary et al., 2024).

6 Policy Recommendations

6.1 Biodiversity as Part of the Urban Planning Process

Rather than keeping cities stuck to models that are overloaded with infrastructure, the idea of smart cities should instead focus on integrating biodiversity into cities. Not only will green belts, ecological corridors, and restored wetlands protect species on urban lands, but also provide resilience against urban heat islands and flooding. All these actions enhance recreational areas and human (Ahmed et al., 2024).

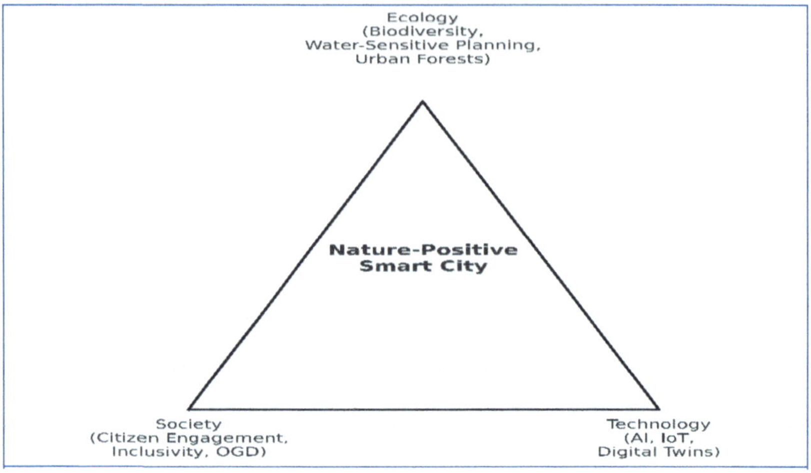

Fig. 3. STES Model for Nature-Positive Smart Cities (Bandauko & Nutifafa Arku, 2023).

6.2 Advancing Smart Governance

Smart city governance must be participatory, transparent and pro-citizen. Trust between citizens and governments can be built by expanding the utilization of Open Government Data (OGD) portal as well as enhancing its usability and accessibility. The idea needs to be institutionalised with member citizens not being mere beneficiaries of the future development of cities but pocket architects (Shuang & Zheng, 2024).

6.3 Sustainable Resources Management

Cities are confronted with increasing problems concerning water and the degradation of soil. As a result, rainwater harvesting, wastewater reuse, treated effluent recycling, and groundwater augmentation need to be the hotspots in the field of intelligent cities. This means that compact cities should be encouraged to reverse the sprawl trend and take agricultural lands away from expansive development, while also providing unlimited expansion of artificial habitat to ecosystems (Suryawan et al., 2024).

6.4 The Conversion to Renewable Energy and Green Mobility

A reduction in the use of fossil energy due to ecological and overall health concerns is of the utmost importance. Expansion in the field of investment in the systems of renewable energy (photovoltaic, wind energy, biomass) and the development of technological solutions in the field of green mobility (electric public transport, non-moto trajectory concepts, energy-harmonious infrastructure) will reduce emissions, clean the air and enable smart cities to fulfil international environmental climate and biodiversity policy obligations (Wu et al., 2024). A phased implementation strategy is essential for achieving nature-positive smart cities, with recommended actions and policy objectives categorized across short-, medium-, and long-term planning horizons, as presented in Table 2.

Table 2. Pathway of policy Action in Nature-Positive Smart Cities (Demirel, 2023).

Time Horizon	Actions	Policy Objective
Short-term (1–2 years)	Protect remaining urban green spaces; incorporate water-sensitive urban design (WSUD) principles into ongoing smart city projects	Immediate reduction of ecological and hydrological risks
Medium-term (3–5 years)	Strengthen participatory governance mechanisms; restore urban ecosystems to enhance biodiversity	Institutional fortification and inclusive environmental governance
Long-term (>5 years)	Implement integrated urban planning based on Socio-Technical-Ecological Systems (STES); align smart city initiatives with climate adaptation and biodiversity goals	Systemic and long-term ecological resilience

7 Challenges and Limitations

7.1 Digital Divide and Governance Gaps.

The nature-positive smart cities are hindered by the presence of governance gaps and digital divides, which can be easily measured. Internet access in majority of the European cities is above 85% whereas, in areas of Punjab and the Gulf, the figure is under 50–60 percent, restricting citizen use of digital governance platforms. The same applies to use of open government data portals and use of participatory planning tools which are also not equally distributed at the regions. They directly inhibit such disparities limit the inclusive decision-making process and undermine the ecological monitoring and accountability mechanisms (Bandauko & Nutifafa Arku, 2023; Mortaheb & Jankowski, 2023).

7.2 Data Limitations

The research draws mostly from secondary datasets and published reports, useful as that information may be, but is not able to capture dynamics at local levels. This is limited by the fact that primary field data, which could be a household survey or even an ecological survey conducted on the ground, is not used to establish the patterns that have been observed using secondary sources (Su & Fan, 2023; Vettriselvan & Ramya, 2024). Figure 4 shows the four strategic priorities: biodiversity, governance, resource management and renewable energy.

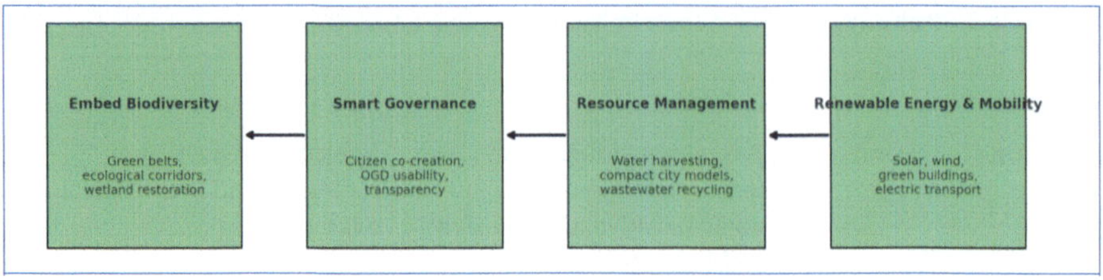

Fig. 4. Policy Roadmap for Nature-Positive Smart Cities (Ionescu et al., 2025).

7.3 Regional Adoption Variation

There are no universal city models: the scale of their implementation differs according to the capacities of governments around the world, available financial resources, and cultural factors (Azizi & Kouddane, 2024; Dhiman & Alghamdi, 2024). While cities in most parts of Europe show encouraging signs of ecological integration, other cities such as Punjab and the Gulf are increasingly under pressure from sustained unregulated ecological stress and administrative strain (Braga et al., 2021; Ramu et al., 2022). The interpretations of the results above are also restricted by the context since remember that interpretation cannot happen independently from the context (Dahmane et al., 2025; Xia et al., 2021).

8 Conclusion

On a conceptual basis, this review asserts an urgent need for rethinking smart cities not so much as techno-oriented places or nodes but rather as nature-regal cities. The comparative analysis between Punjab, Europe, China and the Gulf region teases out a continuum of integration-from Europe's highest ecological convergence to the continued vulnerabilities between Punjab and the Gulf. Digitalisation and ICT infrastructure has led to large increases in efficiency, yet without adequate regulations it often shifts ecological factors to the side, with resulting effects like dewatering, loss of biodiversity and climate stress.

The STES framework (Society, Technology, Ecology) suggested here provides a symmetrical basis for new urbanization in the future. He indicates that each dimension of the 'relocalization eco-economy-project-or one of the three required elements-cannot be successfully deployed unless citizen participation, technological ingenuity, and ecological renewal are deployed at the same time, if a resilient situation is ever to be obtained. Implementation paths identified as relevant and practical include actions that have already been established in cities, such as the integration of gardens and sustainable agriculture, the incorporation of biodiversity in urban planning, participatory governance, and water-sensitive resource allocation and speeding up renewable energy transitions.

However, issues surrounding the digital divide, gaps in governance and the use of secondary data restrict the scope of findings to be globally relevant. Identified as follows, there is once more to support the thesis, that there is no blueprint for smart cities; rather each context will determine, and do determine, the structure and levels of society's commitment towards their development. Instead, cities need to apply the STES framework in ways that match local capacity, cultural contexts and economic priorities for the city's ecology, the road to Nature-Positive Smart Cities involves reconciling society, technology, and ecology. Ideas that will enhance resilience and sustainability as well as guarantee that the urbanisation of the future is for the welfare of people, the well-being of the environment and of future generations.

9 Future Directions

While this review presents the potential of the STES framework to inform nature-positive building of smart cities, spaces for further research along several paths are open:

- Integration of Data from Primary Data Sources and Field Surveys-More importantly, future studies need to combine field surveys, ecological monitoring and citizenship perceptions studies to be able to help ground-truth planetary data from derivative data sources and inform what a top-down smart city model often misses.
- Luis Menendez, Boulder, CO: The use of AI, IoT, remote sensing, and digital twins should be further tested in the context of biodiversity monitoring, mitigation of urban heat island effects, and predictive climate modeling in cities.
- Cross-Regional Comparative Studies: In addition to the more developed case studies in Punjab, Europe, China and the Gulf, case studies drawing on African, Latin American and Southeast Asian cities can broaden the learning on the different governance systems and challenges for ecological protection.
- Policy Experimentation and Governance Models: PILOTs that employ concepts of citizen co-production, open government information portals, and tools for ecological budgeting could experiment with the ways that the speed of ecological integration can be strongly influenced by governance experimentation.
- Socio-Ecological Equity: Practical studies concerning digital inequalities, participatory and equitable approaches to the production and distribution of smart technologies, and ecological justice concerning digital connections, open systems, and green laboratories; equitable smart cities for marginalized populations
- Climate Resilience and Global Frameworks: Further work is needed to bring urban models into closer understanding of the Kunming-Montreal GBF, SDGs and COP commitments where the impact of global policy may be estimated down to the minutiae in local conditions and contingencies.
- Longitudinal and Predictive Studies: Cities can track performance of the urban system at the ecological scale for decades, to observe the outcome of implementing smart city interventions, as opposed to displacing people.

References

Ahmed, K., Kumar Dubey, M., Kumar, A., Dubey, S.: Artificial intelligence and IoT driven system architecture for municipality waste management in smart cities: a review. Meas.: Sens. **36**, 101395 (2024). https://doi.org/10.1016/j.measen.2024.101395

Alizadeh, T.: Global trends of smart cities : a comparative analysis of geography, city size, governance, and urban planning. Academic Press (2021)

Azizi, L., Kouddane, N.: The green city as a driver of sustainable development. J. Umm Al-Qura Univ. Eng. Archit. **15**(4), 384–397 (2024). https://doi.org/10.1007/s43995-024-00077-x

Bandauko, E., Nutifafa Arku, R.: A critical analysis of 'smart cities' as an urban development strategy in Africa. Int. Plan. Stud. **28**(1), 69–86 (2023). https://doi.org/10.1080/13563475.2022.2137112

Benites, A.J., Simões, A.F.: Assessing the urban sustainable development strategy: an application of a smart city services sustainability taxonomy. Ecol. Ind. **127**, 107734 (2021). https://doi.org/10.1016/j.ecolind.2021.107734

Bibri, S.E.: Data-driven smart eco-cities and sustainable integrated districts: a best-evidence synthesis approach to an extensive literature review. Eur. J. Futures Res. **9**(1), 16 (2021). https://doi.org/10.1186/s40309-021-00181-4

Bibri, S.E.: The underlying components of data-driven smart sustainable cities of the future: a case study approach to an applied theoretical framework. Eur. J. Futures Res. **9**(1), 13 (2021). https://doi.org/10.1186/s40309-021-00182-3

Bibri, S.E.: Data-Driven smart eco-cities of the future: an empirically informed integrated model for strategic sustainable urban development. World Futures **79**(7–8), 703–746 (2023). https://doi.org/10.1080/02604027.2021.1969877

Bibri, S.E., Alexandre, A., Sharifi, A., Krogstie, J.: Environmentally sustainable smart cities and their converging AI, IoT, and big data technologies and solutions: an integrated approach to an extensive literature review. Energy Inform. **6**(1), 9 (2023). https://doi.org/10.1186/s42162-023-00259-2

Braga, I.F.B., Ferreira, F.A.F., Ferreira, J.J.M., Correia, R.J.C., Pereira, L.F., Falcão, P.F.: A DEMATEL analysis of smart city determinants. Technol. Soc. **66**, 101687 (2021). https://doi.org/10.1016/j.techsoc.2021.101687

Chatti, W., Majeed, M.T.: Information communication technology (ICT), smart urbanization, and environmental quality: evidence from a panel of developing and developed economies. J. Clean. Prod. **366**, 132925 (2022). https://doi.org/10.1016/j.jclepro.2022.132925

Dahmane, W.M., Ouchani, S., Bouarfa, H.: Smart cities services and solutions: a systematic review. Data Inf. Manage. **9**(2), 100087 (2025). https://doi.org/10.1016/j.dim.2024.100087

Demirel, D.: The impact of managing diversity on building the smart city a comparison of smart city strategies: cases from Europe, America, and Asia. Sage Open **13**(3) (2023). https://doi.org/10.1177/21582440231184971

Deng, T., Zhang, K., Shen, Z.-J.: A systematic review of a digital twin city: a new pattern of urban governance toward smart cities. J. Manage. Sci. Eng. **6**(2), 125–134 (2021). https://doi.org/10.1016/j.jmse.2021.03.003

Dhiman, G., Alghamdi, N.S.: SMoSE: artificial intelligence-based smart city framework using multi-objective and IoT approach for consumer electronics application. IEEE Trans. Consum. Electron. **70**(1), 3848–3855 (2024). https://doi.org/10.1109/TCE.2024.3363720

Gerasopoulos, E., et al.: Earth observation: an integral part of a smart and sustainable city. Environ. Sci. Policy **132**, 296–307 (2022). https://doi.org/10.1016/j.envsci.2022.02.033

Goel, R.K., Yadav, C.S., Vishnoi, S.: Self-sustainable smart cities: socio-spatial society using participative bottom-up and cognitive top-down approach. Cities **118**, 103370 (2021). https://doi.org/10.1016/j.cities.2021.103370

Ionescu, R.-V., Zlati, M.L., Antohi, V.-M.: Smart cities from low cost to expensive solutions under an optimal analysis. In: Blockchain, Crypto Assets, and Financial Innovation (pp. 483–522). Springer Nature Singapore (2025). https://doi.org/10.1007/978-981-96-6839-7_18

Mortaheb, R., Jankowski, P.: Smart city re-imagined: city planning and GeoAI in the age of big data. J. Urban Manage. **12**(1), 4–15 (2023). https://doi.org/10.1016/j.jum.2022.08.001

Nikki Han, M.J., Kim, M.J.: A critical review of the smart city in relation to citizen adoption towards sustainable smart living. Habitat Int. **108**, 102312 (2021). https://doi.org/10.1016/j.habitatint.2021.102312

Padhiary, M., Roy, P., Roy, D.: The future of urban connectivity (pp. 33–66) (2024). https://doi.org/10.4018/979-8-3693-6740-7.ch002

Park, J., Yoo, S.: Evolution of the smart city: three extensions to governance, sustainability, and decent urbanisation from an ICT-based urban solution. Int. J. Urban Sci. **27**(sup1), 10–28 (2023). https://doi.org/10.1080/12265934.2022.2110143

Patel, U.R., Ghaffarianhoseini, A., Ghaffarianhoseini, A., Burgess, A.: Digital Twin Technology for sustainable urban development: a review of its potential impact on SDG 11 in New Zealand. Cities **155**, 105484 (2024). https://doi.org/10.1016/j.cities.2024.105484

R., A.R.N., Laishram, T., Bal, S.: Energy transition: paving the way for a greener future. Nanomaterials Energy **12**(3), 100–109 (2023). https://doi.org/10.1680/jnaen.23.00057

Ramu, S.P., et al.: Federated learning enabled digital twins for smart cities: concepts, recent advances, and future directions. Sustain. Cities Soc. **79**, 103663 (2022). https://doi.org/10.1016/j.scs.2021.103663

Shamsuzzoha, A., Nieminen, J., Piya, S., Rutledge, K.: Smart city for sustainable environment: a comparison of participatory strategies from Helsinki Singapore and London. Cities **114**, 103194 (2021). https://doi.org/10.1016/j.cities.2021.103194

Shuang, Q., Zheng, Z.: Analysis on the impact of smart city construction on urban greenness in China's megacities. J. Environ. Manage. **355**, 120568 (2024). https://doi.org/10.1016/j.jenvman.2024.120568

Su, Y., Fan, D.: Smart cities and sustainable development. Reg. Stud. **57**(4), 722–738 (2023). https://doi.org/10.1080/00343404.2022.2106360

Suryawan, I.W.K., et al.: Smart urbanism, citizen-centric approaches and integrated environmental services in transit-oriented development in Jakarta Indonesia. Res. Globalization **8**, 100181 (2024). https://doi.org/10.1016/j.resglo.2023.100181

Vettriselvan, R., Ramya, R.: Sustainable curriculum design and development (pp. 471–486) (2024). https://doi.org/10.4018/979-8-3693-7723-9.ch027

Wirtz, B.W., Müller, W.M.: An integrative collaborative ecosystem for smart Cities—A framework for organizational governance. Int. J. Public Adm. **46**(7), 499–518 (2023). https://doi.org/10.1080/01900692.2021.2001014

Wu, P., Zhang, Z., Peng, X., Wang, R.: Deep learning solutions for smart city challenges in urban development. Sci. Rep. **14**(1), 5176 (2024). https://doi.org/10.1038/s41598-024-55928-3

Xia, X., Wu, X., BalaMurugan, S., Karuppiah, M.: Effect of environmental and social responsibility in energy-efficient management models for smart cities infrastructure. Sustain. Energy Technol. Assess. **47**, 101525 (2021). https://doi.org/10.1016/j.seta.2021.101525

Zhu, J., Gianoli, A., Noori, N., de Jong, M., Edelenbos, J.: How different can smart cities be? A typology of smart cities in China. Cities **149**, 104992 (2024). https://doi.org/10.1016/j.cities.2024.104992

Author Index

L. Choudhary et al. (Eds.): TIMES 2025, SCI, pp. 317–318, 2027.
https://doi.org/10.1007/978-3-032-34636-0

The manufacturer's authorised representative in the EU is Springer Nature Customer Service Centre GmbH, Europaplatz 3, 69115 Heidelberg, Germany. If you have any concerns regarding our products, please contact ProductSafety@springernature.com

Printed and bound by CPI Group (UK) Ltd, Croydon, CR0 4YY
30/10/2026
02242258-0002